CLASSIFICATION OF THE "TRUE SEALS"

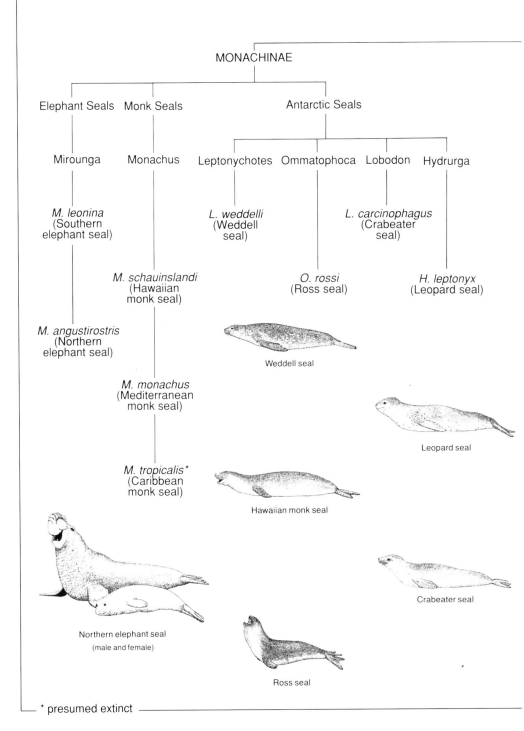

MONACHINAE

Elephant Seals | Monk Seals | Antarctic Seals

Mirounga | Monachus | Leptonychotes | Ommatophoca | Lobodon | Hydrurga

M. leonina
(Southern
elephant seal)

L. weddelli
(Weddell
seal)

L. carcinophagus
(Crabeater
seal)

M. schauinslandi
(Hawaiian
monk seal)

O. rossi
(Ross seal)

H. leptonyx
(Leopard seal)

Weddell seal

M. angustirostris
(Northern
elephant seal)

Leopard seal

M. monachus
(Mediterranean
monk seal)

*M. tropicalis**
(Caribbean
monk seal)

Hawaiian monk seal

Crabeater seal

Northern elephant seal
(male and female)

Ross seal

* presumed extinct

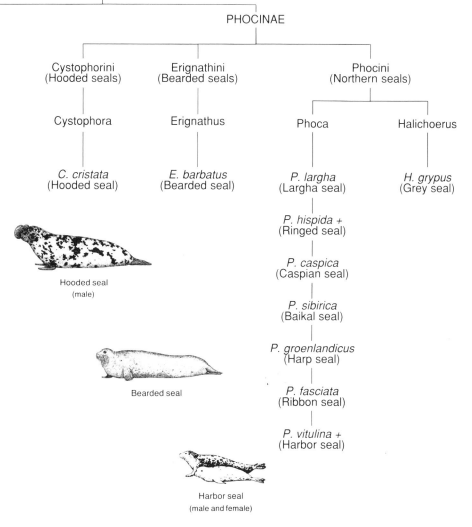

PHOCOIDEA
|
PHOCIDAE
(earless seals)

PHOCINAE

| Cystophorini (Hooded seals) | Erignathini (Bearded seals) | Phocini (Northern seals) |

Cystophora   Erignathus   Phoca   Halichoerus

C. cristata (Hooded seal)

E. barbatus (Bearded seal)

P. largha (Largha seal)

H. grypus (Grey seal)

P. hispida + (Ringed seal)

P. caspica (Caspian seal)

P. sibirica (Baikal seal)

P. groenlandicus (Harp seal)

P. fasciata (Ribbon seal)

P. vitulina + (Harbor seal)

Hooded seal
(male)

Bearded seal

Harbor seal
(male and female)

Grey seal
(male)

+ there are 5 recognized subspecies of the
harbor seal and ringed seal (see Table 2)

The Pinnipeds

SEALS, SEA LIONS, AND WALRUSES

The
Pinnipeds

Seals, Sea Lions, and Walruses

Marianne Riedman

University of California Press
Berkeley / Los Angeles
Oxford

Contribution number 12
of the Monterey Bay Aquarium

University of California Press
Berkeley and Los Angeles
University of California Press, Ltd.
Oxford, England
© 1990 by
The Regents of the University of California
Printed in the United States of America
2 3 4 5 6 7 8 9

Library of Congress Cataloging-in-Publication Data

Riedman, Marianne.
    The pinnipeds : seals, sea lions, and walruses / Marianne Riedman.
        p.   cm.
    Bibliography: p.
    Includes index.
    ISBN 0-520-06497-6.
    ISBN 0-520-06498-4 (ppb.).
    1. Pinnipedia.   I. Title.
QL737.P6R54   1990
599.74'.5–dc20   89-31542

Endpapers: The taxonomy of the pinnipeds, with illustrations of the various species. *Drawings by Pieter Folkens.*

To my grandparents
Fred and Veda Riedman
and
my father
Fred L. Riedman

CONTENTS

ILLUSTRATIONS

Plates
(Following Page 168)

Figures

Maps

TABLES

ACKNOWLEDGMENTS

I am deeply grateful to the many friends, colleagues, and family members who contributed in numerous ways to the preparation of this manuscript. Without their generous help and support, it would not have been possible to produce this book. I would first like to acknowledge my father, who initially sparked my interest in marine biology. He taught me to skin-dive when I was seven years old and later to scuba dive in the clear waters off Catalina Island in southern California, where I caught my first glimpse of seals, dolphins, and other marine life. I am truly indebted to him for this early start as well as his continued support and encouragement throughout my years of graduate school. I would also like to thank my grandparents, mother, sisters, brother, and the rest of my family and friends for their much-appreciated love and support during the writing of this book.

My sincerest thanks are also extended to Jim Estes, Bob Gisiner, Ron Schusterman, and especially Roger Gentry and Burney Le Boeuf for carefully reviewing the manuscript and providing many helpful ideas and comments. The time they took from their busy schedules to read through this lengthy manuscript is much appreciated. I am particularly grateful to Burney Le Boeuf for his guidance and encouragement and for his continuing friendship and support. Burney first introduced me not only to the theories of behavioral biology but also to the amazing elephant seals. I was fortunate to find such a generous, astute, and enjoyable teacher and advisor.

I also wish to acknowledge my friends, fellow students, and professors at the University of California at Santa Cruz who provided me with many interesting hours of discussion and stories about seals as well as much-needed companionship on Año Nuevo Island and at other isolated field research stations, especially Joanne Reiter, Bob Gisiner, Anne Hoover,

Mark Pierson, Kathy Panken, Leo Ortiz, Breck Tyler, Keith Skaug, and Steve Davenport.

I appreciate the help and information kindly provided for the book by Tom Loughlin, John Francis, Jerold Lowenstein, Chip Deutsch, Alan Baldridge, Tim Gerrodette, John Ling, Sheila Anderson, Nigil Bonner, Leo Ortiz, Dan Costa, Kathy Ralls, Pat Fiorelli, Frans Lanting, Mark Pierson, Bob Byrne, Anne Hoover, Gerry Kooyman, Kathy Frost, David Lavigne, Don Siniff, Steve Leatherwood, Francis Fay, Michael Bigg, John Burns, Carolyn Heath, Ian Stirling, and many others. In particular, this book could not have been written without the encouragement and substantial help provided by Roger Luckenbach, an outstanding teacher from whom I learned a great deal during the many times he patiently read the manuscript and offered advice. I also wish to thank Charles Mertz, my high school marine biology teacher, who inspired me to continue in the field.

Many of my friends and colleagues at the Monterey Bay Aquarium provided invaluable assistance, and I am grateful to them for their help and patience with me as I was writing this book: Michelle Staedler, Jim Watanabe, Peter Barnett, Chris Harrold, Sue Lisin, Roger Phillips, Julie Graef, Patty Anderson, Barbara Hrabrich, Anna Crabtree, Drew Anderson, Bill Townsend, Dave Rasco, and Chuck Farwell.

I am grateful for the generous support of the Banbury Foundation, the Monterey Bay Aquarium, and the Institute of Marine Sciences at UCSC, as well as Bob and Kate Ernst, Julie Packard, Bill Doyle, Gene Haderlie, and the other members of the Monterey Bay Aquarium Research Board. I'd like to express my appreciation to Debra Christofferson, Janis Warschauer, and especially Patty Anderson for typing the bibliography and to the Monterey Bay Aquarium Graphics Department for providing all the maps and figures for the manuscript: Bill Thomason, Cindy Sagen, Kathy Kopp, Don Hughes, and especially Sean Cummings. Pieter Folkens kindly supplied most of the drawings, and I am grateful to him for his assistance. Irene Campagna also generously provided some of her pinniped illustrations, and Brad Decaussin supplied the white shark illustration.

Numerous individuals generously contributed photographs for the book: Steve Amstrup, Sheila Anderson, Nancy Burnett, Victoria Corliss of the New England Aquarium, Dan Costa, W. Curtsinger of the International Fund for Animal Welfare, Jorge de Castro, Susan Dudenhoffer, Jim Estes, Brian Fadley, Michael and Lauren Farley, Pat Fiorelli, Pieter Folkens, Jeff Foott, John Francis, R. Frank, Kathy Frost, Roger Gentry, Bob Gisiner, Anne Hoover, R. Itther, Brendan Kelly, Gerry Kooyman, Kit

Kovacs, Frans Lanting, David Lavigne, Steve Leatherwood, Burney Le Boeuf, John Ling, Lloyd Lowry, Rob Mattlin, Larry Minden, Flip Nicklin, Dave Olsen, Tom O'Shea, Mark Pierson, Galen Rathbun, Michelle Staedler, William Stortz, Jeanette Thomas, Phil Thorson, Frank Todd, Bill Townsend, Breck Tyler, Kennan Ward, Jim Watanabe, Steve Webster, Doc White, Randy Wells, and Graham Worthy.

Finally, I'd like to express my deepest thanks to the University of California Press editors Ernest Callenbach and Stephanie Fay for the enormous amount of time and effort they devoted to the production and design of the book. I very much appreciate Stephanie's excellent and thorough editorial skills; her patience in dealing with complex scientific jargon has made the book much more readable for nonspecialists. I'm especially grateful to Ernest Callenbach, who patiently guided the manuscript to completion; his creative and helpful advice, support, and encouragement are deeply appreciated.

PREFACE

I am a man upon the land
I am a selchie on the sea
CELTIC POEM

Celts from the Orkney and Hebrides islands of Scotland to western Ireland have long told legends about selchies, or silkies — seals that could take off their skins and walk upon the land as men or women. In one often-told story, an isolated fisherman happens upon a beautiful woman strolling on a beach. The fisherman takes her seal skin, which he finds hidden among the rocks, knowing that in so doing the silkie woman must become his wife. After marrying her, the fisherman is careful to hide away the seal skin so that his wife will stay with him always on land. After bearing many earthly children, the silkie wife accidentally discovers her skin. Donning it, she takes her changeling children — who are silkies too — back into the sea, not only leaving her husband heartbroken but also populating the ocean with more seals.

Countless stories of seals and humans can be found throughout the world — in the mythology of the ancient Greeks, the culture of aboriginal Norwegians, and the legends of Eskimos. As these stories indicate, the lives of seals and humans have been closely intertwined for many thousands of years. Even in the Stone Age these abundant and valuable mammals were hunted with relative ease; seals resting on land were vulnerable and easily killed. Especially in the Northern Hemisphere, seals provided primitive humans with food, clothing, shelter, and fuel. Some Arctic coastal communities, such as those of the Inuit Eskimos, were almost entirely dependent on seal hunting for survival. The importance of sealing is reflected in the Inuits' religion and culture, in which seals play a prominent role. Inuits even make use of the three-foot-long walrus baculum — the penis bone — which serves as a walking staff called an *osik*.

In modern times, man's relationship with seals changed dramatically with the development of commercial sealing in the eighteenth and nineteenth centuries when a great many species were hunted almost to extinction. The lucrative trade for seal products resulted in the intensive slaughter of many species for their velvety fur, rich oil, and ivory tusks. In China there was even a market for Steller sea lion whiskers—used for cleaning opium pipes—and the genitalia of breeding bulls, which were dried, pressed into a powder, and used as an aphrodisiac. Paradoxically, the adaptive features that have allowed pinnipeds (from the Latin meaning "wing-footed" or "fin-footed") to survive in the marine environment—their thick blubber and dense fur—are the very attributes that have made them attractive to commercial sealers, who brought about the near extinction of many species.

Seals are familiar to most of us now as exhibit or performing animals in zoos and aquariums. Nearly everyone has seen the California sea lion balancing a ball on its nose, although most people are unaware that sea lions can learn much more difficult and complex tasks than balancing balls or applauding their own performances by clapping their foreflippers together. In our Western culture, the popularity of seals is evident in stuffed animals, toys, sculpture, and children's books, in which seals appear as characters. Public concern and compassion have increased for seals and other wildlife that are still slaughtered for their fur. The practice of clubbing harp seal pups, or "whitecoats," only recently prohibited by the Canadian government, was well publicized and protested by many. Fur coats made from the skins of seals and other fur-bearing mammals have lost their appeal. Growing public awareness of the need to conserve pinnipeds and other marine mammals in the United States was in large part responsible for the passage of the Marine Mammal Protection Act of 1972.

A visit to almost any Pacific or Atlantic shoreline is likely to yield a glimpse of these unique sea mammals, the pinnipeds: seals, sea lions, fur seals, and walruses. (I use the words *pinniped* and *seal* interchangeably throughout the book.) Whether surfing waves, perching on a pier, or lounging on mudflats, seals seem equally at ease on shore and in the ocean. In fact, one of the most remarkable attributes of this group of marine mammals is their ability to live in these two very different worlds. Even within each realm the various species of seals have adapted to an amazing diversity of environments—from the polar ice caps to the desertlike Galápagos Islands. In the water, seals are able to inhabit shallow coastal areas, landlocked freshwater lakes, or the open ocean, where they can dive to depths of over four thousand feet.

Current research is generating a great deal of exciting information on many species of pinnipeds, such as the polar seals, which have been nearly impossible to study in the past even on land because they inhabit inaccessible and remote locations. Although it has also been difficult to observe seals underwater in the past, the Time-Depth Recorders that have come into use relatively recently have produced fascinating information on diving patterns and foraging behavior in free-ranging pinnipeds. In addition, long-term research on tagged individuals is yielding a new understanding of how reproductive strategies and breeding success in species such as grey seals, elephant seals, and northern fur seals change during these animals' lifetimes. Such continuous longitudinal field studies are indeed rare in any species. The seal's habit of returning year after year to the rookery of its birth has helped make it possible for biologists to observe individuals over many years. Researchers are also currently discovering how sea lions "think": these intelligent animals can comprehend complex commands and, like dolphins and chimpanzees, can even learn a simplified sort of language. Examples discussed in the text are often drawn from the better-studied species for which more information is available. I admit, however, that northern elephant seals take up more than their fair share of space in the book because I have spent many years studying this species myself.

Although everyone knows what seals look like, their lives are in many ways still a mystery. In this book I review much of what is currently known of the thirty-three species of pinnipeds and, in so doing, hope to expand the reader's appreciation of these diverse creatures and their fascinating ways of adapting to an amphibious existence. This book provides a broad overview and synthesis of the biology of pinnipeds, emphasizing behavior, reproduction, ecology, and specialized adaptive trends for living in both marine and terrestrial environments. It is not an account of species or a comprehensive review of pinniped biology. Instead, it was written as a general introduction to the lives of seals for the student and interested general reader. Therefore it focuses on overall patterns and trends rather than detailed discussions of the life history and biology of each species. The book also reviews some topics of current interest among scientists and summarizes information in a way that is designed to be useful to specialists. I hope that anyone who is intrigued by the pinnipeds will find here something new and interesting to learn about these versatile and appealing marine mammals.

1

Adaptations for a Marine Existence

From the study of seals — those queer amphibious animals
living at the interface of land and sea — we gain new respect
for the far-limit possibilities of mammalian adaptation.
VICTOR SCHEFFER

As the marine mammalogist Victor Scheffer notes in the epigraph to this chapter, the pinnipeds are remarkable mammals whose amphibious nature has necessitated a wide range of adaptations to life both in the water and on land or ice. The ocean, although an extraordinarily vast, rich, and diversified environment, presents a special collection of challenges and problems for its air-breathing, warm-blooded inhabitants. Over millions of years the marine environment has dramatically shaped their evolution, affecting the anatomy, physiology, and behavior of both the pinnipeds and the other marine mammals — cetaceans (whales, porpoises, and dolphins), sirenians (manatees and dugongs), and sea otters.* In fact, some marine mammals, like porpoises, look so fishlike that the layperson sometimes even thinks of them as large fish.

The price that pinnipeds have paid for evolving such specialized marine adaptations has been a loss of the attributes that allowed them to live safely on land. Perhaps the most costly loss is their lack of efficient mobility on land; pinnipeds are at a disadvantage onshore, where they are especially vulnerable to land predators. Unlike other marine mammals, however, pinnipeds can live for long periods both in the ocean and on

* The scientific names of all species mentioned in the text are given in the Index.

land. Seals see and hear relatively well both in the air and underwater. They are able to tolerate the frigid temperatures of oceans depths as well as the blistering heat they are sometimes subjected to while basking on land. Seals stay at sea for weeks and sometimes months at a time, yet they also remain on land during the breeding and molting seasons, when many species do not eat or drink for weeks. Although cetaceans are in many ways the most highly specialized mammals for a marine existence, the plasticity of adaptation to vastly different environments among the pinnipeds is truly remarkable.

This chapter presents a synopsis of the adaptive trends that help the pinnipeds as well as the other three groups of marine mammals survive and reproduce in a marine environment. I include very brief discussions of marine adaptations in the cetaceans, sirenians, and sea otters for comparative purposes. Some adaptations associated in part with living in the ocean, such as those related to feeding and reproduction, are discussed in subsequent chapters. Chapter 1 focuses on the general morphological and physiological adaptations that are clearly associated with living in the ocean. The problems of determining what represents an adaptation exclusively to aquatic living are discussed by Estes (in press), who reviews adaptations for aquatic living in carnivores. I have organized the adaptive trends discussed in this chapter into five broad categories: (1) locomotion and aquatic propulsion, (2) temperature regulation, (3) breathing and diving, (4) osmotic adaptations, and (5) sensory adaptations.

Locomotion and Aquatic Propulsion

Laurence Irving (1966) wrote that "seals and whales excite our interest and perhaps a trace of envy, as we watch them move swiftly through the water by propulsion that appears so much less laborious than our own." Yet seals have had to overcome the difficulties involved in moving efficiently through a dense aquatic medium. Seawater is eight hundred times denser than air and far more difficult to move through. Because marine mammals have developed sleek, streamlined bodies shaped somewhat like torpedoes, they are able to move through water without a great expenditure of energy and without much "drag," or resistance (Fig. 1.1). Appendages such as limbs and feet have degenerated and become modified by degrees into fins and flippers that can propel the animals through the water with great force. Sex organs and mammary glands are retracted inside the body within slits or pockets beneath the skin. In the phocids

FIG. 1.1. Ringed seal, showing streamlined shape. *Photo by Brendan Kelly.*

("true" seals like the harbor seal) and odobenids (walruses), the testes are inguinal, or inside the body. Otariids (sea lions and fur seals), however, have externally visible, scrotal testes. Marine mammal ears are located internally or are reduced in size when external. The otariid ear is tiny, and phocid seals have no external ears at all—only a hole visible on each side of the head. Even eyelashes have vanished and eyelids have become modified as well. To compensate for this lack of eye protection, marine mammals constantly produce copious amounts of eye mucus.

The hair or fur has largely disappeared in walruses as well as cetaceans and sirenians, further enhancing the streamlining effect, although a close examination of the skin reveals a sparse covering of coarse hairs in some species. Phocid and otariid seals have retained a fur coat that is plainly visible, although the coats of sea lions and some phocids are more sparse and streamlined than those of most land mammals. The thick layer of blubber under the skin also promotes a smooth, streamlined body shape.

The fur seals, however, are well known for their thick fur and were even hunted specifically for their luxuriant pelage (Fig. 1.2). Surprisingly, a fur seal's fur may actually reduce drag in some cases by thickening the water boundary layer around the animal (Romanenko et al. 1973). In one experiment, the drag on a seal model was greater when the fur was shaved than when it was present (Mordvinov and Kurbatov 1972). In addition, because seals lack erector pili muscles in the skin, their fur lies flat.

FIG. 1.2. New Zealand fur seal. *Photo by Rob Mattlin.*

Erector pili are present in many mammals, including humans. Everyone has seen an angry cat, for instance, with its fur standing on end.

It has been suggested that the network of smooth-muscle bundles in the skin of pinnipeds may alter the configuration of the skin and thereby reduce drag (Sokolov 1982). Such conformation increases laminar flow over the body and produces important propelling vortices. When a marine mammal (or any object) moves through the water, the fluid particles near it are slowed and adhere to its body. The fluid layers farther from the body are inhibited less and less until a region of steady flow is attained. Laminar flow occurs when there is little drag on the fluid particles closest to the body and the outer fluid layers glide over one another, allowing the animal to slip easily through the water. In the streamlined dolphins, for instance, the flow is more highly laminar than it is even for seals. When a human swims through the water, however, drag is high and the fluid particles near the body are slowed so much that they cannot glide within the outer layers. The speed of the fluid particles is referred to as the critical velocity when laminar flow becomes turbulent flow, in which eddies form and the body's motion greatly slows.

Although pinnipeds are well adapted for moving efficiently through the water, they tend to swim slower than cetaceans, cruising at speeds of perhaps 5–15 knots. A number of dolphins are able to swim at a top velocity of 20 knots for many hours (Fig. 1.3), whereas killer whales can sprint through the water at speeds of up to 60 knots. The need to accelerate rapidly to high speeds is important to both odontocete cetaceans and many pinnipeds, which forage on swift-moving schools of fish, on penguins, or on other marine mammals. Pinniped bodies are not as streamlined or stiff as the fusiform and tapered bodies of cetaceans, but seals, lacking the clavicle, or shoulder bone, have greater flexibility of movement. The pinniped foreflipper can dexterously swivel for scratching most parts of the body as well as for maneuvering underwater (Fig. 1.4). The body of a sea lion is so pliable, in fact, that it can practically bend over backward and "touch its toes" (Fig. 1.5). Cetaceans, however, because their cervical neck vertebrae have become compressed and fused together, are stiffly streamlined and move with great speed. Since pinnipeds lack this cervical compression, their necks are flexible, even snakelike in otariids, making them versatile predators (Fig. 1.6).

The otariid seal uses its long foreflippers, or pectoral fins, for balance and propulsion, so that it moves through the water, its long neck extended, almost as a bird flies through the air. Its rear flippers, which act as

FIG. 1.3. Bottlenose dolphin leaping out of water. Dolphins, which have a stiff, fusiform shape, swim at high speeds. *Photo by Randy Wells.*

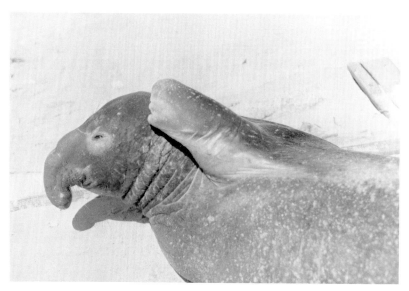

FIG. 1.4. Northern elephant seal scratching itself with its foreflippers; note five digits and flexibility of foreflipper. *Photo by author.*

FIG. 1.5. Galápagos sea lion bending backward; note extremely pliable neck and torso. *Photo by Steve Webster.*

Fig. 1.6. Snakelike neck and head of a California sea lion, shown swimming underwater. *Photo by Doc White.*

stabilizers, are not used much in swimming. Fur seals swim at speeds of about 16 knots, while sea lions can reach bursts of speed of 25–30 knots (C. Ray 1963). Sea lions sometimes "porpoise," or leap out of the water, while playing together or when traveling at high speeds; from a distance they look somewhat like a school of dolphins (Fig. 1.7). Harp seals also leap out of the water while swimming long distances during migrations.

FIG. 1.7. California sea lions "porpoising." *Photo by Doc White.*

This method of swimming may allow them to swim faster than if they moved solidly through the water, since there is much less drag in air. Moreover, young sea lions are often seen "riding waves" along the shoreline, like human bodysurfers, in a form of apparent play. Sea lions may even "bow ride" the wake of large whales, substantially cutting down on the energy cost of locomotion.

Phocids swim by moving their rear flippers and lower body in a lateral, or side-to-side, sculling motion. The hindflippers may either be held together and spread or used alternately in a series of strokes. Cresswell (1985) presents evidence that immature gray seals may move both hindflippers together rather than separately. The foreflippers are generally held immobile against the body, although they may be used for turning and maneuvering at slow speeds. T. M. Williams (1985) has measured the cost of aquatic locomotion in a harbor seal and found it to be low relative to that of similar movement in aquatic mustelids such as mink and sea otters. Phocids may travel at speeds of 12–20 knots, but they generally cruise at 5 knots or less. The neck of phocids, and to a lesser degree walruses, is telescoped or compressed, and the body is shortened and compacted during swimming to produce the drag-reducing torpedo shape characteristic of cetaceans (Fig. 1.8).

The odobenids are the slowest-swimming pinnipeds; they generally cruise at speeds of 5 knots or less. Because walruses feed on sessile, bottom-dwelling shellfish, they do not seem to require the speed and agility of the seals and sea lions, which prey on fish and squid. When they are pursued by hunters or predators such as polar bears, however, they can reach speeds of 16 knots in short bursts (King 1983). Walruses swim somewhat like phocids, using their hindflippers in alternating strokes. They use their foreflippers in tandem for maneuvering as well as for stabilizing themselves.

The dense-boned and sluggish sirenians usually swim slowly in comparison with most pinnipeds, at a speed of less than 5 knots (Anderson 1981), although dugongs (Marsh et al. 1981) and manatees (Hartman 1979) can reach 12 and 14 knots, respectively, in short bursts when being pursued by a boat. Despite their large size and bulky appearance, sirenians are surprisingly agile underwater. While the more streamlined dugongs possess flukes similar to those of cetaceans, a manatee's broad

FIG. 1.8. Harbor seal swimming underwater. *Photo by Steve Webster.*

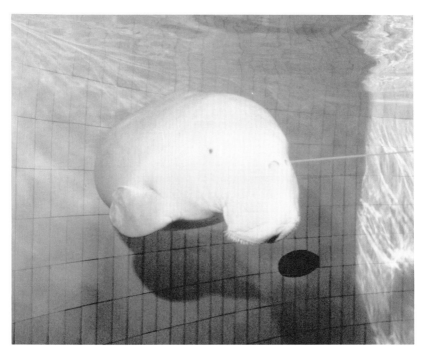

Fig. 1.9. Dugong. *Photo by Galen Rathbun, Sirenia Project, U.S. Fish and Wildlife Service.*

Fig. 1.10. Manatee mother and calf underwater. *Photo by Galen Rathbun, Sirenia Project, U.S. Fish and Wildlife Service.*

tail is flattened into a rounded horizontal paddle (Figs. 1.9, 1.10). Dugongs and manatees use powerful sweeps of their flukes or broad tail paddle to move through the water. The shortened and rounded sirenian foreflippers, although not used much for propulsion, may be used for balance and stability and to direct marine plants toward the animal's mouth.

Sea otters move slowly through the water in relation to the pinnipeds, typically cruising at speeds of 2–3 knots, although they are able to reach swimming velocities of 9 kilometers per hour, that is, 5 knots (Kenyon 1969; Riedman and Estes MS). Because sea otters feed mainly on sessile or very slow-moving invertebrates or on bottom fish, they have less need than the pinnipeds to swim at high speed. Sea otters usually swim on their backs, propelling themselves with their webbed hindflippers, which they move with a bicyclist's pumping action. For greater speed, otters swim on their stomachs, using undulating motions of their torsos and hindflippers (Fig. 1.11).

Because seals, sea lions, and walruses spend part of their lives on land, they have retained the ability to move about out of water, although less efficiently than terrestrial mammals. All otariids have hindflippers that can be turned under the body for "walking" on land. An otariid swings its

FIG. 1.11. Resting sea otter keeping hindflippers above the surface and dry. *Photo by Dick Buchich.*

heavy neck and head from side to side, creating momentum, while alternately moving its foreflippers and hindflippers. In fact, sea lions and fur seals are the most agile of the pinnipeds on land. On Año Nuevo Island, many California sea lions have made their way into an old two-story house (Fig. 1.12), climbing up the stairs and into every room. Some sea lions have even been found lounging in the bathtub on the second floor. One California sea lion with an apparently misguided sense of direction twice made its way up city streets in San Francisco, actually visiting the men's restroom (fortunately, it was a male sea lion) of the same building each time (Vanderbrook, pers. comm.).

Phocid seals cannot turn their hindflippers under their bodies and are somewhat less agile on shore than otariids. On land, phocids lunge, bounce, and wriggle along, using their foreflippers for balance. Some species, such as grey seals, also use their foreflippers to help pull them forward. Walruses move like otariids, but they keep their bodies closer to the ground, progressing by lunging forward, as phocids do. Besides the pinnipeds, the only other marine mammals that can haul out on land are the sea otter and the marine otter, which "walk" differently from any pinniped. Sea otters move by arching their backs and quickly bounding along, somewhat like loping rabbits.

One would expect that certain forms of ice or snow would be a much easier substrate for seals to move about on than sandy or rocky areas of land. Terrestrial locomotion on ice is in fact easier for many of the pagophilic, "ice-loving," phocids, which sled across the ice at high speeds. Antarctic penguins similarly "toboggan" across the ice on their bellies when in a hurry or traversing long distances. (The problems ice presents for phocids, however, are discussed in chapter 3.) The Antarctic crabeater seal is the fastest sprinter on ice, reaching reported speeds of up to 25 kilometers per hour when chased (O'Gorman 1963). At this speed, the seal could easily overtake a man running at top speed. As the crabeater seal sprints, it swings its head and hind body from side to side while pulling alternately with its foreflippers. The Arctic ribbon seal is also quite a good sledder on ice, attaining velocities comparable to those of a man sprinting at high speed (Burns 1981b).

Temperature Regulation

Everyone knows that ocean water is cold, rapidly drawing heat out of the body. More precisely, the thermal conductivity of water is 25 times that of

FIG. 1.12. Sea lions in house on Año Nuevo Island. *Photo by Frans Lanting.*

air; that is, water absorbs heat from the body of a mammal about 25 times more rapidly than air does. A person immersed even in tropically warm waters of 80 degrees eventually becomes chilled, while the same person remains comfortable indefinitely in 80-degree air. Seals inhabiting near-freezing polar waters must contend with the ever-present cold, as do most deep-diving pinnipeds and other marine mammals that encounter frigid vertical thermoclines, or abrupt changes in temperature. Heat loss is therefore a problem for pinnipeds and most other marine mammals. A seal must maintain a constant core body temperature of close to 100 degrees F (38 degrees C) in ocean water that may be only 30–40 degrees F (0–5 degrees C). Conversely, overheating may also pose difficulties both for seals hauled out on land during hot days and for large whales engaged in prolonged or strenuous activity, especially during seasonal visits to warm waters.

The following adaptations allow pinnipeds and other marine mammals to regulate their temperature and to conserve heat: larged-sized and compact body; blubber or dense fur; a high rate of metabolism, or internal heat production, in some cases; and countercurrent heat exchange systems, all of which are discussed later in this chapter. Irving (1969) and Whittow (1987) review such adaptations in all marine mammals, as do Blix and Steen (1979) for pinniped pups in cold polar environments. Some of the same adaptations that help pinnipeds retain heat in the water, such as thick blubber and dense fur, may also promote overheating on land, again showing that the pinnipeds must make continual adjustments because of their amphibious way of life. More is known at the present time about thermoregulation in pinnipeds than in cetaceans, especially the large whales.

SIZE AND SHAPE OF BODY

All marine mammals are relatively large sized in comparison with terrestrial mammals. Even the smallest seal, the ringed seal, weighs 50 kilograms, and the largest pinnipeds, such as elephant seals and walruses, weigh up to 4 tons. As Le Boeuf (1978) points out, many species of land carnivores are comparatively small sized. Least weasels, for instance, weigh as little as 25 grams. Large size helps to conserve warmth; large-bodied mammals chill less quickly than small mammals because the surface area where heat is lost is minimal in comparison with the bulk of the entire animal. As an animal grows, the proportion of surface area to bulk comparatively decreases. And, as the immensity of some mysticete whales illustrates, large size does not pose a gravitational problem in the

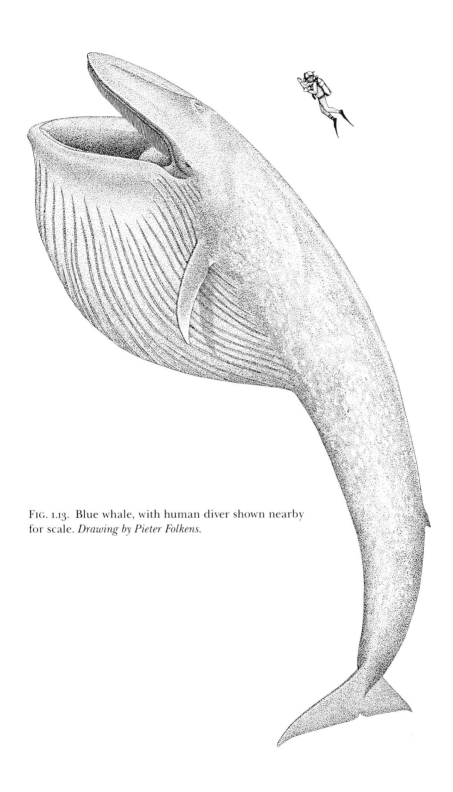

FIG. 1.13. Blue whale, with human diver shown nearby for scale. *Drawing by Pieter Folkens.*

ocean, where a heavy body is easily supported by the buoyancy of salt water. Blue whales, in fact, are the largest animals in the world, reaching lengths of up to 100 feet (30 meters) and weights of nearly 200 tons (Fig. 1.13). In comparison, the largest terrestrial mammal—the African elephant—weighs a mere 2–8 tons. In addition to large size, pinnipeds and other marine mammals have a compact body shape with no long limbs, other than fins and flippers, from which heat can dissipate rapidly.

BLUBBER AND FUR

The dense fur or blubber—the subcutaneous layer of fat beneath the skin—covering marine mammals minimizes heat loss in the water. Fat not only insulates the animals from cold but also serves as an energy reservoir, providing nourishment during long periods of fasting. Some pinniped males do not eat or drink for nearly three months during the breeding season. A fasting migratory mysticete whale theoretically could survive on its fat stores for as long as 4–6 months. During certain times of the year—at the beginning of the breeding season, for example—up to 50 percent of the total body weight of a whale or seal may be made up of blubber. In fact, in walruses and elephant seals the layer of blubber can reach a thickness of several inches, and some of the larger whales are insulated by a 2-foot layer of fat.

In contrast, most pinniped pups are born with very little blubber and must quickly develop a covering of fatty insulation to survive. In Weddell seals, for instance, the newborn's hair, or lanugo, is several times longer than the adult's fur, which helps to conserve warmth (Elsner et al. 1977). Harp seal pups are born onto the cold Arctic ice with virtually no protective fat. Sunlight, among other things, may help to keep these pups warm: their white coat soaks up solar radiation especially well. Sunlight is reflected efficiently from the individual hairs onto the surface of the skin, where it is absorbed. The pup's coat traps this warmth between hair and skin, producing a "greenhouse" effect that prevents heat loss (T. Øritsland 1970b; N. A. Øritsland 1971). In addition, although the newborns lack a layer of blubber and have fur of poor insulative quality, they do possess a layer of brown adipose, or fatty, tissue (also known as BAT). This specialized brown fat, also found in human babies and hibernating mammals, helps keep the pups warm by means of "nonshivering thermogenesis" (Grav et al. 1974; Grav and Blix 1975; Blix et al. 1975). Instead of shivering to produce heat, seals metabolize fat during this process almost exclusively to produce heat instead of storable energy (or ATP); that is, the fat is not oxidized via normal cellular respiration. Harp seal adults also have deposits of brown fat around veins transporting cool blood from the flippers back to the heart. Just as this brown fat helps newborn pups to generate heat, it can produce heat to warm returning blood.

Some newborn pinnipeds also seem to cope with heat loss by maintaining a high level of heat production (Blix and Steen 1979; Whittow 1987). For instance, although newborn northern fur seals are nearly blubberless, their high resting metabolic rate, along with their thick fur, allows them to keep warm (Blix et al. 1979). Newborn harbor seals swimming for several hours in water of only 41 degrees F (5 degrees C) were able to avoid hypothermia despite their lack of blubber or fur. Their heat production, normally 2.6 times that of an adult terrestrial mammal, increased even more in the frigid water (K. Miller and Irving 1975).

While phocid seals, walruses, and sea lions conserve heat by means of a heavy layer of blubber, fur seals and sea otters have a thick coat of fur to protect them against the cold. (Although fur seals are also insulated by a thick layer of blubber, sea otters have a comparatively thin layer.) This fur consists of two layers: the flattened, protective outer guard hairs and the extremely fine, thick underfur, which remains dry by trapping tiny air bubbles in its oily fibers, insulating the skin from cold water. Sea otter underfur contains up to an amazing one million hairs per square inch, although the number varies from 170,364 to 1,062,070 hairs per square inch, depending on the part of the body examined (T. D. Williams et al., in press). Fur seal underfur, less dense than that of sea otters, contains about 300,000 hairs per square inch. (In comparison, the human head is covered with an average of only about 100,000 hairs.) Both fur seals and sea otters are especially vulnerable to environmental contaminants such as oil; they chill and soon die if part of their fur becomes dirty or contaminated (Kooyman and Costa 1979; Geraci and St. Aubin 1980). Harp seal fur, which also has some insulative value, consists of primary hairs as well as the secondary hairs that retain a layer of warm water next to the skin (Tarasoff et al. 1972; Frisch et al. 1974).

METABOLIC ADJUSTMENTS

Many biologists believe that under natural conditions pinnipeds have a high rate of metabolism (relative to that of similarly sized land mammals), which helps them to generate heat in cold ocean water or on polar ice by burning calories at a rapid rate (e.g., Irving and Hart 1957; Andersen 1966; Irving 1969, 1973; Ridgway 1972; Iversen and Krog 1973). Yet recent research has led others to disagree; this research suggests that pinniped metabolic rates are not all that different from those of land mammals.

Several studies of pinniped metabolism (as measured by oxygen consumption) have shown that pinnipeds' resting metabolic rates are about 1.5 to 3 times greater than those of land mammals of similar sizes (Kooyman 1981c). Species whose resting metabolic rates have measured higher than those of similarly sized terrestrial mammals include captive harbor seals, 1.7–2.2 times (Hart and Irving 1959; K. Miller et al. 1976); Weddell seals, 1.8–2.2 times (Kooyman et al. 1973; Elsner et al. 1977; Kooyman 1981d); noncaptive northern fur seals, 3.2 times (Costa and Gentry 1986); and noncaptive northern elephant seals, 2.4 times (Costa et al. 1986).

Yet Lavigne et al. (1986) believe that the metabolic rates of pinnipeds are not significantly different from those of land mammals and suggest that previous studies did not always make comparisons under standardized conditions. When resting metabolism was measured by oxygen consumption rates in captive harp seals, grey seals, and ringed seals (H. S. Innes and Ronald 1985) and in captive hooded seals (Rea 1987), their metabolic rates were found to be only slightly higher (1.1–1.2 times) than those typical of a terrestrial mammal of similar size. In seals the resting metabolic rates of juveniles were higher than those of adults, but then the same difference exists in land mammals. Innes and Ronald (1985) concluded that pinnipeds do not need an elevated metabolic rate for thermoregulation in the ocean. Their findings further underscore the sophistication of pinniped heat conserving adaptations and indicate the need for further research.

But what about seals that spend weeks at a time fasting on land? At this time, it pays to actually reduce metabolic costs. For example, northern elephant seal females tend to sleep a great deal and move around very little during the breeding season (J. Reiter, pers. comm.), and fasting weaned elephant seal pups periodically experience apnea (cessation of breathing) during sleep, thus reducing their metabolic rates by nearly 50 percent (Huntley and Costa 1983; Huntley 1984). We now know that even human metabolic rates decline during and following periods of prolonged fasting, paradoxically making it more difficult for us to lose weight and maintain weight loss after resuming normal eating patterns.

COUNTERCURRENT HEAT EXCHANGE SYSTEM

Overheating, rather than heat retention, can be a problem for pinnipeds hauled out on land during very warm days — especially when fat stores are seasonally high — as well as for tropical cetaceans and the larger whales when they are highly active. A marine mammal is unable to release heat by sweating or panting (the northern fur seal, which sometimes pants, is an exception). To dissipate heat and cool their bodies, as well as to conserve warmth when necessary, pinnipeds and other marine mammals have developed an internal thermostat known as a counter-current heat exchange blood vessel system. This versatile heat exchange network is made up of intertwining arteries and veins in the flippers, flukes, and dorsal fins, where there is little blubber or fur. The veins that carry blood from the flippers back to the heart initially lie beside arteries that circulate blood back to the flippers. The arterial blood passes on warmth to the cooler venous blood, so that the seal retains heat. In pinnipeds, this network of arteries and veins, or arteriovenous anastomosis, is also found in the superficial layers of skin throughout the body (Tarasoff and Fisher 1970; Irving et al. 1962; Bryden 1978; Molyneux and Bryden 1978). The capillary networks become highly elaborated, and in some cases arterioles may join venules by direct connections via one or more arteriovenous anastomoses. These anastomosing vascular beds allow blood to flood the skin much more rapidly than ordinary capillary networks. Therefore, more rapid heat loss is possible when the animal needs to dissipate heat, but the seal can also shunt the blood away from the skin to maintain its core body temperature.

If a seal is heat stressed, it is able to pump hot blood into the flipper, which it may wave about in the air or dip into a tidepool to cool the circulating blood. The skin of many tropical cetaceans is pink, in fact, because of numerous tiny capillary vessels close to the skin's surface. When walruses hauled out on land become warm, their skin also becomes suffused with blood and turns a rosy pink color. Seals in cold environments still need to supply blood to their flippers, but when that blood is shunted through the anastomosing beds, it conducts heat to the veins to maintain a normal inner body temperature. Phocids resting on ice may be quite comfortable maintaining a skin temperature of only 1 degree C—just above freezing—while their internal body temperature remains normal. These highly elaborated vascular networks, also present in cetaceans and sirenians, function in countercurrent heat exchange in addition to serving as blood reservoirs during deep dives. Such *retia miribilia*, from the Latin meaning "miraculous nets," are found in a wide variety of other animals, including sharks and many other fish, certain wading birds, sloths, anteaters, and lemurs. Walruses have extensive retia in their lower foreflippers and hind limbs. Some of the most important and better-studied retia are found in the mammalian kidney and fish swim bladder, which helps the fish maintain a neutral buoyancy.

FIG. 1.14. Northern elephant seal flipping sand on its back. *Photo by W. E. Townsend, Jr.*

THERMOREGULATORY BEHAVIOR

Pinnipeds and other marine animals employ behavioral as well as physiological means of thermoregulating. Pinnipeds hauled out on land may slowly wave their flippers in the air or immerse them in water or damp sand to relieve heat stress. Or they may flip sand onto their backs with their foreflippers to gain some protection against hot temperatures and direct sunlight. Sand flipping is common among species such as elephant seals, New Zealand sea lions, and southern sea lions (Fig. 1.14). In the New Zealand sea lion, an increased rate of sand flipping occurs among pregnant females and among both males and females at higher temperatures (Marlow 1975). During the hot dry breeding season of the southern sea lion, the animals cool themselves by constantly shifting position, waving their foreflippers, flipping pebbles onto their backs, burying their rear flippers and foreflippers in wet sand, and resting in tidepools when these are available (Campagna 1987; Campagna and Le Boeuf 1988). The northern fur seal, as already mentioned, actually pants like a dog to dissipate heat (Gentry 1981a).

Fur seals and sea lions inhabiting tropical or dry, hot areas, such as the Guadalupe and Galápagos fur seals, may seek shelter under large boulders or inside caves or may cool off frequently in nearby tidepools and shallow surf. Even otariids that live in more temperate areas, such as the Steller sea lion and many populations of California sea lion, enter the water or rest in tidepool areas when the air temperature rises. Studies of thermoregulatory behavior in the California sea lion have shown that this species cannot effectively regulate body temperature on extremely warm days; to cool off, these animals must enter the sea and immerse themselves (see, e.g., Whittow et al. 1971; Odell 1974).

The monk seal is the only phocid seal that inhabits tropical areas. As might be expected, overheating sometimes poses a problem for this seal when it hauls out on land. Like sea lions, the Hawaiian monk seal cools off on warm days by resting near the water in damp sand, often digging holes to expose the cool deeper layers of earth (Fig. 1.15). Monk seals rest in the dry upper areas of beaches only on cool or windy days. While resting, they often position themselves so as to expose their pale ventral fur, which has a significantly lower temperature than the darker dorsal fur (Whittow 1978).

When temperatures are cool, a number of sea lions, including New Zealand sea lions (Walker and Ling 1981b), southern sea lions (Vaz-Ferreira and Palerm 1961), and Steller sea lions (Gentry 1973), lie on their bellies with their flippers tucked under their bodies. During windy or rainy weather, Australian sea lions seek shelter inland among the sand

FIG. 1.15. Hawaiian monk seal cooling off in a sand wallow. *Photo by E. Kridler.*

dunes and coastal vegetation whereas New Zealand sea lions enter the ocean (Walker and Ling 1981b). In the Arctic pack ice on especially cold days and at night, harp seal pups search for refuge from wind and rain under overhanging ice hummocks (Lavigne 1982b).

Sea lions and elephant seals often gather in tight groups to conserve heat; most fur seals, however, do not huddle together. Walruses may also cluster close together during cold or stormy weather, particularly during the pupping season. The young calves, which have little blubber, are probably more vulnerable to heat loss than the adult animals. But walruses also haul out in tight groups during warm weather, dispersing only when temperatures become very hot, so such aggregations have other functions as well (Taggart and Zabel 1987).

To conserve heat when resting on the water's surface, fur seals and sea lions extend both hindflippers and one foreflipper above the surface. Because of the arc formed by the flippers, sealers used to call this the jug-handle position. Today, pinnipeds resting in this characteristic position are said to be jugging (Fig. 1.16). Sea otters also conserve heat while floating in the water by stiffly extending their less densely furred forepaws as well as their hind feet above the water in a rigid position. In fact, otters

FIG. 1.16. California sea lions resting on the surface in a "jug-handle" position, with both hindflippers and one foreflipper extended above the surface. *Photo by W. E. Townsend, Jr.*

are often so reluctant to wet their warm, dry paws that when disturbed, they carefully swim away or completely roll in the water without wetting their extended paws and feet.

Breathing and Diving

Air-breathing mammals like pinnipeds that live in the ocean must be able to take in an adequate amount of oxygen while swimming or diving for food. The various species of marine mammals, which find most of the food they eat beneath the ocean's surface, have evolved to differing degrees the adaptations associated with deep diving. Because sea otters forage in shallow waters near shore and can capture their prey quickly, their dives usually last only a minute or so. Similarly, manatees and dugongs do not need to dive deeply; they feed in shallow coastal or riverine waters on aquatic plants that grow only near the surface where sufficient light can penetrate. Most otariid seals and a number of phocids make brief dives to relatively shallow depths of less than 150–200 meters. A few pinnipeds and many cetaceans, however, make long, deep dives to obtain their prey. For instance, sperm whales remain underwater for up to 90 minutes and have become entangled in deep-sea cables at depths of at least 1,000 meters (about 3,000 feet). Pilot whales and white whales are able to dive to 600 or more meters. Bottlenose whales may be able to stay submerged for up to two hours.

Among the pinnipeds, phocid seals seem to be especially well adapted for deep diving. The Weddell seal can dive for up to 73 minutes to a maximum depth of 600 meters (about 2,000 feet) (Kooyman 1981c). Ridgway (1985), comparing diving patterns in the Weddell seal and bottlenose dolphin, notes that although the dolphin holds its breath only 10–15 percent as long as the seal, dolphins dive much more rapidly. For instance, in one hour two dolphins made as many as 25 dives to 100 meters. The seals, in contrast, did not make as many dives in such rapid succession.

Perhaps the most remarkable diving behavior observed in a pinniped so far has been documented in the northern elephant seal (Le Boeuf et al. 1985; 1986a; 1988; in press; Le Boeuf, pers. comm.). Elephant seal adult females are able to dive as deep as 1,250 meters (4,100 feet) and can remain submerged for up to 62 minutes. Despite these incredibly deep, long, and almost continuous dives, the seals spend surprisingly little time resting on the surface between dives—less than 3 minutes (Fig. 1.17).

A number of marine mammals, in fact, can dive to 600 meters or more,

FIG. 1.17. Northern elephant seal blowing bubbles out of its nose during a dive. *Photo by Doc White.*

and most are able to descend to 200 meters (Ridgway 1985). These superb divers have evolved the ability to conserve oxygen while holding their breath for prolonged periods and can withstand the pressure of thousands of tons exerted by the surrounding seawater. Table 1 summarizes information on dives for various pinniped species. Diving behavior in pinnipeds is discussed in more detail in chapter 5.

PRESSURE

Although pinnipeds and other marine mammals are practically weightless at the water's surface, their bodies are subjected to enormous pressure during deep dives. Since there is one additional atmosphere for each 33 feet descended, at a depth of 600 meters the underwater pressure exerted on a seal is equivalent to that of about 60 atmospheres, that is, sixty times the pressure at the surface. Deep-diving mammals, with their lungs and other hollow air spaces, like the inner ear, have to contend with such tremendous underwater pressures. Unlike humans, however, these mammals do not suffer from the bends, a painful and sometimes fatal disorder that can injure or kill a human scuba diver who surfaces too quickly. Underwater pressure forces nitrogen from a diver's lung air into the blood. When a diver surfaces too quickly, reduced pressure causes nitrogen to bubble out of the blood and collect in joints and nerve tissue, blocking circulation and causing pain and paralysis.

Yet most seals and cetaceans can ascend quickly following deep dives. Why are they not subject to the bends? Marine mammals, unlike scuba divers, dive with only a lungful of air and hold their breath for the duration of the dive. Pinnipeds, in fact, exhale before diving, so that their lungs are only partially filled and they are able to submerge more easily. Nitrogen cannot accumulate in their blood because the amount of air in their lungs is reduced and they do not continuously ventilate their lungs, (humans, of course, carry with them a constant air supply in a pressurized scuba tank). Moreover, unlike human divers, who sometimes feel painful pressure in air-filled sinus passageways, pinnipeds have air spaces only in their lungs and ears. In addition, seals and whales have flexible ribs that allow the lungs and alveoli (tiny air pockets within the lungs) to collapse, sending air into rigid, nonabsorptive pockets in the upper body that are less susceptible to the effects of pressure. According to Ridgway (1985), the collapse of lung alveoli takes place at a comparatively shallow depth of 25–50 meters in seals and at a depth of about 70 meters in dolphins, which dive with their lungs filled. The somewhat flattened shape of a seal's heart facilitates the collapse of the lungs under pressure. In pinnipeds, the long, elastic trachea can also collapse in response to increasing pressure. And during dives the middle-ear sinus of pinnipeds (and cetaceans) swells with a network of thick tissue and venous channels so that the air space in the ear is filled.

Scientists have not yet solved the puzzle of high-pressure nervous syndrome (HPNS) and deep-diving seals. Experimental studies have shown that the nervous tissue of some terrestrial mammals stops functioning properly at certain depths. Under high pressure, the nerves actually become hypersensitive, firing too readily and inappropriately (see, e.g., Rostain et al. 1986; Jannasch et al. 1987). In addition, certain enzymes may cease to work under high hydrostatic pressure, at least in some deep-water fishes (Siebenaller and Somero 1978). Yet a number of pinnipeds dive to depths beyond 25 atmospheres, and northern elephant seals are able to descend to nearly 4 times this depth. How are they able to do it without suffering from HPNS? Biologists hope to answer this intriguing question in the near future.

OXYGEN CONSERVATION

A person would pass out within three minutes of being deprived of oxygen; how can a seal or whale hold its breath for such long periods without asphyxiating, or suffocating? First, a diving pinniped or cetacean

TABLE 1
Depth and Duration of Dives for Selected Pinniped Species

SPECIES	DEPTH (meters)		DURATION (minutes)		CONDITIONS*	REFERENCES
	Mean	Maximum	Mean	Maximum		
Hawaiian monk seal	10–40	121–175			TDR	Kooyman et al. 1983
Northern elephant seal	450	1,250	20	62	TDR	Le Boeuf et al. 1985, 1986a, 1988, in press; Le Boeuf, pers. comm.
Southern elephant seal	350–650+	>30–175			Noncaptive	Matthews 1952
Weddell seal	130	600	8–15	73	TDR	Kooyman 1966, 1975, 1981b,c
Hooded seal	75				Noncaptive (pup)	Scholander 1940
				18	?	Lavigne & Kovacs 1988
Bearded seal		200		20	Experimental	Scholander 1940
		275			?	Lavigne & Kovacs 1988
Harp seal					Fish. gear	Nansen 1925
			2.5	10	Noncaptive	Øritsland & Ronald 1975
Baikal seal				43	Noncaptive	Pastukhov 1969
		90		68	Experimental	Pastukhov 1969
Ringed seal					Stomach con.	McLaren 1958
				18	Experimental	Ferren & Elsner 1979
				21	Fish. gear	Freuchen 1935
Harbor seal		206		30	Experimental	Harrison & Kooyman 1968; Kooyman et al. 1972
	~3		~3.2		Noncaptive	Carl 1964

Species					Condition*	Reference
Grey seal		70		16	Telemetry	Thompson et al. 1987
		146			Fish. gear	Collett 1881
		225			Experimental	Scronce & Ridgway, unpubl. data
				23	?	Lavigne & Kovacs 1988
California sea lion		274			TDR	Feldkamp et al. 1983
Galápagos sea lion	37	186	<2.0	6.0	TDR	Kooyman & Trillmich 1986b
Steller sea lion		180			Stomach con.	Fiscus & Baines 1966
		183			Fish. gear	Kenyon 1952
Australian sea lion		37			Fish. gear (pup)	Walker & Ling 1981a
New Zealand sea lion	200	460		12	TDR	Gentry, Roberts, & Cawthorn 1987
Northern fur seal	68	207	2.2	7.6	TDR	Gentry et al. 1986
Galápagos fur seal	26	115	<2.0	5.0	TDR	Kooyman & Trillmich 1986a
South American fur seal	34	170	2.5	7.1	TDR	Trillmich et al. 1986
Antarctic fur seal	30	101	<2.0	4.9	TDR	Kooyman et al. 1986
South African fur seal	45	204	2.1	7.5	TDR	Kooyman & Gentry 1986
Australian fur seal		200			Fish. gear	Shaughnessy & Warneke 1987
Walrus		80		10	Noncaptive	Vibe 1950
		91			Stomach con.	Buckley 1958

NOTE: Tilde (~) indicates approximation.

* Conditions are indicated as follows: TDR (Time-Depth Recorder attached to noncaptive animal); Noncaptive (observations made on free-ranging pinnipeds); Experimental (captive conditions); Fish. gear (animal caught or entangled in fishing gear); Stomach con. (estimation of depth based on species of prey consumed).

† Represents the modal or preferred dive depth for all females (95% of all dives were within this range).

is able to reduce oxygen consumption by shunting blood to only essential organs and tissues: the brain, the heart, and a few other vital organs. Body temperature and metabolic rate decrease. The heartbeat also slows considerably. This reduction in heartbeat, to as little as one-tenth of its normal rate, is called bradycardia. In fact, the heartbeat slows during diving in all air-breathing vertebrates, including many species of diving ducks, cormorants, penguins, snakes, alligators, marine iguanas, and even humans. The marine iguana's heart may actually stop beating for over an hour (Dawson et al. 1977).

Bradycardia and the channeling of blood to critical organs save oxygen. Since muscles are deprived of their usual supply of blood and oxygen, they compensate by containing large amounts of a special compound called myoglobin, an iron-bonding pigment related to hemoglobin that stores oxygen. Myoglobin helps the seal to tolerate the large accumulation of carbon dioxide in its bloodstream that prompts breathing during periods of prolonged breath-holding. Myoglobin therefore helps the animal to conserve oxygen during high-speed movement or deep dives. In fact, cetacean and pinniped muscles appear nearly black when they are exposed to the air. This dark coloration is due to the extremely dense concentrations of deep red myoglobin in the muscle tissues.

A seal's muscles are also able to handle the high amount of lactic acid that accumulates in its system during periods of heavy or prolonged activity (such as diving), which causes muscle exhaustion in land mammals. The muscles of marine mammals can function with insufficient oxygen for many hours. Most of the time, the muscles function aerobically (with oxygen), even though the seal may be diving and exercising heavily. They function anaerobically when oxygen supplies are depleted by prolonged submergence. When a seal surfaces from a long dive, it takes several rapid and deep breaths until the body's oxygen supply is replenished; at the same time, at least in harbor seals, a pronounced tachycardia (extremely rapid heartbeat) occurs (Fedak et al. 1985). Seals and whales even have an automatic mechanism that cuts off breathing if they are knocked unconscious, so that their lungs will not fill with water.

In pinnipeds, the oxygen-transporting circulatory system is very large. A seal's total blood volume in relation to its body weight is 1.5–2 times greater than that of other mammals. According to Scheffer (1976), "In a full-grown walrus the forked veins which drain the lower body are so enormous that a man can draw them over his legs like a pair of pants!" Wickham et al. (1985) and others (e.g., Lenfant et al. 1970) have identified a number of characteristics of phocid seal blood that appear to help the seal to cope with the hypoxia, or oxygen deficiency, of long dives. Phocid blood has a greater capacity than the blood of terrestrial mammals to store oxygen because phocids have fewer and larger red blood cells with a higher concentration of hemoglobin, which stores and carries oxygen. And despite the high viscosity, or thickness, of phocid blood, the rate of blood flow is high.

SLEEPING IN WATER

In addition to foraging underwater, pinnipeds often must sleep in the water. Fur seals, for example, spend months at a time at sea, where they sleep in open water. And although harbor seals prefer to sleep on land, they are often forced to sleep in the water during high tides, when hauling grounds are scarce or unavailable. But can seals actually sleep underwater? Yes—although when sleeping underwater most seals wake up frequently and regularly to surface and breathe. Northern elephant seals, however, may possibly sleep hundreds of meters underwater; this possibility is discussed in chapter 5.

REM (rapid-eye-movement) sleep is the lightest stage of sleep, in which brain-wave activity is highest and dreaming takes place. It does not occur in all animals, and it does not seem to occur in pinnipeds during underwater sleep. But it does take place when pinnipeds sleep at the surface of the water and on land, despite frequent awakenings (Ridgway et al. 1975b; Almon and Renouf 1985). Sleep in grey seals (and possibly other pinnipeds) differs from that in land mammals in that REM sleep occurs before slow-wave sleep, and rapid and regular heart rate and respiration are maintained during REM sleep (Ridgway et al. 1975b).

Harbor seals sleeping on the surface of the water often assume a posture known as bottling: most of the seal's body remains submerged, but the animal's face pokes above the surface like a snorkel, allowing the animal to breathe regularly while sleeping or resting. Elephant seals sometimes rest in the water in a similar manner. The long proboscis of the adult male is conspicuous on the water's surface, and the seal can actually be heard snorting and gurgling with each expiration. Respiratory modifications for easier breathing are most pronounced in the cetaceans. The nostrils, or blowholes, of whales and dolphins have gradually migrated to the top of the head, allowing the animals to breathe more easily while swimming or floating at the surface—somewhat as a snorkel helps a skin diver to breathe (Fig. 1.18).

Fɪɢ. 1.18. Grey whale blowing at the surface. *Photo by Doc White.*

Water Balance:
Adaptations for Living in Salt Water and Fasting

ELIMINATION OF SALT

Although humans adrift at sea might lament, like Coleridge's Ancient Mariner, "Water, water, every where. / Nor any drop to drink," pinnipeds have managed to adapt to their salty environment. Nevertheless, living in salt water poses osmoregulatory problems for seals and other marine mammals, since they must maintain a proper concentration of body fluids despite great differences or changes in their external aquatic environment. Most marine mammals are hypo-osmotic; that is, they are constantly in danger of losing water to the more concentrated, or salty, ocean water in which they live. As human swimmers and divers know, they tend to become dehydrated after long periods in the ocean.

Water-balance mechanisms appear to be highly versatile in some species, such as harbor seals, belugas, manatees, and various porpoises that can inhabit either the ocean or freshwater rivers. Harbor seals, for example, are often seen traveling many miles upstream in rivers of the Pacific Northwest. Their kidneys appear to adapt to differences in the salinity of the water. A number of landlocked phocid seals actually spend their entire lives in freshwater lakes. The Baikal seal, Caspian seal, and some subspecies of ringed seal inhabit large freshwater lakes in Russia (Fig. 1.19).

Pinnipeds conserve water and maintain a proper water balance in several ways. The heavily lobulated kidneys of seals and most marine mammals are extremely efficient at concentrating urine; they absorb water and eliminate excess salt in urine that has a concentration equal to or greater than that of seawater (see, e.g., Irving et al. 1935; Pilson 1970; Keyes et al. 1971; Tarasoff and Toews 1972; Bester 1975; Vardy and Bryden 1981). The chloride concentration in seal urine is therefore as high as or higher than that of ocean water (which contains an average of 535 millimoles of chloride per liter). A seal that drinks a liter of seawater, after excreting the salts, ends up with a slight net gain of pure water. A human, however, experiences a net water loss of about a third of a liter of water after drinking the same amount of seawater. Human kidneys are less powerful than those of seals or whales, and even the most concentrated urine they can produce is always more dilute than salt water. Water is therefore wasted in eliminating the excess salt in urine. Consequently, someone adrift in a life raft and in need of water only promotes dehydration by drinking seawater.

FIG. 1.19. Baikal seal. *Photo by Jeanette Thomas.*

Many pinnipeds and other marine mammals appear to restrict their intake of seawater and derive fluids from the foods they consume. Harbor seals derive about 90 percent of their fresh water from water present in fish they eat and most of the remainder from metabolically produced water and inspired water vapor (Depocas et al. 1971). Northern fur seal juveniles held in captivity obtained 65 percent of their water intake from preformed water in the fish they ate and the remainder mainly from metabolically produced and inspired water; they also ingested seawater, which supplied a small amount of water as well (Fadely 1987).

The salt and fluid content of ingested prey and the resultant salt load vary according to the diet of a particular group or species. Pinnipeds and odontocete whales that eat fish ingest food with a salt content of less than 1 percent. Walruses and crabeater seals, as well as mysticete whales, eat foods with an elevated salt content: plankton, crustaceans, and molluscs. Most marine invertebrates are either hyperosmotic or isosmotic (in osmotic equilibrium) with seawater and therefore have a high salt content. The herbiverous sirenians also take in high levels of salt, since they feed on marine seagrasses and plants that are isosmotic with the surrounding seawater, although manatees also feed on freshwater vegetation.

On occasion pinnipeds have been observed drinking seawater, a practice called mariposia. Gentry (1981b) has observed various sex and age classes of California sea lions, Steller sea lions, and northern fur seals drinking seawater during the nonbreeding season. A female leopard seal was also observed drinking seawater by Brown (1957). Weddell seal females have fresh water available in the form of snow, which they sometimes eat during the breeding season when they are fasting and lactating (Kooyman and Drabek 1968).

Mariposia is especially common among adult male otariids. The following species of sea lions and fur seals have been observed ingesting water while fasting on land during the breeding season: California sea lions, Steller sea lions (the California population but not the Alaska population), northern fur seals in California as well as St. George Island, New Zealand fur seals (populations on the warmer South Neptune Islands only), South American fur seals, and South African fur seals (Rand 1959; Gentry 1981b, unpubl. data; Vaz-Ferreira, unpubl. data).

Why do otariid males and other pinnipeds drink seawater? Although the answer is still unclear, Gentry (1981b) points out that most of the otariid species or populations observed ingesting seawater live in warmer climates and lose water by urination, panting, and sweating. He suggests that such water loss, along with prolonged fasting by territorial males, may be severe enough to promote the drinking of seawater, which may play a role in nitrogen excretion and help conserve oxidatively produced water from fat reserves.

FASTING: CONSERVING WATER AND BURNING FAT

Many pinnipeds fast for long periods during their breeding seasons (see Table 17). Adult male pinnipeds may go without food or water for up to three months while maintaining a territory or a position in the dominance hierarchy at the breeding rookery. In many phocid species, females do not eat or drink for over one month while expending tremendous amounts of energy nursing their pups. (Fasting and feeding strategies in female pinnipeds in relation to lactation and pup dependency periods are discussed in chapter 8.) A number of fasting pinnipeds and whales are known to obtain fluids from metabolically produced water, that is, fluid yielded by the oxidation of stored fat. Breeding pinnipeds that haul out on land and fast for periods of up to three months certainly must produce and conserve water; these animals depend on their fat stores for oxidatively produced water during this period.

The energetics of fasting has been particularly well studied in the northern elephant seal by Ortiz et al. (1978), Costa and Ortiz (1983), and Condit and Ortiz (1985), who have shown that elephant seals have the unique ability to use mainly fat stores rather than protein during their long fasts. In addition, Ortiz et al. (1978) demonstrated a long-term positive water balance in fasting elephant seal weaned pups, which means that all forms of water loss must be less than or equivalent to metabolic water production. The weaners, which fasted for periods of 32–52 days, obtained their necessary water as well as energy from oxidation of fat stores. Elephant seals also minimize water loss by producing only a small amount of concentrated urine. To synthesize protein from a fixed store of amino acids during long fasts, elephant seals appear to have developed extremely efficient mechanisms for recycling amino acids (Pernia et al. 1980).

Huntley et al. (1984) have found that elephant seals can substantially reduce the loss of water through evaporation by means of a countercurrent heat exchange within the large and complex nasal passage (a method first suggested by Ortiz et al. 1978). Water conserved in this manner is important to the seal during long fasts. The amount of water recovered upon exhalation depends on the structure and temperature of the animal's nasal surfaces; by condensing water, the complex nasal turbinates of elephant seals reduce its loss during respiration.

Yet another water-conserving feature is evident in lactating female seals. The milk of seals and other marine mammals is extremely high in fat and protein and contains comparatively little water. Among other things, this high fat–low water ratio allows the mother to provide nutrients for her young with a minimal expenditure of water. Pinnipeds are able to transfer nourishment to their young many times more efficiently than a nursing land mammal. In some species of phocid seals, the water content of the milk decreases strikingly during lactation. The gradually rising milk fat content and declining water content throughout the nursing period reflect the fasting mother's increasing physiological stress as well as her pup's changing metabolic requirements.

Sensory Adaptations

SOUND PRODUCTION

The development and refinement of various sensory systems have enhanced the ability of the amphibious pinnipeds to navigate, forage, and communicate successfully both in the hydrosphere and the atmosphere. A number of recent reviews of sensory systems are available for pinnipeds (e.g., Schusterman 1981a), as well as delphinids (e.g., Popper 1980) and marine mammals as a group (e.g., Fobes and Smock 1981; Watkins and Wartzok 1985). Many marine mammals produce two forms of sound, one for communication and another, called echolocation, for navigation and foraging. Sound production for communication purposes is discussed at greater length in chapter 9.

Sound waves travel exceedingly well in an aquatic medium, and the speed of sound is increased five times in water. Low-frequency sounds in particular can travel hundreds and even thousands of miles in the ocean. This characteristic of ocean acoustics has probably promoted the evolution of the long-range signaling and communication systems among some mysticete whales; individual whales may be in acoustic contact with one another over an entire ocean basin (R. S. Payne and McVay 1971). The production of elaborate "songs" by humpback whales has been well documented (see, e.g., R. S. Payne 1978; Winn and Winn 1978).

The evolution of a truly remarkable sound production system referred to as echolocation has dramatically improved the ability of at least the odontocete cetaceans to navigate and gather food underwater. Echolocating sounds consist of a series of broad-band (including the frequencies between 20 and 220 kilohertz) clicks emitted from the odontocete skull in brief, intense bursts or trains. The main frequency ranges from 40–50 kilohertz to about 125 kilohertz, depending on background noise. The click trains travel through the water, bouncing back or reflecting off objects such as schools of fish. The returning sound waves, or echos, provide the animal with information about the objects within its environment, working like our artificially produced sonar (Fig. 1.20) (see, e.g., Schevill 1956, 1964; Kellogg 1958; Norris et al. 1961; Moore 1980; Murchison 1980; Nachtigall 1980; Schusterman 1980).

At times, odontocetes even produce amazingly intense low-frequency sounds after emitting their echolocation clicks, evidently to acoustically stun or disorient prey like fish and cephalopods so that they can be captured more easily (Berzin 1972; Norris and Møhl 1983). This intriguing idea, called the "big bang" theory by Ken Norris, is supported by an abundance of indirect evidence. For example, it would help to explain how sperm whales, which are not particularly well equipped to capture large swift-swimming squid, can use these prey as a major food resource. Additional evidence is presented by Marten et al. (1987), who found ten cases in which wild odontocetes produced these loud impulse sounds when foraging for fish. The impulse sounds are louder and last longer than echolocation clicks. Experiments to test the "big bang" theory are currently being conducted by Ken Marten at the Long Marine Lab, using playbacks of artificial and real "bang" sounds recorded in the field to see how fish react to the sounds. Simulated "bangs" produced in the Santa Cruz harbor were actually intense enough to kill anchovies that were exposed to the noise.

Besides cetaceans, bats are the only mammals known to use echolocation as a principal means of locating prey as they forage under conditions of near-zero visibility. However, other terrestrial mammals, including voles, shrews, tenrecs, and birds such as the oil bird and cave swiftlet also appear to use echolocation at times. In addition, marine mammals such as baleen whales and pinnipeds may possess echolocation-like abilities to a limited degree. Baleen whales seem not to actively echolocate, although they are capable of producing short pulses of sound. At the present time there is a difference of opinion among researchers as to whether pinnipeds produce echolocation sounds.

Clearly, pinnipeds do not have either the well-developed underwater auditory capabilities or the sound production system associated with the odontocetes' highly developed and sophisticated echolocation abilities. Seals generally depend on visual and tactile senses to locate prey, at least when sufficient light is present. Although no pinnipeds have been found conclusively to use active echolocation in their natural surroundings, free-ranging leopard seals have been recorded producing high-frequency

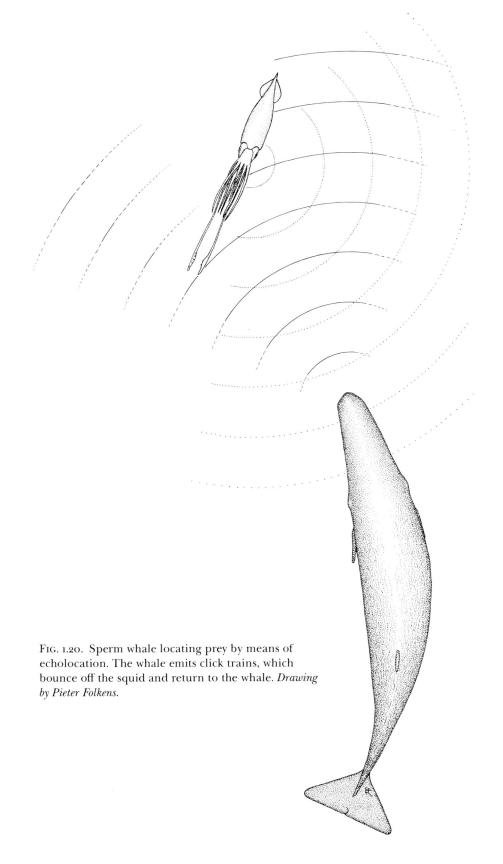

FIG. 1.20. Sperm whale locating prey by means of
echolocation. The whale emits click trains, which
bounce off the squid and return to the whale. *Drawing
by Pieter Folkens.*

ultrasonic (that is, beyond the range of human hearing ability) sound (Thomas et al. 1983), which could have value in echolocation. In addition, Renouf (1981) has found that noncaptive harbor seals produce various clicking sounds, especially at night. The harbor seals studied in San Francisco Bay (Fancher 1977) and the ringed seals (*Phoca hispida*) of Lake Saimaa, Finland (Hyvarinen 1987), are able to navigate and capture fish in extremely murky waters.

In captivity, seals and sea lions sometimes emit underwater clicking sounds when visibility is poor or virtually nonexistent, suggesting an echolocation function for these sound emissions. Oliver (1978) demonstrated that a captive grey seal learned to use sound to navigate and orient itself, although he did not find that the seal echolocated. Yet Poulter (1967, 1969) and Poulter and Del Carlo (1971) have claimed that California sea lions and Steller sea lions use sophisticated sonar systems to orient, navigate, and find food underwater, and Renouf and Davis (1982) have similarly speculated that the same may be true of harbor seals. These ideas, however, have been called into question by several investigators who not only have pointed out the lack of appropriate control procedures in the Poulter and Renouf studies but also have conducted carefully controlled studies of their own that did not show that seals echolocated (see Schevill 1968; Schusterman 1967, 1975, 1981a; Scronce and Ridgway 1980; and Wartzok et al. 1984). Schusterman (1981a) suggests that even though apparently healthy blind seals and sea lions are successfully able to locate and capture food, we cannot assume that these seals have evolved echolocation abilities. He speculates that sightless seals have instead learned to use sound localization and other nonvisual, perhaps tactile, cues to locate food. So at least for now, the question whether pinnipeds truly echolocate remains open.

HEARING UNDERWATER AND IN THE AIR

Underwater audition, or hearing sensitivity, is undoubtedly an extremely useful sensory system to pinnipeds as well as other marine mammals. A seal must be able to distinguish important sounds from the considerable amount of ambient noise in the ocean. In fact, seals are able to hear better underwater than in the air. Well-developed underwater directional hearing in both pinnipeds and odontocete cetaceans helps these animals to locate prey as well as conspecifics by means of vibrations transmitted through the water (Møhl 1964; Gentry 1967). In the air, sounds reach the ear on either side of the head at a slightly different time and with varying intensities, allowing an animal to pinpoint a sound source with great accuracy. A land mammal underwater, however, loses not only directional hearing but also hearing sensitivity and hears mainly low-frequency sounds. In water, sound waves cannot enter the ear in the usual way— through the auditory canal to the inner ear. Instead, waterborne sound waves reach the inner ear through the skull from all directions at once. Any diver who has tried to talk underwater to a diving buddy knows that it is nearly impossible to make words understandable. The seal ear has been modified to improve hearing sensitivity and directional hearing under- water, with the inner ear located so that it does not touch most other bones in the skull, thus diminishing the nondirectional sound waves pelting it through the skull. Further structural modification of the skull and inner ear enhances sound reception and directionality (Repenning 1972).

In the water pinnipeds can hear a wide range of frequencies to over 70,000 hertz (cycles per second). In comparison, humans cannot hear sound frequencies over 20,000 hertz or below 20 hertz. Underwater, pinnipeds gain about 20 decibels, according to Fobes and Smock (1981), which is similar to a human's loss of 18 decibels in moving in the opposite direction, from air to water. Schusterman and Moore (1978, 1980), how- ever, found only a 5 decibel gain in pinniped hearing sensitivity under- water. While it had been thought that phocids hear better underwater than otariids, Moore and Schusterman (1987) have demonstrated that at least one otariid—the northern fur seal—is as sensitive as a phocid to underwater sound. Phocids may differ from otariids only in their ability to detect higher frequencies underwater. (Variation in the audiograms of various pinniped species tested, however, may also have resulted from variation in the quality of testing environments.) Phocids that have been tested for underwater and aerial hearing capabilities include harp seals (Terhune and Ronald 1971, 1972) and harbor seals (Møhl 1968); phocids tested only for underwater hearing include the grey seal (Ridgway 1973) and the ringed seal (Terhune and Ronald 1971, 1972). The only otariid tested for aerial and aquatic hearing prior to the work of Moore and Schusterman (1987) was the California sea lion (Schusterman et al. 1972; Schusterman 1974).

On land, pinnipeds are thought to hear slightly less well than most terrestrial carnivores. Although otariids and some phocid seals hear a wide range of sound frequencies in air as well as or perhaps slightly better than humans do, their hearing overall is less sensitive than ours. Because pinnipeds lack a large external ear, their hearing sensitivity in air is naturally reduced. Moreover, the heavy wax coating in their auditory canal probably inhibits sound perception to some degree. According to

King (1983), the northern phocids and perhaps the walrus hear better in air than other pinnipeds. Moore and Schusterman (1987), however, found that aerial hearing in two otariids — the northern fur seal and the California sea lion — is greater in absolute sensitivity as well as in the range of frequencies detected than that of phocids tested. In fact, the northern fur seal is the most sensitive to airborne sounds (ranging from 500 hertz to 32 kilohertz) of all pinnipeds that have been tested and is equally or more sensitive to waterborne pure tones (ranging from 2 to 28 kilohertz). A seal also has good directional hearing in air — an important aid for a mother in searching for her pup, a breeding bull in locating "the competition," and all seals in detecting predators.

For a variety of theoretical as well as practical reasons it is important to measure the impact of man-made noise (such as that produced by oil and gas development operations) on marine mammals and to understand the aerial and hearing abilities of pinnipeds and other marine mammals. Unlike cetaceans, seals may be vulnerable to disturbances resulting from aerial sounds — another consequence of their amphibious existence. For instance, pinnipeds breeding on the California Channel Islands may experience disturbances as a result of frequent sonic booms and overhead aircraft (C. F. Cooper and Jehl 1980).

CHEMORECEPTIVE SENSES

Underwater, pinnipeds and most other marine mammals have virtually no sense of smell (their nostrils are usually closed tight anyway) and, as far as we know, a limited sense of taste. Chemoreceptive (olfactory and taste) sensory abilities have progressively degenerated in sirenians and cetaceans, especially in the odontocetes. Comprehensive reviews of chemoreception in marine mammals are provided by Lowell and Flanigan (1980) and Watkins and Wartzok (1985).

Marine mammals that spend much time in the air — pinnipeds and sea otters — appear to have an acute sense of smell out of water (Riedman and Estes MS). The sense of smell plays an especially important role in social and reproductive events that take place on land among the pinnipeds. Biologists are well aware that many pinnipeds, by scent, can detect the presence of humans hundreds of feet away. Many sea lions stampede into the water if they catch the scent of a person upwind. No doubt seals are able to discern the presence of other predators by their keen sense of smell. During the breeding season, adult males often investigate a female's anogenital area to determine, presumably by chemoreceptive means, whether she is in estrus. The frequent practice of nose-to-nose nuzzling of mothers and pups is also an important means of mutual recognition and of conveying and receiving information via chemoreception (see, e.g., Ross 1972). The importance of chemoreception in the bonding of a mother and her newborn pup is discussed in chapter 8.

TACTILE SENSES

Tactile sensory abilities are well developed among most marine mammals, especially pinnipeds. The most obvious sensory structure is the skin. Many colonial species of pinnipeds, such as walruses, sea lions, and elephant seals, group together densely on land—a tendency called thigmotaxis. In fact, sometimes the animals clump so tightly that they are literally piled on top of one another. Burney Le Boeuf calls such aggregations the downtown area of an elephant seal rookery (Fig. 1.21). Other pinnipeds, such as harbor seals, are less thigmotactic and rarely touch one another while resting together.

All pinnipeds have an abundance of vibrissae, or whiskers, that contain many sensitive nerve fibers. A single vibrissa of the ringed seal (*Phoca hispida saimensis*) contains ten times the number of nerve fibers typically found in one vibrissa of a land mammal. Moreover, vibrissae of ringed

Fɪɢ. 1.21. Tightly clumped group of northern elephant seal weaned pups, with subadult male in weaner pod, at Año Nuevo Island. *Photo by author.*

seals are structurally distinctive from those of land mammals (Hyvarinen 1987). Such vibrissae, which serve as sense organs, are absent for the most part in cetaceans but are present in sea otters. The vibrissae along either side of the snout, called the mystacial whiskers, vary in length, number, position, and structure among the seal species. The whiskers have a smooth appearance in the walrus, monk seal, bearded seal, and all otariids. The remaining species have whiskers that are beaded, having a corrugated appearance. Some seals also have whiskers over the eyes (superciliary whiskers) and near the nose (rhinal whiskers). Like other carnivores, pinnipeds on land may use their whiskers when socially interacting with other animals. The position of the seal's whiskers—held back or stiffly forward—conveys such states as attentiveness or aggression.

Long and sensitive vibrissae appear to help pinnipeds detect vibrations of prey in the water, enhancing their ability to forage, especially in murky depths where visibility is poor (Stephens et al. 1973). The ringed seals (*P.h. saimensis*) of Lake Saimaa, Finland, have exceptionally well developed vibrissae, which appear to help them find their way in the dark and often cloudy waters beneath the ice (Hyvarinen 1987). Some healthy blind seals even inhabit the lake. Hyvarinen believes that the ringed seals can sense compressional waves as well as sounds with their vibrissae. Sound waves could be received by way of the blood sinuses and by tissue conduction through the vibrissae.

Foraging walruses probably depend heavily on their sensitive vibrissae to locate molluscs on the muddy sea floor. Each walrus possesses 600–700 vibrissae, many more than both other seals and most land carnivores. The whiskers of wild walruses wear down to a length of only 1–2 inches, although captive walruses that do not have to forage for themselves may have whiskers as long as 5 inches (Fig. 1.22). While fur seals have fewer mystacial whiskers than walruses—perhaps 40–60—their vibrissae are quite long, reaching lengths of over 18 inches (Bonner 1968).

Pinniped vibrissae even appear to aid in navigation. For instance, when a largha seal was blindfolded in an experiment but its vibrissae were left alone, it was able to surface in the center of a breathing hole in the ice. When the blindfolded seal had its vibrissae restricted, however, it bumped into ice near the hole several times before locating it (Sonafrank et al. 1983). Experimental observations of captive ringed seals also showed that vibrissae helped the seals locate ice holes drilled in a frozen pond; visual and acoustic cues were even more important for such navigation, however (Wartzok et al. 1987). Montagna (1967) has suggested that the whiskers might also function to gauge the speed at which a seal swims, although this has not yet been proved experimentally.

FIG. 1.22. Walrus, showing numerous vibrissae. *Photo by Kathy Frost.*

VISION

Seals, like most other marine mammals (with the notable exception of river dolphins), are able to see well both in air and underwater. Pinnipeds must not only see clearly in water and on land but also under conditions of extremely variable light intensities — while foraging in deep, dimly lit waters or breeding on ice or along bright, sandy beaches. Seals have greatly enlarged orbits and eyes that are large in relation to body size (Fig. 1.23). Walruses are the exception among pinnipeds in that their eyes are relatively small and placed toward each side of their heads, more like fish eyes. Because walruses feed on sessile bottom molluscs, they do not require the acute vision of other pinnipeds for capturing moving prey (Fig. 1.24).

Most marine mammals can see well in unclear or dim light — an important ability since much of the available light is absorbed underwater. Under low-light conditions in air or underwater, the pinniped pupil dilates in a large circle to let in more light. In bright sunlight the pupil constricts to a narrow vertical slit (Lavigne and Ronald 1972; Jamieson and Fisher 1972; Lavigne 1973). Seals as well as whales have well-developed ciliary muscles for contracting and dilating the pupil.

The pinniped and cetacean eye is particularly light sensitive and contains high numbers of rods (so named for the rod-shaped photorecep-

FIG. 1.23. California sea lion underwater, showing large eyes. *Photo by Nancy Burnett.*

FIG. 1.24. Male walrus; note small-sized eyes. *Photo by Anne Hoover.*

tor cells in the retina that respond to low light), which help them to see during the night or at great depths where light penetrates poorly. Although the seal can perceive light in dark environments, its ability to distinguish color and fine detail diminishes in darkness. Moreover, pinnipeds have an especially well developed *tapetum lucidum* (Lavigne et al. 1977), a specialized layer behind the retina that contains large amounts of guanine crystals, allowing the retina to reflect light like a mirror and giving it a metallic appearance. The tapetum functions to send the light that passes through the retina back through it a second time, essentially doubling the light-gathering work of the rod cells. Many land mammals (especially nocturnal carnivores) as well as other marine mammals that need to see well in the dark possess a tapetum, which makes the eyes of pinnipeds, like those of a cat, seem to glow if a light shines on them at night.

But seals must also contend with extremely intense levels of ultraviolet rays that reflect off snow and ice and to a lesser degree off water and light sandy beaches. These ultraviolet rays can damage the cornea (the transparent convex structure covering the eye) and in humans, at least, cause snow blindness. Unlike humans, ice-inhabiting seals have corneas that tolerate the abundant ultraviolet light of polar and subpolar environments (Hemmingsen and Douglas 1970; Lavigne et al. 1977). Because the human cornea lacks the natural protection of the seal's, people who spend time outdoors near water or snow often wear sunglasses or contact lenses specially coated to shield the eyes from damaging ultraviolet rays.

Well-controlled experiments on captive pinnipeds have shown that under good to moderate lighting conditions, pinnipeds—like the cetaceans—can see almost as well in air as in water (Schusterman and Balliet 1971; Schusterman 1972). The seal's eye, however, appears to be somewhat better adapted for underwater than for aerial vision. A seal's underwater visual acuity is thought to be comparable to that of a cat on land, according to Schusterman (1972). Seals have overcome the problem encountered by all mammals that try to see clearly underwater: that water and the cornea have the same refractive index. (The refractive index is a measure of the refraction, or bending, of light in a particular substance; because the refractive index of water is higher than that of air, light travels more slowly in water than in air.) Consequently, underwater the cornea can no longer be used as a lens to focus images as it does in air. In air, the cornea as well as the inner eye lens focus light on the retina at the back of the eye. When the cornea's ability to focus vanishes underwater, a land mammal's eye is unable to refract light rays enough to see clearly and the mammal is farsighted, or hypermetropic. Swimmers experience this blurry vision when trying to see underwater without face masks (which provide an air barrier that allows the eye to focus as on land). To solve this problem, both pinnipeds and whales have evolved an extremely rounded fish-eye lens, which refracts light appropriately for underwater vision (see, e.g., Jamieson and Fisher 1972).

This enlarged fish-eye lens, however, sometimes makes it more difficult for a seal to see clearly in air. Because the lens as well as the cornea bends the light rays at the wrong angle in air, the seal becomes nearsighted, or myopic, as well as astigmatic at times (Fig. 1.25). Yet the pinniped eye is able to make adjustments in air that allow relatively clear vision, as Schusterman (1972) elegantly demonstrated with experiments on captive California sea lions. As light levels were decreased, the sea lion's visual

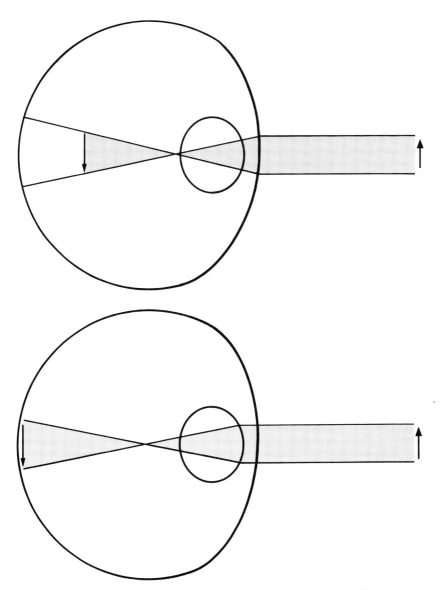

Fig. 1.25. The pinniped eye (adapted from Lavigne and Kovacs 1988).
Monterey Bay Aquarium Graphics.

a. (*top*) Optics of the pinniped eye in air. The image is focused in front of the
   retina, so that vision becomes blurry and nearsighted.

b. Optics of the pinniped eye underwater. The image is focused on the
   retina, so that vision is relatively clear.

acuity in air became quite poor in comparison with its underwater visual acuity, which remained relatively good—similar to that of a baboon tested under the same lighting conditions. In other words, while both aerial and underwater visual acuity declined with decreasing levels of background light, visual acuity dropped much less sharply underwater than in air (Fig. 1.26).

On land a seal sees most clearly in bright sunlight when the pupil contracts to a thin slit and a tiny pinhole. This contraction helps to alleviate the seal's nearsightedness since the lens and cornea are unable to bend light as efficiently when only a small amount of it passes through the pinhole. In addition, blurry vision caused by astigmatism is improved since little focusing is required through a pinhole lens (Lavigne et al. 1977).

Although color vision has never been demonstrated for pinnipeds, seals may be able to see color to a limited degree, since their eyes contain at least some cones, which allow eyes to perceive color as well as fine detail in bright light. In one experiment a spotted seal was able to distinguish

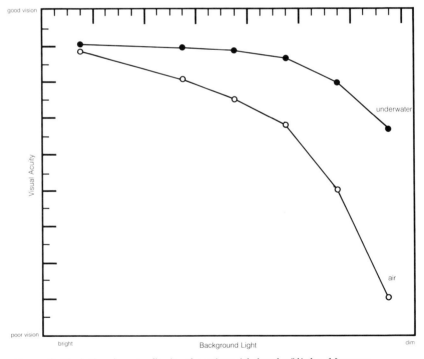

FIG. 1.26. Variation in a seal's visual acuity with level of light. *Monterey Bay Aquarium Graphics.*

between two objects that were identical except in color (Wartzog and McCormick 1978). Experiments with California sea lions, however, indicate that they do not see in color. Instead, they perceive colored objects in varying shades of black, white, and grey (Gisiner, pers. comm.). A seal underwater cannot see much color anyway, since only certain wavelengths of light or color can penetrate beneath the surface. A seal's underwater world, therefore, is mostly blue or green. Various pinniped species are visually sensitive to the blue or green light wavelengths that predominate in the particular marine environment they inhabit (Lavigne et al. 1977). For instance, southern elephant seals, which dive to great depths in open waters characterized by an abundance of blue color wavelengths, possess visual pigments in their rods most sensitive to blue light. Seals living in greener coastal or polar waters, such as harp seals, possess more green-sensitive visual pigments. This correlation also holds true for many whales and fishes.

The pinniped eye is protected on land and underwater by the strongly keratinized corneal epithelium. Keratin is a tough, fibrous protein that also forms the outer layer of fingernails, hair, hoofs, and horns. A thick sclera (the white, fibrous outer tissue covering all of the eyeball except the cornea) also forms a protective coating that safeguards the eye against the underwater pressure experienced by a seal submerged at great depths. The lacrimal glands steadily lubricate the seal's eye with tears to help protect it from salty seawater and sand. A seal's eye is also protected by the nictitating membrane, which is like an inner eyelid that effectively wipes away sand and debris. This membrane is also present in a number of other mammals as well as birds.

In conclusion, pinnipeds are extremely well adapted to survive in an extraordinary diversity of aquatic and terrestrial environments. Many species, such as hooded seals and ringed seals, are able to give birth and breed on polar ice under freezing conditions. Other pinnipeds, such as the Hawaiian monk seal and Galápagos seals, must tolerate blistering temperatures while hauled out on land. Northern elephant seals and Weddell seals can dive to depths of 600–1,250 meters — deeper than many cetaceans are known to dive. Some seals are able to live in freshwater lakes, while others move freely between rivers and the ocean. With their sensory abilities, most pinnipeds can see, hear, and communicate both underwater and on land. In fact, one can only imagine what other fascinating adaptations of the amphibious pinnipeds researchers will discover in the future.

2

Evolution, Classification, and Distribution of Pinnipeds

Evolution

Millions of years after the first mammal-like animals appeared on earth, several groups of land mammals began to live in the ocean, probably to avoid land predators and to take advantage of the abundant food in an immense and richly productive marine environment. Today, a glimpse of a possible transitional process from land to sea may be found on the Tokelau Islands of the South Pacific, where pigs have developed the habit of wandering widely over coral reefs during low tide in search of food. According to the *Bangkok Post* (May 5, 1983), about three hundred pigs living on the island of Fakaofo forage along the shallow coral reefs for small molluscs, sea slugs, and fish. Apparently the pigs are excellent swimmers; they spend much of their time wading with heads submerged as they search for food.

On the other side of the world along the coasts of Scotland are domestic sheep that survive by grazing on algae in the intertidal zone. Described by Thomson (1954) as "scraggly creature[s], long in the leg and neck, goatlike and wild, but with fine wool like the Shetland," these "marine sheep" have foraged on kelp and seaweeds for hundreds of years. They seem at home in the sea and often venture far out on surf-pounded promontories and swim in shallow waters to reach a new patch of sea-weed. According to a native Orkney Islander, "If ye study the sheep, ye'll see how they follow the way o' the sea. They'll shift round to the lee side o' the island afore a storm, and they'll ken the run o' the tides afore ye'll ken it yersel'." The sheep actually share the beaches with grey seals; the seals tend to ignore the sheep, who in turn seem to go out of their way to avoid the seals.

Although we may have to wait a few million years to see truly marine pigs and sheep, at the present time we can observe the evolutionary result of millions of years of adaptation to the ocean in four groups of mammals: cetaceans, sirenians, pinnipeds, and sea otters. The different length of time each group has inhabited the ocean is generally reflected in its degree of aquatic specialization and dependency on land. Thus among the marine mammals, in relation to cetaceans and sirenians, the amphibious pinnipeds have evolved more recently. To provide a basis of comparison with the pinnipeds, I give a brief evolutionary history and basic description of the other marine mammal groups in the following paragraphs.

The cetaceans (from the Latin *cetus*, "whale or sea monster") have lived longer in the sea than any other mammal. Accordingly, they show the most pronounced adaptation to an exclusively oceanic existence, as is evident in their complete lack of dependence on land and in the design of their bodies, which are sleekly streamlined, with rigid foreflippers and no hind limbs. Cetaceans are believed by many scientists to have descended from a primitive artiodactyl, or even-toed, ungulate (hoofed mammal) that lived some sixty million years ago during the Paleocene following the extinction of the dinosaurs at the end of the Cretaceous period. Another line of evidence based on molecular studies, however, suggests that cetaceans evolved much more recently (Lowenstein 1985a). The earliest fossils of toothed whales (odontocetes) and baleen whales (mysticetes) appeared roughly 27–30 million years ago in the Oligocene period. According to molecular evidence presented by Lowenstein, the odontocete and mysticete whales diverged some 20–25 million years ago.

Sirenians (from the word *siren*, meaning "mermaid") are believed to have descended from the same protoungulate ancestor that gave rise to elephants, aardvarks, and hyraxes 50–55 million years ago in the early Eocene. Present-day sirenians—dugongs and manatees—have also lost their hind limbs, yet their foreflippers have remained pliable for maneuvering underwater among sea grasses. They live in shallow coastal waters and freshwater rivers. According to the molecular studies of Rainey et al. (1984), dugongs and manatees probably diverged some twenty million years ago. The fossil record, however, leads other scientists to believe the two groups separated as long ago as 45 million years. Like cetaceans, the sirenians spend their entire lives in the water, giving birth aquatically and never hauling out on land.

Sea otters are the most recently evolved group of marine mammals. They appear to have descended from an otterlike creature—*Enhydritherium*—5–7 million years ago during the Miocene/Pliocene. The separation of this lineage from primitive lutrines, however, may have occurred considerably earlier (Berta and Morgan 1985). Sea otters are the smallest, slowest-swimming, and least streamlined group of marine mammals. While their hind feet have become modified flippers, otters have retained the front paws of a terrestrial carnivore. Sea otters inhabit nearshore waters and may haul out on land, but unlike pinnipeds, they retain essentially no ties to land because they are able to mate and raise their pups entirely at sea (Fig. 2.1). The more ancient pinnipeds, which appear to have inhabited the ocean 20–25 million years longer than sea otters, are still highly dependent on land and must return to shore each year to breed. So although otters are not as morphologically modified as other marine mammals, they have evolved a completely marine existence.

The pinnipeds evolved from arctoid, "bearlike," carnivore ancestors roughly thirty million years ago during the late Oligocene/early Miocene, apparently entering the ocean to take advantage of the new and abundant

FIG. 2.1. Sea otter mother and pup. *Photo by Fred Bavendam.*

FIG. 2.2. Walruses. *Photo by Brian Fadely.*

food resources created by an upwelling of colder, nutrient-rich waters along the coasts of Europe and North America. The upwelling resulted from a major global change in climate about thirty-six million years ago that caused a relatively abrupt cooling of ocean waters and altered oceanic circulation patterns (Repenning 1980). According to one view of pinniped evolution, phocid seals (earless, or true, seals) and the otarioid seals (eared seals and walruses) diverged from a common ancestor about twenty-five million years ago in the early Miocene. According to the biphyletic (meaning "descended from separate ancestors") view, phocids originated independently of the older otarioids perhaps 20 million years ago. Some scientists believe that the walruses diverged from the eared seals about twenty million years ago, but many others now think that walruses are actually more closely related to phocids than to otariids (see, e.g., Wyss 1987) (Fig. 2.2).

The origin of the pinnipeds is obscure as well as controversial. Are pinnipeds biphyletic or monophyletic (originating from a single common ancestor)? Although there is no clear answer to this question and many still favor the biphyletic view, the most recent evidence supports the theory that pinnipeds are monophyletic. In fact, pinniped biologists and evolutionary systematists have been arguing about pinniped relations for a hundred years. Appropriately enough, Lowenstein (1986) calls them "another quarrelsome tribe," alluding to Linnaeus's description of pinnipeds as a "dirty, curious, quarrelsome tribe, easily tamed and polygamous." (Whether pinniped biologists are also dirty, curious, and polygamous has not yet been the subject of scientific scrutiny.)

According to those who believe that pinnipeds had a biphyletic origin (e.g., McLaren 1960; King 1964a; C. E. Ray 1976; Repenning 1976, 1980; Tedford 1976), the two pinniped superfamilies evolved independently of one another. Basing their judgment on the fossil record and comparative morphology, scientists believe that the phocid seals descended from an otterlike carnivore in the North Atlantic in the middle Miocene, whereas the eared seals and walruses evolved during the late Oligocene from a bearlike or doglike ancestor in the North Pacific. (Interestingly, the skulls of sea lions, bears, and such canids as large dogs and wolves are nearly indistinguishable at first glance.)

The monophyletic view is supported by studies of pinniped evolution at the molecular level, which involve comparing serum proteins or DNA or RNA of different living species. In addition, Berta (1987) hypothesizes a monophyletic origin based on a nearly complete fossil skeleton of the ancient pinniped *Enaliarctidae*. Comparative studies of hindflipper structure in the three families, among other things, also lends support to the monophyletic view (Wyss 1987).

The molecular structure of even extinct and fossilized plants and animals can be compared. The timing of evolutionary divergence and the degree of genetic relatedness between two species can therefore be determined by comparisons based on the similarities of their proteins or DNA rather than on the basis of morphology or appearance. Evolutionary and taxonomic relationships based on the similarity of one animal's appearance to another's can be misleading for two reasons. First, two unrelated animals, such as dolphins and fish, may look somewhat alike simply because they have evolved in similar environments—a process called convergent evolution. Second, it is difficult to tell by their appearance when two species diverged from a common ancestor, since some features, such as teeth or bones, may evolve more quickly in one lineage than in another (Lowenstein 1985b).

If pinnipeds originated from a common ancestor, the proteins and DNA molecules of both otariid and phocid seals should resemble each other more closely than they resemble those of other carnivores. If pinnipeds are biphyletic, however, the phocids would be genetically more like otters, whereas the otariids would be more like bears or dogs. So far, the molecular evidence shows that both groups of pinnipeds are more like each other and therefore monophyletic. Sarich (1969) found that albumin proteins of phocids and otariids were more similar to each other than to the albumin proteins of doglike carnivores. Using eye lens alphacrystalline protein sequences, de Jong (1982) showed that the proteins of phocids and otariids were more like each other than like those of other mammals. Finally, Arnason and Widegren (1986) compared the DNA of phocids and otariids and found a closer resemblance between the genetic material of the two groups than between that of either group and other mammals. So the molecular evidence, at least, supports a monophyletic origin for the pinnipeds, while some interpretations of the fossil record and the comparative anatomy of living pinnipeds suggest a biphyletic origin.

Although scientists may disagree about the evolution of pinnipeds, how seals evolved has always been perfectly clear to the Inuit Eskimos. The story they tell of the creation of seals revolves around Nuliajuk (pronounced Nooli AH juk), also known as Sedna the Eskimo girl. The Netsilik Eskimos worshipped Nuliajuk as their most powerful spirit — the mother of all the animals and the mistress of the sea and land. As one version of the story goes, Sedna and her father were at sea when some seabirds whom they had offended created a terrible storm. The father panicked and flung his daughter overboard as an offering to the birds. As Sedna clung to the edge of the boat, her father cut off her fingers, which, falling into the ocean, were transformed into seals and whales. Ever since,

FIG. 2.3. Sedna the Eskimo seal spirit. *Drawing by Roger Luckenbach.*

Sedna, or the spirit Nuliajuk, has controlled the weather and the seals, taking the animals away from the hunters when she is angry (Fig. 2.3). The Netsilik Eskimos even have a song about the creation of seals—"magic words for hunting seals":

> O sea goddess Nuliajuk
> When you were a little unwanted orphan girl
> we let you drown.
> You fell in the water
> and when you hung onto the kayaks crying
> we cut off your fingers.
> So you sank into the sea
> and your fingers turned into
> the innumerable seals.
>
> You sweet orphan Nuliajuk,
> I beg you now
> bring me a gift,
> not anything from the land
> but a gift from the sea,
> something that will make a nice soup.
> Dare I say it right out?
> I want a seal!
>
> You dear little orphan
> creep out of the water
> panting on this beautiful shore,
> puh, puh, like this, puh, puh.
> O welcome gift!
> in the shape of a seal!

Classification, Distribution, and Abundance

The modern seals, sea lions, fur seals, and walruses make up a distinctive yet diverse group of mammals: the order Pinnipedia. Some taxonomists (e.g., King 1983) do not consider the pinnipeds as a separate order but instead include them as a suborder under the order Carnivora within the arctoid group of carnivores. Still others (e.g., Eisenburg 1981; Nowak and Paradiso 1983) feel that because of the pinnipeds' uniqueness as a group, it is convenient to regard them as a separate order, as I will do here. At least everyone seems to agree that the Pinnipedia is made up of two superfamilies: the Otarioidea and the Phocoidea. Otarioids include the families Otariidae (eared seals) and the Odobenidae (walruses). The phocoids consist of the family Phocidae (true seals).

Thirty-three different species of pinnipeds are found throughout the world today: eighteen phocids, fourteen otariids, and one odobenid (Table 2). The Caribbean monk seal is not included in the total number of species since, unfortunately, it appears to be extinct (Le Boeuf et al.

TABLE 2

Classification of the Pinnipeds
(Order Pinnipedia)

Superfamily Phocoidea
  Family Phocidae
    Subfamily Monachinae (Southern phocids:* 8 species)

      Hawaiian monk seal (*Monachus schauinslandi*)
      Mediterranean monk seal (*Monachus monachus*)
      [Caribbean monk seal (*Monachus tropicalis*)]†

      Northern elephant seal (*Mirounga angustirostris*)
      Southern elephant seal (*Mirounga leonina*)

      Weddell seal (*Leptonychotes weddelli*)
      Ross seal (*Ommatophoca rossi*)
      Crabeater seal (*Lobodon carcinophagus*)
      Leopard seal (*Hydrurga leptonyx*)

    Subfamily Phocinae (Northern phocids: 10 species)

      Hooded seal (*Cystophora cristata*)
      Bearded seal (*Erignathus barbatus*)
      Grey seal (*Halichoerus grypus*)
      Harp seal (*Phoca groenlandica*)
      Ribbon seal (*Phoca fasciata*)
      Largha seal (*Phoca largha*)
      Caspian seal (*Phoca caspica*)
      Baikal seal (*Phoca sibirica*)
      Ringed seal (*Phoca hispida*)
          P.h. hispida (Arctic Basin)
          P.h. ochotensis (Northern Japan)
          P.h. botnica (Baltic Sea)
          P.h. ladogensis (Lake Ladoga)
          P.h. saimensis (Lake Saimaa)

      Harbor seal (*Phoca vitulina*)
          P.v. vitulina (eastern Atlantic)
          P.v. richardsi (eastern Pacific)
          P.v. stejnegeri (Kuril harbor seal)
          P.v. concolor (western Atlantic)
          P.v. mellonae (Seal Lake)

Continued on next page

* The terms *southern* and *northern* refer only generally to the distribution of phocids; northern elephant seals, for instance, occur in the Northern Hemisphere, although not in the Arctic.

† The Caribbean monk seal appears to be extinct and is not included in the total of 33 pinniped species.

TABLE 2, *continued*

Superfamily Otarioidea
Family Otariidae
Subfamily Otariinae (Sea lions: 5 species)
California sea lion (*Zalophus californianus*)
        Z.c. californianus (California sea lion)
        Z.c. wollebaeki (Galápagos sea lion)
        Z.c. japonicus (Japanese sea lion)
Steller sea lion (*Eumetopias jubatus*)
Southern sea lion (*Otaria byronia*)‡
Australian sea lion (*Neophoca cinerea*)
New Zealand sea lion (*Phocartos hookeri*)
Subfamily Arctocephalinae (Fur seals: 9 species)
Northern fur seal (*Callorhinus ursinus*)
Guadalupe fur seal (*Arctocephalus townsendi*)
Juan Fernández fur seal (*Arctocephalus philippii*)
Galápagos fur seal (*Arctocephalus galapagoensis*)
South American fur seal (*Arctocephalus australis*)
New Zealand fur seal (*Arctocephalus forsteri*)
Antarctic fur seal (*Arctocephalus gazella*)
Subantarctic fur seal (*Arctocephalus tropicalis*)
Arctocephalus pusillus
    A.p. pusillus (South African fur seal)
    A.p. doriferus (Australian fur seal)
Family Odobenidae (1 species)
Walrus (*Odobenus rosmarus*)
      O.r. rosmarus (Atlantic walrus)
      O.r. divergens (Pacific walrus)

‡ Also known as the South American sea lion *Otaria flavescens* (Shaw).

1986b). Pinnipeds make up over one quarter (28 percent) of the 116 species of marine mammals, the majority of which are cetaceans. An estimated 50 million pinnipeds exist throughout the world today. Roughly 90 percent of them are phocid seals; the remaining 10 percent are otariids and odobenids.

Fewer species of pinnipeds as well as whales inhabit the Antarctic than the Arctic, but the Antarctic seal populations are larger. In fact, there are substantially more phocid seals worldwide than otariids, largely because there are so many crabeater seals in the Antarctic — 30 million or so (Laws 1984). Antarctic seals generally have larger bodies than other phocids, a characteristic that provides them with insulation against the cold on or under the frigid polar ice where they live. Laws (1977b) also suggests that their large size may be related to the more plentiful food supply — such as the abundant krill — in the oceans of the southern hemisphere.

TABLE 3
Estimated World Population of Each Pinniped Species

SPECIES	ESTIMATED POPULATION	STATUS	REFERENCES
Hawaiian monk seal	1,500	Variable; some colonies stable, some declining, some increasing	Gilmartin et al. 1987; T. Gerrodette, pers. comm.
Mediterranean monk seal	500	Declining	Sergeant et al. 1978; de Visscher 1985
Caribbean monk seal	0	Extinct	Kenyon 1977; Le Boeuf et al. 1986b
Northern elephant seal	120,000	Increasing	Le Boeuf & Bonnell 1980; Le Boeuf, unpubl. data
Southern elephant seal	550,000–750,000	Variable; some colonies stable, others declining	Laws 1984; J. Ling, pers. comm.
Weddell seal	800,000	Stable, but some colonies declining	Laws 1984
Ross seal	200,000	Unknown	Laws 1984
Crabeater seal	30,000,000	Increasing	Laws 1984
Leopard seal	440,000	Increasing	Laws 1984
Hooded seal	500,000	Unknown, but probably variable	Øritsland 1960
Bearded seal	500,000	Unknown, but probably variable	Stirling et al. 1977; Stirling & Archibald 1979
Harp seal	2,250,000–3,000,000	Variable	Ronald & Dougan 1982
Ribbon seal	240,000	Variable	Burns 1981b
Caspian seal	600,000	Stable	Popov 1979a; King 1983
Baikal seal	50,000	Probably stable	Popov 1979b; King 1983
Ringed seal	6,000,000–7,000,000	Variable	Stirling et al. 1977; Stirling & Calvert 1979; Bigg 1981
Largha seal	400,000	Probably stable	Bonner 1979
Harbor seal	400,000–500,000	Variable	Bigg 1981
Grey seal	120,000	Increasing	Bonner 1981a
California sea lion	145,000	Probably increasing	Le Boeuf et al. 1983
Galápagos sea lion	40,000	Unknown	Gentry & Kooyman 1986

Continued on next page

SPECIES	ESTIMATED POPULATION	STATUS	REFERENCES
Steller sea lion	245,000–290,000	Stable, but some colonies declining, others increasing	Loughlin et al. 1984, 1987; Merrick et al. 1987
Southern sea lion	275,000	Declining	Vaz-Ferreira 1981; Vaz-Ferreira & Ponce de Leon 1987
Australian sea lion	3,000–5,000	Unknown	Walker & Ling 1981a; Ling, pers. comm.
New Zealand sea lion	4,000–6,000	Unknown	Walker & Ling 1981b
Northern fur seal	1,215,000	Decreasing	Gentry & Kooyman 1986
Guadalupe fur seal	2,500	Increasing	M. Pierson, pers. comm.
Juan Fernández fur seal	6,300	Increasing	Torres 1987
Galápagos fur seal	30,000	Probably increasing	Trillmich 1987a
South American fur seal	500,000	Increasing	Vaz-Ferreira & Ponce de Leon 1987; Majluf 1987; Croxall & Gentry 1987
New Zealand fur seal	55,000	Probably increasing	G. J. Wilson 1981; Mattlin 1987; Ling 1987
Antarctic fur seal	1,200,000	Increasing	McCann & Doidge 1987
Subantarctic fur seal	300,000	Increasing	Roux 1987a; Bester 1987; Kerley 1987; Gentry & Croxall 1987
South African fur seal	1,200,000	Increasing, but some colonies declining	Gentry & Kooyman 1986; Shaughnessy 1987; Cressie & Shaughnessy 1987; David 1987a; Butterworth et al. 1987
Australian fur seal	25,000	Stable	King 1983; Warneke & Shaughnessy 1985
Walrus	250,000	Fluctuating	Reeves 1978; Estes & Goltsev 1981; Fay et al. 1989

The global distribution patterns of pinnipeds reveal that certain groups or species tend to be restricted to particular regions of the world. For instance, there are no otariids in the extreme polar regions. Only phocid seals live in the Antarctic and Arctic, with the exception of walruses, which inhabit northern circumpolar waters. In contrast, the only phocids found in the tropics are the endangered Hawaiian and Mediterranean monk seals, whose populations are extremely small. The Hawaiian monk seal population numbers about fifteen hundred, and only an estimated five hundred Mediterranean monk seals still survive along the coasts of Greece, Turkey, and North Africa (Table 3; Maps 1–3).

A number of fur seals and sea lions live in tropical or subtropical areas, although northern and Antarctic fur seals range widely into areas with colder climates as well. All of the fur seals except the northern and Guadalupe fur seals are found in the Southern Hemisphere, but sea lions are common in both hemispheres. Although many otariids live in areas characterized by warm or tropical climates, ocean waters in these areas are often cool and rich in nutrients because of current patterns. No pinnipeds at all are currently found in some parts of the world, such as Asia and India, presumably because the lack of nutrient-rich coastal waters and upwelling in these areas keeps primary production low, so that there are few food resources for seals. Repenning et al. (1979) believe that fossil pinnipeds, however, inhabited warmer waters than most of the modern pinnipeds. They also speculate that the joining of Africa and

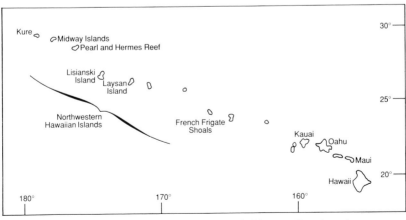

MAP 1. Distribution of the Hawaiian monk seal. *Monterey Bay Aquarium Graphics.*

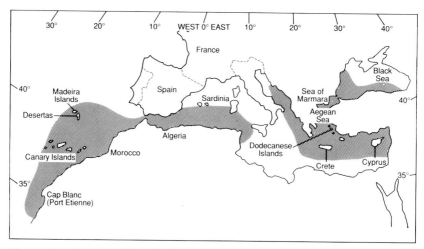

MAP 2. Distribution of the Mediterranean monk seal. *Monterey Bay Aquarium Graphics.*

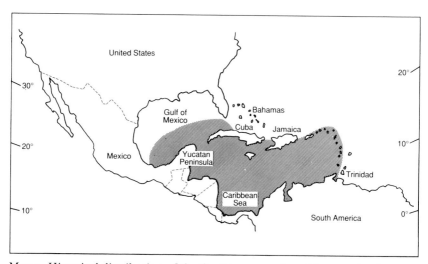

MAP 3. Historical distribution of the Caribbean monk seal (now extinct). *Monterey Bay Aquarium Graphics.*

Asia substantially reduced or closed the Tethyan connection before the primitive pinnipeds were able to disperse into the Indian Ocean.

DIFFERENCES BETWEEN PHOCIDS, OTARIIDS, AND ODOBENIDS

There are a number of fundamental morphological, behavioral, and reproductive differences between the three pinniped families: the Phocidae, the Otariidae, and the Odobenidae (Table 4). Although morphology has often been emphasized in distinguishing the families in

TABLE 4

Differences between Phocid, Otariid, and Odobenid Seals

PHOCIDS	OTARIIDS	ODOBENIDS
No external ears	Visible external ears	No external ears
Furred hindflippers that cannot be turned forward under the body	Hairless hindflippers that can be turned forward under the body and used for movement on land	Hairless hindflippers that can be turned under the body and used for movement on land
Mammae with 2 teats (except monk and bearded seals, which have 4)	Mammae with 4 teats	Mammae with 4 teats
Mostly marine; also freshwater and estuarine	Marine; occasionally ascend freshwater rivers	Exclusively marine
Variable breeding systems	Uniform breeding systems; all polygynous	Polygynous breeding system (resembling "mobile" lekking system)
Variable sexual dimorphism (monomorphism, pronounced sexual dimorphism, and slight reverse sexual dimorphism)	Pronounced sexual dimorphism	Pronounced sexual dimorphism
Time of mating variable (during lactation or after pup is weaned)	Mating takes place a few days after birth	Mating takes place during lactation
Lactation period generally brief, from 4 days to several weeks	Lactation period relatively long, from several months to over 2 years	Protracted lactation, lasting 2 or more years

Continued on next page

TABLE 4, *continued*

PHOCIDS	OTARIIDS	ODOBENIDS
Annual parturition	Parturition usually annual; sometimes biennial	Parturition biennial; longer interval between births elapses in older females
Females and males of many species fast or feed little during breeding and lactation	Females of all species feed during lactation; territorial males often fast during breeding	Females and males feed throughout breeding and lactation; feeding is reduced during northward migration
Milk tends to be rich in fat	Milk tends to be less fatty	Milk tends to be less fatty
Rapid growth rate of pups	Slower growth rate of pups	Slower growth rate of pups

previous reviews, each group has distinctive behavioral and reproductive attributes that sets it apart. King (1983) provides a detailed list of anatomical features characterizing each of the three families. One obvious feature distinguishing otariids from phocids is the pinna, or small, furry earflaps on the otariids. Appropriately enough, otariids are often called eared seals, while phocids are sometimes called earless seals, since their ears are not plainly visible (Figs. 2.4, 2.5).

The other major morphological difference between the two groups is reflected in their movement on land and in water. All otariids have hindflippers that can be turned under the body and used to move about on land (Fig. 2.6). The hindflippers of phocids, in contrast, cannot be turned forward under the body for "walking," so that phocid seals usually move more slowly and awkwardly on shore, bouncing and lurching. In the water, however, even the pinnipeds most cumbersome on land become swift and agile swimmers. Otariids swim differently from phocids, using their long front flippers to propel themselves through the water (Fig. 2.7). Phocids, in contrast, swim by moving their rear flippers and lower body in a lateral, or side-to-side, sculling motion.

One also finds distinctive biological and behavioral differences between the two groups having to do with their diverse systems of reproduction. Otariid seals have similar breeding systems, while those of phocids are much more variable. All sea lions and fur seals are highly polygynous, and adult males are usually territorial. The territorial males generally fast during the breeding season. Sexual dimorphism is pronounced, with adult males two to four times larger than adult females. Pregnant otariid

FIG. 2.4. Ringed seal; note characteristic phocid ear. *Photo by Lori Quakenbush.*

FIG. 2.5. Guadalupe fur seal female, showing typical otariid ear. *Photo by Phil Thorson.*

FIG. 2.6. California sea lion, moving on land in typical otariid fashion. *Photo by Frans Lanting.*

FIG. 2.7. California sea lion swimming underwater, using long foreflippers. *Photo by Jim Watanabe.*

females haul out onto a male's territory, give birth soon after arriving, and mate a few days after parturition. Females then alternate periods of feeding at sea for several days with shorter bouts of nursing their pup on land. This pattern of feeding at sea and nursing on land continues for several months until the pup is weaned.

The breeding behavior of phocids, in contrast, is less easily categorized. Mating systems, as well as the degree of sexual dimorphism, are variable. Some phocids, such as the Arctic ringed seal, breed in solitary family groups, with an adult male and a female-pup pair making up one such unit. Named after the distinctive grey-white rings on the dark backs of adults, ringed seals are monomorphic; that is, females and males are of equal size. Large size helps a mother to provide her pup with milk rich in fat and protects her against the polar cold. Large size in the male is not necessary for successful reproduction, since the mating system tends toward monogamy and there is really no need for a male to dominate others (Fig. 2.8). In contrast, other phocids, like the elephant seal, are highly polygynous and extremely sexually dimorphic, with males forming dominance hierarchies to compete for females. Elephant seals and other phocids, such as grey seals, breed in large groups that may contain hundreds of animals.

FIG. 2.8. Adult female and male harp seals of monomorphic, or similar, size. The male is attempting to mate with the female. *Photo by Kit Kovacs.*

Most phocid females, unlike otariids, nurse their pups for short, intensive periods and often fast or feed very little during lactation. While all otariids copulate soon after parturition, most phocids mate during late lactation or after weaning their pup. The length of lactation varies from less than a week in hooded seals to as long as six weeks in some populations of ringed seals. The milk of fasting phocids tends to increase in fat content during lactation, whereas the composition of otariid milk, which appears to contain somewhat less fat than phocid milk, probably varies little over the long nursing period. These divergent patterns concerning variation in milk composition during lactation reflect the respective nursing strategies (fasting versus feeding) and the length of lactation in each family.

OTARIIDS: FUR SEALS AND SEA LIONS

Otariids are divided into two subfamilies: the Arctocephalinae (fur seals) and the Otariinae (sea lions). The two groups probably diverged roughly two million years ago. In fact, all fur seals used to be called sea-bears; the genus name *Arctocephalus* derives from two Greek words, *arktos*, "bear," and *kephalē*, "head." The five species of sea lions have a rounded snout, and their pelage is short and coarse (Fig. 2.9; Maps 4–6). Fur seals are characterized by a more pointed muzzle, longer foreflippers, and a thick,

FIG. 2.9. Australian sea lions. *Photo by Dan Costa.*

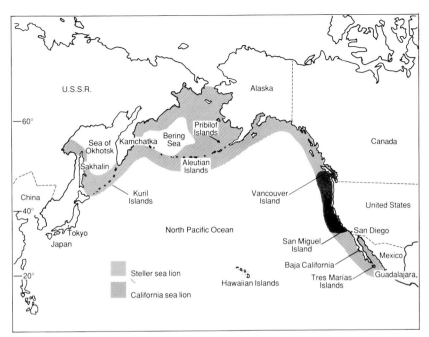

MAP 4. Distribution of the Steller sea lion and the California sea lion. See Map 6 for the distribution of the Galápagos sea lion (*Z.c. wollebaeki*), a subspecies of California sea lion. *Monterey Bay Aquarium Graphics.*

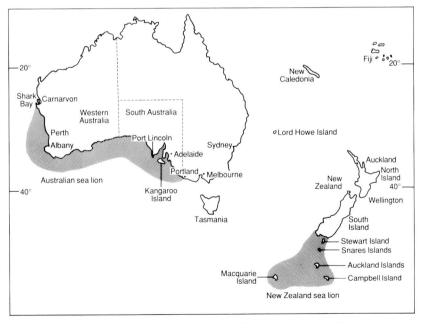

MAP 5. Distribution of the New Zealand sea lion and the Australian sea lion. *Monterey Bay Aquarium Graphics.*

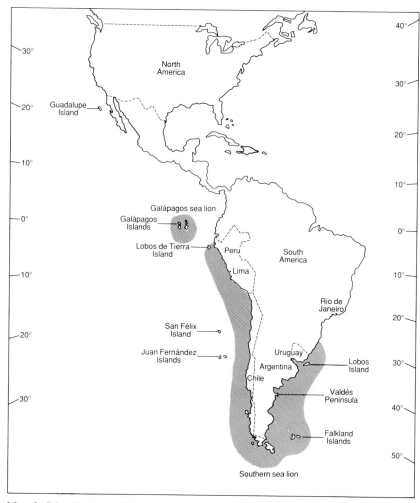

MAP 6. Distribution of the southern sea lion and Galápagos sea lion. *Monterey Bay Aquarium Graphics.*

luxuriant coat. Their waterproof underfur is covered with longer "guard" hairs, often giving them a somewhat grizzled appearance. In addition, most fur seals are smaller than sea lions. The similar social and reproductive behavior of sea lions and fur seals has even led some taxonomists to suggest that the differences between these otariids are not great enough to justify separating them into the subfamilies Otariinae and Arctocephalinae (Repenning 1976; Repenning and Tedford 1977). Moreover, several instances of hybrid offspring produced by matings of *Zalophus* (California sea lions) and *Arctocephalus* have been reported (see, e.g., Mitchell 1968).

FIG. 2.10. Northern fur seals; note distinctive flattened profile and shorter snout. *Photo by Dan Costa.*

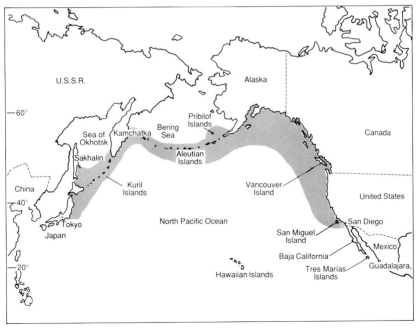

MAP 7. Distribution of the northern fur seal. *Monterey Bay Aquarium Graphics.*

Because there still are behavioral and biological differences between the two subfamilies, however, I will subdivide the otariids into the fur seals and sea lions.

The northern fur seal (*Callorhinus*) differs in several respects from the other eight species of fur seals (*Arctocephalus*), and the two genera have probably evolved separately for a relatively long time (Fig. 2.10). *Callorhinus* is found in subarctic waters of the North Pacific (with the exception of a small population on San Miguel Island off California) whereas

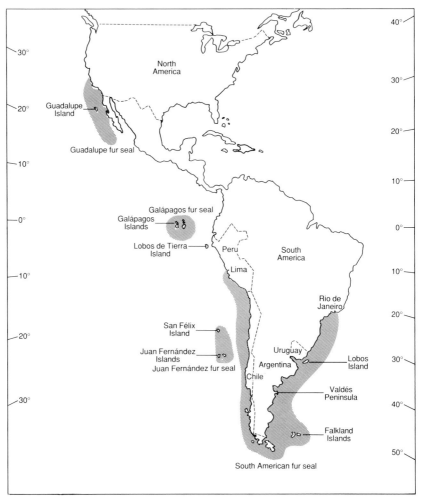

MAP 8. Distribution of the South American fur seal, Juan Fernández fur seal, Galápagos fur seal, and Guadalupe fur seal. *Monterey Bay Aquarium Graphics.*

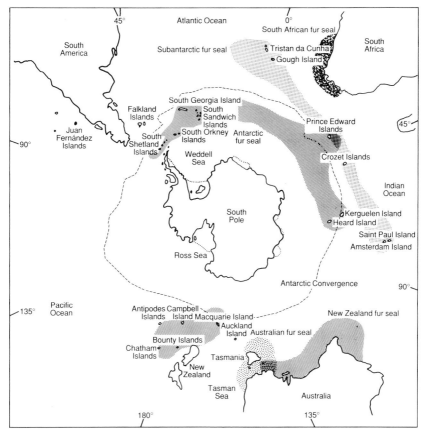

MAP 9. Distribution of the Antarctic fur seal, subantarctic fur seal, South African fur seal, Australian fur seal, and New Zealand fur seal. *Monterey Bay Aquarium Graphics.*

most other fur seals live in the southern hemisphere (Maps 7–9). Moreover, northern fur seals are more pelagic than most other species of fur seals (although the Antarctic fur seal also spends a great deal of time at sea). In fact, the northern fur seal spends 9–10 months of the year at sea. In addition, a number of morphological features of the skull and foreflippers are unique to northern fur seals. For instance, they have a shorter rostrum that curves downwards, creating a "squashed in" profile. The northern fur seal is also characterized by longer ear pinna, elongated hindflippers, and distinctive foreflippers in which the fur ends abruptly along a line across the wrist; in other fur seals the fur descends over the foreflipper metacarpals (Gentry 1981a).

PHOCID SEALS

Phocids used to be split into three subfamilies: the Cystophorinae, Mon-achinae, and Phocinae. The Cystophorinae included the hooded seals (*Cystophora*) and elephant seals (*Mirounga*). The adult males of both genera are characterized by a long elephantine proboscis. Hooded seal males (also called bladdernose seals) can blow from one nostril a strange-looking red balloonlike sac that may attain the size of a baseball (Fig. 2.11). Most males are left-nostril balloon blowers, although a few males have been seen with the red balloon blown from the right nostril (D. M. Lavigne, pers. comm.). More recent classification schemes eliminate the subfamily Cystophorinae, placing *Mirounga* in the Monachinae and *Cystophora* in the Phocinae on the basis of skull and skeletal characteristics (King 1983). Therefore, the two phocid subfamilies include only the Monachinae and Phocinae. Monachines consist of the two species of tropical monk seals, the four species of Antarctic seals, and the two species of elephant seals that generally inhabit temperate waters (Maps 10, 11) (although some southern elephant seal populations also live on ice or snow in the subantarctic).

The monachines used to be divided into two tribes, the Monachini (monk seals) and the Lobodontini (Antarctic seals), although the differ-

FIG. 2.11. Hooded seal adult male inflating its hood. *Photo by Kit Kovacs.*

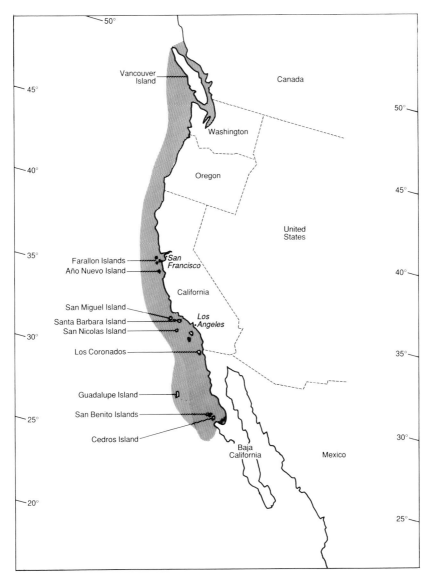

MAP 10. Distribution of the northern elephant seal. *Monterey Bay Aquarium Graphics.*

ences between the tribes are small and fairly insignificant. With the addition of elephant seals to the Monachinae, however, many no longer recognize tribes within the subfamily (Hendey and Repenning 1972; King 1983). Accordingly, I will not subdivide the monachines in this review.

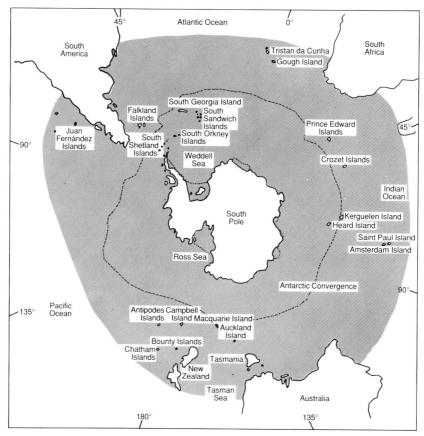

MAP 11. Distribution of the southern elephant seal. *Monterey Bay Aquarium Graphics.*

The Hawaiian monk seal of the Monachinae appears to be the most primitive phocid seal and has been aptly termed a "living fossil" (Repenning and Ray 1977). Muizon and Hendey (1980) have separated the four genera of Antarctic seals into two groups on the basis of dental and cranial features: the Weddell-Ross seal group and the crabeater-leopard seal group. The latter group has more specialized cheek teeth, with well-developed accessory cusps that seem to aid in straining plankton; better developed molars; and a longer snout, among other characteristics, than the Weddell and Ross seals (Map 12).

The Phocinae were previously split into only two tribes: the Erigna-thini, which contained just one species (the bearded seal), and the Pho-

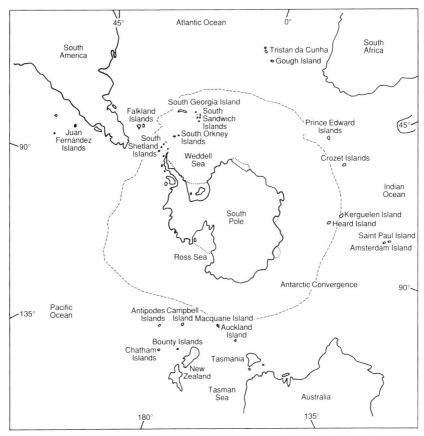

MAP 12. Distribution of the Weddell seal, crabeater seal, Ross seal, and leopard seal. All the Antarctic seals occur primarily inside the Antarctic convergence, although there have been numerous sightings north of the convergence. *Monterey Bay Aquarium Graphics.*

cini. With the addition of hooded seals to the phocines, a third tribe is now recognized: the Cystophorini (Burns and Fay 1970). King (1983) lists morphological differences between the three tribes in detail. All phocines live in the Northern Hemisphere, and many live in Arctic or subarctic waters (Map 13). The bearded seal is the largest of the Arctic phocids. As its name suggests, the bearded seal sports a remarkable abundance of long sensitive whiskers on its snout (Fig. 2.12).

The Phocini consist of the grey seals (Map 14) which are found in the North Atlantic and represent the largest seals of the Phocinae, and the widespread genera *Phoca*, whose members are made up of several diverse species found mainly in the Arctic and temperate areas of the North

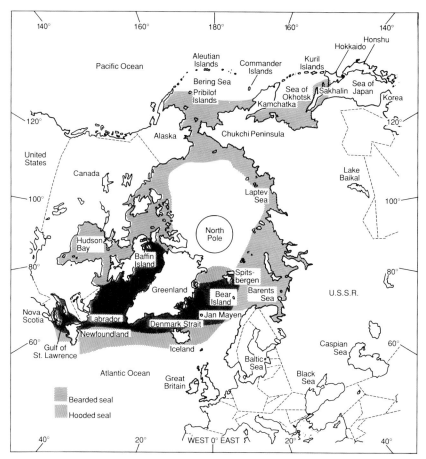

MAP 13. Distribution of the bearded seal and hooded seal. *Monterey Bay Aquarium Graphics.*

Pacific and North Atlantic (Maps 15, 16). Appropriately enough, the Greek word *phōkē* is derived from a Sanskrit term meaning "swollen or plump animal." The grey seal's scientific name *Halichoerus* comes from Greek words meaning "sea pig." Canadians also call grey seals horseheads. Several species of *Phoca* live exclusively in large freshwater lakes in Russia and Canada. The ringed, Baikal, and Caspian seals are the most abundant and widely distributed of the Arctic phocids (Map 17). The three species have in common a relatively small size, a delicate skull, and an attachment to ice; they are often included in the subgenus *Pusa* (Frost and Lowry 1981). Moreover, all three are important in subsistence and commercial fishing.

Fig. 2.12. Large bearded seal pup. *Photo by Lloyd Lowry.*

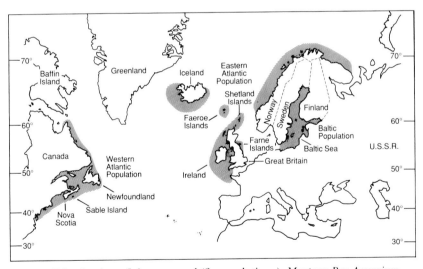

Map 14. Distribution of the grey seal (3 populations). *Monterey Bay Aquarium Graphics.*

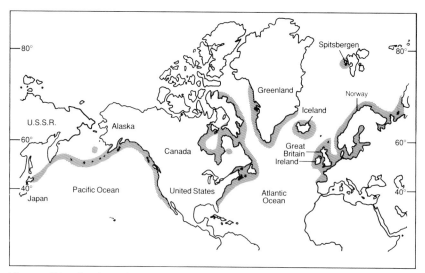

MAP 15. Distribution of the harbor seal (5 subspecies). *Monterey Bay Aquarium Graphics.*

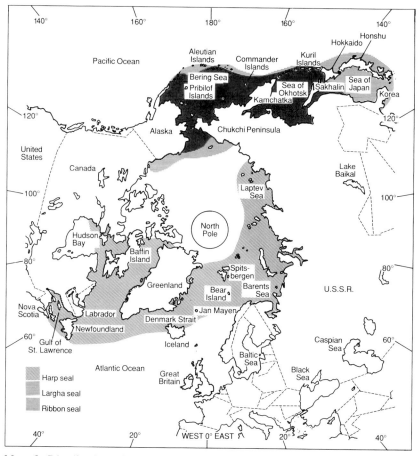

MAP 16. Distribution of the harp seal, largha seal, and ribbon seal. *Monterey Bay Aquarium Graphics.*

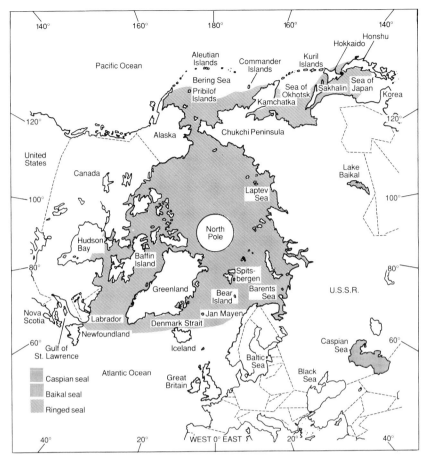

MAP 17. Distribution of the Caspian seal, Baikal seal, and ringed seal.
Monterey Bay Aquarium Graphics.

THE ODOBENIDS

Today, walruses live only in circumarctic waters near the pack ice (Map 18). The walrus family contains only one extant species, which is split into two subspecies: the North Atlantic walrus (*Odobenus rosmarus rosmarus*) and the North Pacific walrus (*O.r. divergens*). The scientific name *Odobenus rosmarus*, which means "tooth-walking seahorse," was probably derived from the walrus's habit of using its long tusks to haul out on ice. The North Pacific walrus is larger sized, longer tusked, and more abundant than its Atlantic counterpart. The more sedentary Atlantic walruses are patchily distributed along the Canadian archipelago east to Greenland and the Soviet Union as far as the Laptev Sea; Pacific walruses are most numerous in the Chukchi and Bering seas.

The odobenids share morphological and behavioral features with both the phocids and otariids, but odobenids are also distinguished by several unique characteristics. The largest and heaviest pinnipeds (with the exception of male elephant seals), the odobenids have foreflippers proportionately smaller than those of otariids. Walruses usually swim with their hindflippers like phocids, their foreflippers acting as stabilizers. On land or ice, however, they turn their hindflippers under their bodies and move like otariids. Like phocids, the walruses lack the external ear pinna visible on otariids. The upper canines of the walrus have enlarged to become enormous tusks, which appear to serve many functions, especially in social and breeding contexts (Fig. 2.13). The unique

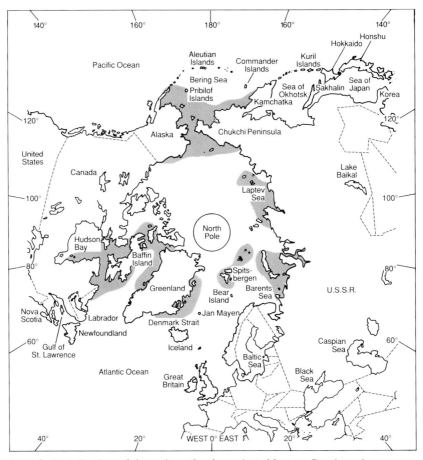

MAP 18. Distribution of the walrus (2 subspecies). *Monterey Bay Aquarium Graphics.*

FIG. 2.13. Walrus, showing distinctive tusks. *Photo by Anne Hoover.*

substrate method of feeding practiced by walruses also distinguishes them from the other two pinniped families. Walruses use their fleshy snout, long vibrissae, and possibly their heavy tusks to probe the ocean mud for clams and other bivalve molluscs in shallow-water areas.

Walruses are the only pinniped that typically gives birth every two years. (Otariids, although some may nurse yearlings or two-year-olds, postponing birth of a new pup, are capable of pupping every year.) In fact, older female walruses may produce a pup only every three or more years. The single calves are nursed for two years or more. The mating system of odobenids appears to be polygynous. During the breeding season, walruses form large aggregations on the pack ice. Because pack ice tends to break up and drift, the breeding animals often move continuously within the vast breeding grounds. Adult males compete for the choice sites near groups of potentially receptive females. Dominant bulls give a "performance" of visual and vocal displays near the female groups, involving a distinctive bell-like sequence of underwater vocalizations. No other pinniped male directs such elaborate displays at breeding females.

3

Ecology

When we take seals from the ocean, what holes are left?
VICTOR SCHEFFER (1975)

At the present time, the answer to Victor Scheffer's question is largely unknown. But there is little doubt that some pinnipeds play an important role in marine and lacustrine, or lake, ecosystems and that, as top consumers, pinnipeds have some degree of impact on various marine communities. Recent studies of walruses, lake-dwelling seals, and other pinnipeds are beginning to shed light on the relationships between seals and their marine environment. In this chapter, I provide a relatively brief and limited overview of the ecology of pinnipeds as well as a description of their various marine and terrestrial habitats. More detailed information on the interactions between pinnipeds and marine communities as well as their population ecology and conservation can be found in Bonner (1982), Hofman and Bonner (1985), McLaren and Smith (1985), Laws (1977a, 1984) and May (1986).

Pinnipeds and the Marine Ecosystem

The role of pinnipeds and most other marine mammals in the organization of pelagic and soft-sediment (inshore sandy or muddy substrates) communities is poorly known and logistically difficult to study. Sea otters and their nearshore environment, in contrast, are relatively easy to observe, and the effect of sea otter predation in structuring rocky inshore communities is dramatic and well documented (Fig. 3.1) (see, e.g., J. A. Estes and Palmisano 1974; VanBlaricom and Estes 1987; Riedman and

FIG. 3.1. Sea otter underwater, holding a sea urchin. *Photo by Richard Mattison.*

Estes 1987). In many areas of California and Alaska, sea otters reduce and effectively limit populations of large invertebrate prey such as sea urchins, abalone, and clams, and this reduction, in turn, influences the structure of nearshore kelp forest communities. According to J. A. Estes (1979), the role of marine mammal predation in pelagic communities may be less important than in rocky nearshore communities. That is, the role of a marine mammal like the sea otter in structuring such marine communities should not necessarily be used as a model for interactions between other marine mammals and their environment. The ecological and evolutionary roles of predators in a number of systems structurally similar to those of pinnipeds and other marine mammals, however, are known to be important.

Although it is extremely difficult to study the relationship between predator and prey among free-ranging pinnipeds in the open ocean, such interactions have been studied among seals inhabiting freshwater lakes in Canada and Russia. For instance, harbor seals (*Phoca vitulina mellonae*) inhabiting freshwater landlocked lakes in Quebec were found to strongly influence freshwater fish communities (Power and Gregoire 1978). Predation by the resident seals of Lower Seal Lake resulted in significantly fewer and smaller lake trout than in similar lakes without seals. The lake trout subject to harbor seal predation not only had a high mortality rate but grew more quickly and reproduced more often than those in lakes without seals. Trout are most vulnerable to seal predation while spawning. Apparently the lake trout, which are aggregating lake spawners, are more vulnerable to seal predation than brook trout, which are dispersed stream spawners and breed in tributary streams where the seals are unable to reach them. The brook trout seem to be more common in lakes with seals, whereas lake trout are the most abundant species in lakes without seals.

As J. A. Estes (1979) points out, however, interactions between marine systems and pinnipeds in the open ocean, especially in deep sea benthic systems, may differ from those between aquatic communities in freshwater lakes, where fish and seals are confined in a more limited space. The effects of walrus predation on shallow marine soft-sediment benthic communities, however, have been studied. Fay et al. (1977) have estimated the impact of Pacific walruses on the food resources of these communities. They calculate that free-ranging walruses consume an average of 6.2 percent of their total body weight per day in benthic invertebrates in terms of net rates of predation. That is, a 1,000-kilogram walrus must actually take 180–240 kilograms of molluscs per day to achieve a net intake of 60 kilograms of molluscan soft parts per day. A single walrus may eat over six thousand individual animals during one feeding bout.

Fay (1982) further estimates that the Pacific walrus population as a whole consumes about 8,900 tons of food each day, which amounts to a net intake of over 3.2 million tons per year. The gross rate of predation on benthic invertebrates therefore amounts to a biomass of 2.5–3 percent of the estimated standing stock of benthos on the Bering-Chukchi shelf (equivalent to 9.5–12.6 million tons per year). But because walruses are concentrated only in certain areas of the Bering-Chukchi shelf, especially during winter, the population no doubt has a much greater effect on food resources in some areas than in others. In fact, the benthic mollusc populations in the walrus wintering areas appear to be subject to the most significant impact, and Fay (1982) believes that the growth of the walrus population may be most severely limited by food resources in these areas.

In plowing through such vast amounts of bottom sediments, walruses not only dramatically influence the structure of macrobenthic assemblages but may also substantially contribute to the productivity in these areas by releasing nutrients that otherwise would have remained trapped in the bottom sediments (J. S. Oliver et al. 1983b, 1985; Fukuyama and Oliver 1985). Grey whales have recently been shown to play a similarly important role in nutrient recycling in the Bering and Arctic seas as well as in British Columbia (J. S. Oliver and Kvitek 1984; J. S. Oliver et al. 1984; J. S. Oliver and Slattery 1985; Kvitek and Oliver 1986; C. H. Nelson and Johnson 1987).

Pinnipeds and other marine mammals for the most part occupy the top trophic positions in marine food webs. (The exclusively herbivorous sirenians are an exception.) While predation on some pinniped populations by nonhuman predators may possibly limit these populations (e.g., Hawaiian monk seals, Galápagos sea lions, and fur seals, all discussed in chapter 4), in the long term most pinniped populations do not appear to be limited by biological or physical disturbances (e.g., by nonhuman predators or changing environmental conditions)—at least for most species there is little evidence to show such effects. More research is needed, however, to answer this question. Pinniped populations appear to be limited by available resources; that is, they are regulated to some degree by food and perhaps by space for breeding (J. A. Estes 1979). Of course, commercial exploitation of some pinnipeds over the past two hundred years has brought some of these species to the brink of extinction. But in this chapter I am referring only to natural predation.

Whether pinniped predation in the ocean seriously reduces the populations of prey remains to be determined. As yet there is no direct evidence to indicate that this happens. It is often difficult to obtain all the information necessary to estimate the impact of a pinniped on its food resources, such as the size and composition of the seal population, its diet, and the daily food requirements of each age class with respect to time of year (Bonner 1982). McLaren and Smith (1985) summarize some of the models developed by May et al. (1979), Laevastu and Larkins (1981), and Laws (1983) relating pinnipeds, their food webs, and the marine ecosystem.

According to Lavigne (1982a), there is no evidence that pinnipeds, or any other marine mammals besides the sea otter, are able to compete effectively with commercial fisheries for food resources. In fact, highly efficient modern fisheries may have an adverse impact on some seal populations (Hofman and Bonner 1985). For instance, female harp seals studied in 1978 after the failure of the North American capelin fishery were found to have depleted energy reserves in comparison to those studied in 1976 (S. R. Innes et al. 1978).

Unfortunately, there are considerable methodological problems with investigating predator-prey relationships among pinnipeds. The ideal approach would be to add the seal to a marine system where it is absent or to remove it from one where it presently occurs. Such methods, however, pose immense logistical as well as political problems. A great deal could be learned from pinniped populations recovering from intensive human exploitation. Although it is already a few decades too late to implement such studies with most of the pinnipeds, they have been implemented with sea otters at San Nicolas Island off southern California. In the summer of 1987 the U.S. Fish and Wildlife Service began the translocation of a small group of mainland California sea otters. The kelp forest communities at San Nicolas Island have been well studied for a number of years, and researchers are excited about the opportunity to document the effects of otter predation in such detail.

Fisheries conflicts with pinnipeds occur throughout the world, yet there is little evidence that pinnipeds either compete with or offer large-scale competition to or damage major commercial fisheries. The localized impact of pinniped predation on fisheries in some areas, however, can be significant. Bonner (1982) discusses three principal ways that seals may affect fisheries: (1) by damaging fishing gear and fish already caught in nets, (2) by preying on commercially exploited fish stocks, and (3) by serving as a definitive host (the one in which a parasite reaches sexual maturity) of parasites that spend their larval stages in food fishes. The destruction of fishing gear caused by seals attempting to feed on captured fish is perhaps the most obvious form of damage, at least from the fisherman's point of view. The impact of seal predation on commercially desirable fish stocks is much less clear. And the role played by seals as a host of parasites that spend part of the life cycle in food fishes represents a limited problem in a few areas. For instance, the parasitic codworm is commonly found in grey seals, which shed the codworm eggs into the water with their feces. After hatching, the larvae encyst in benthic crustaceans, which are then eaten by fish (Platt 1975). Although the parasitic worms are harmless to humans, many people find fish infested with them unappetizing, and infested fish have a lower market value.

Some researchers have attempted to estimate the impact of pinnipeds on fish stocks and on commercial fisheries. The opportunistic feeding behavior of the northern fur seal has been analyzed in detail by Kajimura (1984), who found that the seals feed on sixty-three species of fish and squid, with the dietary composition varying by month and location. Although a few of the fish are commercially valuable, most are not. Summers et al. (1978) estimate that grey seals and harbor seals of the North Sea consume 120–133 million kilograms of fish per year. Although the potential overall impact on fish stocks and fisheries is unknown, the effect on local fisheries can be severe, especially on those using set nets. Harwood and Croxall (1988) review both the interactions between seals (grey seals and harbor seals) and commercial fisheries in the North Atlantic and the feeding ecology of Antarctic pinnipeds (crabeater seals,

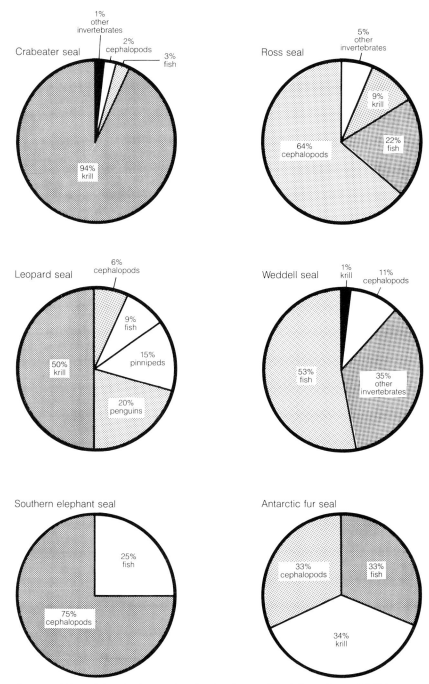

FIG. 3.2 Diet of the six Antarctic pinnipeds (modified from Laws 1977b).
Monterey Bay Aquarium Graphics.

Weddell seals, southern elephant seals, and Antarctic fur seals). They conclude that more information is required on the seals' foraging efforts, the distribution and movements of their prey, and the fisheries' efforts in space and time before the impact on both fisheries and seal populations can be determined. Fortunately, such information may be forthcoming because of recent advances in monitoring pinnipeds using telemetry equipment.

The role of pinnipeds and other vertebrates in the relatively simple Antarctic marine ecosystem has been reviewed by Laws (1977a,b; 1984), T. Øritsland (1977), and Bonner (1982). Each of the four species of Antarctic phocids, in addition to Antarctic fur seals and southern elephant seals in the northern portion of their range, occupies a distinctive position in the Antarctic ecosystem. Two or more species that feed on the same food resource tend to be geographically separated. Conversely, when the range of two or more species overlaps, the seals tend to utilize different food resources. Both the Ross seal and the crabeater seal, for instance, live on the pack ice, yet each species exploits different food resources: the Ross seal eats mainly squid and fish; the crabeater seal consumes krill almost exclusively. Similar divisions of food resources are found north of the pack ice where southern elephant seals and Antarctic fur seals live together. Elephant seals feed on fish and squid; fur seals consume primarily krill. Although the leopard seal has the most extensive distribution of the Antarctic phocids, its range overlaps with that of Weddell seals near the Antarctic continent and over the shelf. While Weddell seals feed on fish, squid, and other invertebrates, leopard seals eat mainly warm-blooded prey and krill. The diet of the six Antarctic species is depicted in Fig. 3.2.

Laws (1977a) roughly estimates the annual food consumption and population biomass of each Antarctic pinniped, relying on the population estimates of Erickson et al. (1971) and Gilbert and Erickson (1977) and on the food consumption data of T. Øritsland (1977). The size of the populations used to estimate food consumption, however, was probably substantially underestimated. Crabeater seals (Fig. 3.3) consume an incredible amount of krill—over 63 million tons each year—a figure that is probably an underestimate. In comparison, Antarctic baleen whales are estimated to consume about 43 million tons of krill every year. As Bonner (1982) points out, the burgeoning commercial interest in harvesting Antarctic krill could threaten pinnipeds as well as other animals in the southern ocean, such as penguins and whales, that depend on krill for their food supply.

Gentry and Kooyman (1986) correlate population sizes in otariids (five fur seals and the Galápagos sea lion) with a number of interrelated

FIG. 3.3. Crabeater seal. *Photo by Gerry Kooyman.*

variables, including the geographic range of each species, the predictability of the environment each inhabits, and the local upwelling and abundance of food resources available to the seals. The tropical otariids have relatively small populations (all tropical species together number only about forty thousand animals), whereas the population of each subpolar otariid exceeds one million. In attempting to explain these respective differences in abundance, Gentry and Kooyman point out that subpolar environments are generally more predictable in productivity (subpolar waters being richer in oxygen and upwelling) than tropical areas. In addition, marine environments nearest the equator are heavily affected by the El Niño Southern Oscillation (ENSO), which causes water temperatures and the availability of food resources to change dramatically from year to year or every few years, making the tropics even less predictable (Map 19a, b). Otariids inhabiting the tropical Galápagos Islands* live nearest to the equator in one of the least predictable environments. Galápagos sea lions depend on a narrow band of upwelling around the islands as well as on seasonal incursions of the cold nutrient-rich Humboldt Current for their food resources (Fig. 3.4) (Barber and Chavez 1983).

* Although situated on the equator, the thirty-seven or so islands that make up the Galápagos Islands undergo seasonal cold water incursions from the Humboldt Current, the Cromwell countercurrents, and localized upwelling. Temperatures on land, however, may be extremely warm.

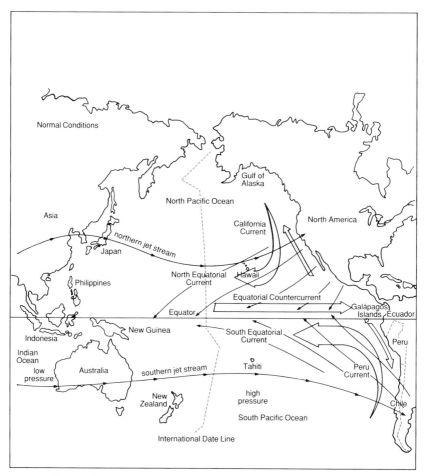

MAP 19. Pacific Ocean currents under (a) normal conditions and (b) El Niño conditions. *Monterey Bay Aquarium Graphics.*

The effects of El Niño events on these rather limited food resources tend to be much more severe in equatorial than in subpolar regions. The greater the distance of the seal populations from the equator, the less the effects of El Niño on their environments (because upwelling continues farther from the equator despite the effects of El Niño).

El Niño events may regulate the population of tropical otariids by causing substantial mortality of young in certain years. With the successive losses of potential breeding females that would result, the population could not recover swiftly from a particular year's loss. The production of young only every two to three years in Galápagos otariids, as opposed to annual pup production in subpolar species, further inhibits recovery

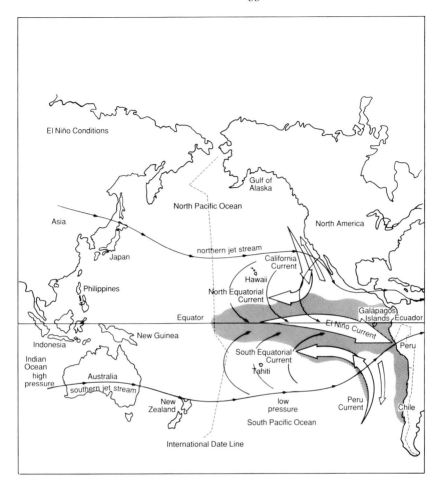

from population declines. Gentry and Kooyman (1986) also suggest that the availability of food resources affects population sizes of otariid seals. Information on foraging behavior in the various fur seals (length of feeding trips, transit time, etc.) indicates that foraging grounds for otariids (and other pinnipeds, for that matter) are much more extensive and food availability is greater in subpolar than in tropical environments. Subpolar waters are therefore able to support much larger otariid populations. The profound effects of ENSO events in the eastern tropical Pacific Ocean on a female's reproductive success, on pup survival, and on population dynamics in a number of pinnipeds are discussed in chapter 8.

FIG. 3.4. Galápagos sea lions. *Photo by Steve Webster.*

Pinnipeds and Their Environment

AQUATIC HABITATS

The most important variables influencing the choice of aquatic habitat seem to be the distribution, abundance, and type of food resources used by a particular species, since seals spend most of their time in the ocean foraging and migrating to and from feeding grounds and breeding rookeries. Very broadly, pinnipeds inhabit five different aquatic habitats, categorized here primarily in terms of major oceanic and freshwater divisions: (1) coastal waters, (2) the open ocean, (3) estuaries and bays, (4) freshwater lakes, and (5) rivers. Although most pinnipeds live in coastal waters, some feed far offshore in deep waters off oceanic islands, such as Guadalupe, Juan Fernández, Hawaii, Auckland, the Falklands, South Georgia, Macquarie, and Kerguelen. These aquatic zones vary dramatically depending on latitude and the area the species lives in — tropical, temperate, or polar.

Many seals, however, inhabit overlapping areas. For instance, harbor seals may live in coastal areas, bays, and estuaries as well as freshwater lakes. Habitat also varies with the time of year in many species that undergo seasonal migrations. The basic ecosystems inhabited by pinnipeds depend on each group's various food resources, which are generally found in certain marine communities. Pinnipeds tend to exploit (1) rocky inshore/kelp forest communities; (2) soft-sediment bottom communities; (3) pelagic communities; and (4) freshwater (often lacustrine) communities.

Perhaps another hallmark of the adaptability of pinnipeds is that some of them are able to live in fresh water exclusively and others can live in both fresh water and seawater (although some cetaceans, such as river dolphins, also inhabit fresh water). Seals of the subgenus *Pusa* inhabit landlocked freshwater lakes in Russia. Both the Caspian Sea and Lake Baikal contain resident species: the Caspian seal and the Baikal seal. Two subspecies of ringed seals (*Phoca hispida saimensis* and *P.h. ladogensis*) also inhabit landlocked lakes in Russia near the Baltic Sea. In some areas of their range, harbor seals may remain in lakes and rivers throughout the year. Iliamna Lake in Alaska is inhabited by harbor seals (*Phoca vitulina richardsi*) all year. Some Canadian harbor seals (*P.v. mellonae*) also live in lakes in Quebec (see the discussion at the beginning of this chapter) (Beck et al. 1970; Power and Gregoire 1978).

Other pinnipeds travel freely between the ocean and freshwater rivers and lakes. Harbor seals (*P.v. richardsi*) in the Pacific Northwest that live along the coast often ascend rivers and freshwater lakes in areas such as the Gulf of Alaska (Pitcher and Calkins 1979) and the southeastern Bering Sea (Everitt and Braham 1980). The eastern Atlantic harbor seal (*P.v. vitulina*) swims up rivers in Europe (King 1983). Juvenile bearded seals also occasionally ascend the rivers that empty into the Bering and Chukchi seas (Burns 1981a). California sea lions generally inhabit coastal areas, but some sea lions also venture into rivers in many areas of the North Pacific (see, e.g., Mate 1973, 1975). Southern sea lions occasionally enter fresh water in the Uruguay and Valdivia rivers (Schlatter 1976).

TERRESTRIAL HABITATS

The amphibious pinnipeds spend part of their lives in a variety of terrestrial habitats on land or ice while breeding, raising young, molting,

FIG. 3.5. California sea lions hauled out on a jetty. *Photo by W. E. Townsend, Jr.*

and resting. The amount of time each species spends on land varies. Northern fur seals spend only about one month on land each year (Gentry, pers. comm.), whereas high-ranking elephant seal bulls may remain ashore for over three months. An especially important variable in the selection of a terrestrial habitat is suitability for breeding activities. In polar and subpolar regions of the world, pinnipeds inhabit many types of sea ice as well as glacial ice. Pinnipeds living in temperate and tropical climates live on a variety of substrates on islands and along the mainland: cobblestone and sandy beaches, sloping rock outcroppings and rock platforms, emergent offshore tidal rocks, rocky shorelines with large boulders and rubble, sandbars and mudflats, sheltered crevices and sea caves, and tidepools. California sea lions and harbor seals often haul out on such man-made structures as piers, jetties, offshore buoys and oil platforms (Fig. 3.5).

Some pinnipeds even venture well inland at times. Australian sea lions move inland to take shelter in coastal vegetation and dunes, especially during rainy or windy weather (Walker and Ling 1981a). These sea lions are also adept at scaling steep cliffs and have been known to climb to heights of 100 feet. Antarctic fur seals sometimes haul out in tussock grass at South Georgia Island (Fig. 3.6). Southern elephant seals also often lie in tussock grass. Northern elephant seals at Año Nuevo may crawl inland and rest in sand dunes, vegetation, or even muddy pools of fresh water. It is generally the subadult and adult males who venture inland, perhaps to escape the ceaseless aggression and battling of the dominant breeding bulls. In some of the more dense and crowded grey seal breeding colonies, such as the Farne Islands and North Rona, females and pups rest far inland from the beach. When female grey seals give birth away from the ocean, they often pup near a stream or freshwater pool (S. S. Anderson et al. 1975; Bonner 1981a). Crabeater seals seem to have traveled both higher and farther inland than any other seal: a live male pup was found 113 kilometers from open water on Crevasse Valley Glacier (Stirling and Rudolph 1968), and a mummified 100-year-old crabeater seal carcass was discovered 1,100 meters above sea level on Ferrar Glacier near McMurdo Sound (E. A. Wilson 1907).

All seals prefer isolated and undisturbed areas, although some species, such as California sea lions and harbor seals, may habituate to human presence. In many species, such as Hawaiian monk seals and grey seals, human disturbance can cause females to abandon their pups, resulting in increased pup mortality. Northern elephant seal females that give birth singly or in small groups in very recently colonized areas are also sensitive to human intrusion. In general, northern elephant seals prefer beaches that are not only isolated from disturbances but also protected from severe winter storms and high surf conditions. Otariids inhabiting tropical or desertlike areas often rest in tidepools, sea caves, and deep crevices, which provide protection from intense sun and heat. In the Galápagos,

FIG. 3.6. Antarctic fur seal female and pup hauled out in tussock grass. *Photo by Dan Costa.*

Fɪɢ. 3.7. Guadalupe fur seal pup resting in the shade of a boulder. *Photo by Mark Pierson.*

for example, fur seals are usually found only on islands with cliff over-hangs and caves that afford protection from the sun. Guadalupe fur seals also prefer to rest either in tidepool areas, deep and shady crevices between boulders, or caves (Fig. 3.7).

Some pinnipeds, such as the widespread harbor seal, tend to be highly adaptable to a wide variety of terrestrial environments. Along the North Pacific coast, harbor seals may haul out on emergent offshore and tidal rocks, sandy and cobble beaches, mudflats, or sandbars on both islands and the mainland. In the northern portion of their range, some harbor seals (*P.v. richardsi*) even haul out on ice. In fact, the social organization, group size, and breeding behavior of harbor seals may vary greatly with geographical location and habitat as well as subspecies (see, e.g., Stirling 1983). The Atlantic grey seals also breed in a variety of habitats, including rocky shores, sandy beaches, sea caves, and ice.

Some ice-inhabiting pinnipeds haul out both on ice and on land, depending on the place and time of year. Most pagophilic, "ice-loving," pinnipeds prefer to live on or near ice but haul out on land where ice is absent. For instance, although walruses prefer to haul out on ice, they also come ashore on small rocky islands when ice is not available (Tsalkin 1937; Nikulin 1947; Popov 1958; Loughrey 1959). In the Arctic, the ice-inhabiting largha seals and bearded seals haul out on land seasonally in some areas as well. Bearded seals routinely haul out on land in areas where ice melts or recedes beyond shallow waters in the summer, such as the White Sea (Heptner 1976), the Sea of Okhotsk (Tikhomirov 1961), and the Laptev Sea (Tavrovskii 1971). The Antarctic Weddell seal and leopard seal also sometimes haul out on land on subantarctic islands, and a small population of Weddell seals breeds on South Georgia Island in places where there is no ice (Bonner 1982). Conversely, pinnipeds that generally live on land sometimes haul out on ice, including Steller sea lions, harbor seals, and grey seals (Table 5).

ICE

Generally, pagophilic pinnipeds inhabit two forms of ice—fast ice (ice attached to land) and pack ice (floating ice)—but both forms have many variations. For instance, ice habitats vary in the degree of attachment of ice to shore, the surface area of the water covered by ice, the depth of the snow cover, and the size and thickness of floes (Burns 1970). Each pinniped species tends to prefer a particular type of ice, often in areas

TABLE 5
Pinnipeds That Inhabit Ice

ANTARCTIC		ARCTIC	
Pack Ice	*Fast Ice*	*Pack Ice*	*Fast Ice*
Ross seal	Weddell seal*	Walrus*	Ringed seal
Crabeater seal	Southern elephant seal†	Harp seal	Caspian seal
Leopard seal*		Hooded seal	Baikal seal
		Bearded seal*‡	Grey seal†§
		Ribbon seal	
		Largha seal*	
		Grey seal†§	
		Harbor seal†‖	
		Steller sea lion†#	

NOTES: Ice is the primary habitat for 13 species; 4 species associate with ice seasonally: southern elephant seal, grey seal, harbor seal, and Steller sea lion. See text for references.

* Pagophilic species that also haul out on land when ice is not available.

† Species in which only certain populations breed on ice or associate with ice for only brief periods (not considered truly pagophilic).

‡ Sometimes found on thin fast ice.

§ Populations in the Gulf of St. Lawrence breed on fast ice; populations in the Baltic breed either on fast ice or pack ice.

‖ *P.v. richardsi* seasonally associates with ice in the Gulf of Alaska and the Bristol Bay area of the Bering Sea.

Steller sea lions in the Bering Sea haul out on pack ice near the ice front during winter.

FIG. 3.8. Aerial view of largha (spotted) seals on ice. *Photo by Kathy Frost.*

that contain suitable food resources. Because many authors writing of pagophilic seals do not define the specialized terms used to distinguish the types of ice, readers unfamiliar with the many varieties of sea ice may be confused. Therefore, a general description and definition of the various types of sea ice used by pinnipeds is provided in the text as well as in the Glossary.

The sea ice inhabited by seals behaves in different, less predictable ways than freshwater ice. For example, sea ice is more elastic, meaning that it bends under a load and breaks less easily. Winds and currents profoundly affect its formation. Where light winds and currents prevail, the ice is relatively smooth; where winds are high and currents strong, it is heavily fractured and rugged, covered with hummocks (rounded hills of ice) and high-pressure ridges. Pressure ridges are formed when thick ice fractured by winds, currents, and tides is compressed into a large pile or ridge, sometimes extending 20–40 feet above and below the ice. Pack ice contantly moves and shifts as a result of winds and surface currents. A band of stable shore-fast ice generally forms along the coast and in other protected areas, such as bays. A flaw lead, or area of open water, usually separates the shore-fast ice from moving pack ice, especially when offshore winds prevail. Pinnipeds and other marine mammals regularly use this flaw zone as a travel corridor or aquatic "highway," as do polar bears and human hunters that exploit polar mammals. A large area of open water that remains open throughout the year is called a polynya. Most pinnipeds that inhabit pack ice live in the areas of temporary or seasonal ice (ice that is generally less than one year old), although a few species venture onto the permanent ice pack of the Arctic Ocean (Figs. 3.8, 3.9). Good general discussions of the dynamics of polar ice are provided by Armstrong et al. (1973) and, in relation to mammals, by Burns (1970), Fay (1974b) and Burns et al. (1981b).

According to both Fay (1974b) and Burns et al. (1981b), ice provides many advantages to seals of the polar regions. In general, both fast and pack ice serve as a solid substrate on which seals can breed, give birth, raise their young, molt, and rest. Other benefits associated with living on seasonal pack ice include an abundant food supply that is easily accessible beneath the ice; shelter in ice ridges and crevices as well as the milder microclimate of the ice pack (where there is less wind and wave action than on smooth exposed ice); isolation from terrestrial predators (although polar bears often venture onto the ice pack); an abundance of space for breeding, resting, and feeding; passive transport with the ice pack to new feeding grounds; cleaner, more sanitary hauling grounds because of the large amount of space and the cleansing properties of the ice and snow; and a variety of habitats within the ice pack from which to

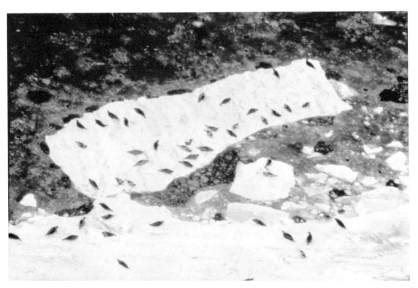

FIG. 3.9. Aerial view of male harp seals on ice. *Photo by R. Frank.*

choose (for instance, floe size varies as well as the movement patterns of ice, its thickness and smoothness, etc.). Yet pinnipeds and other marine mammals living on the fast ice, as well as on certain forms of permanent heavy pack ice, must penetrate the often solid barrier of ice for access to air and food. Only seals that can find or make and maintain breathing holes in the ice are able to live on the stable fast ice. Other pagophilic seals must inhabit the moving pack ice where there are many leads and areas of open water.

Thirteen species of pinnipeds spend most of their lives in areas of polar ice in the Arctic (nine species) and the Antarctic polar ice (four species). In the Antarctic, the Weddell seal inhabits the fast ice, while the Ross seal, crabeater seal, and leopard seal live on pack ice of various forms. The Ross seal seems to prefer the consolidated interior pack ice; the other two species tend to remain on the outer fringes of ice in areas with smaller bergs and ice floes (Bonner 1982). In the Arctic the Caspian seal, Baikal seal, and ringed seal inhabit fast ice, although ringed seals may also molt and rest on moving and permanent pack ice. Six species live on pack ice: walruses, harp seals, hooded seals, bearded seals, ribbon seals, and largha seals. Bearded seals are also sometimes found in areas of the thin fast ice; they prefer to remain in areas where the ice pack is constantly shifting, forming many leads and polynyi (Burns 1981a). Some seals may associate with ice for short periods but are not thought of as true ice-inhabiting pinnipeds because they do not live on ice during most of the year. For instance, some grey seals in the Baltic Sea and the Gulf of St. Lawrence associate with ice seasonally for breeding (Curry-Lindahl 1975; Mansfield and Beck 1977). The Canadian grey seal breeds on fast ice. The Baltic grey seal breeds either on moving pack ice or fast ice, although females pupping on shore-fast ice usually remain close to the edge of the ice for easy access to open water. Steller sea lions that winter in the Bering Sea often haul out on ice floes near the front (Tikhomirov and Kosygin 1966). Harbor seals (*P.v. richardsi*) of the Bristol Bay area of the Bering Sea and the area near the Gulf of Alaska seasonally associate with sea ice to some degree (Burns and Fay 1973). In the southeastern Bering Sea and the northern Gulf of Alaska, harbor seals apparently prefer to use sea ice rather than land when ice is available.

Harbor seals (*P.v. richardsi*) in the northeast Gulf of Alaska and in Prince William Sound come ashore on bergs of glacial ice for breeding, molting, or resting. Harbor seals commonly haul out on glacial ice in Aialik Bay, located on the Kenai Peninsula in south central Alaska. In the Glacier Bay

Fig. 3.10. Harbor seals on glacial ice in Aialik Bay, Alaska. *Photo by Anne Hoover.*

area of southeastern Alaska, glacial ice floes are also used by the resident harbor seals. The ecology and behavior of harbor seals inhabiting glacial ice in Aialik Bay have been studied extensively by Hoover (1983), who found differences in the way seals use glacial ice and sea ice.

In general, sea ice provides seals with more benefits than glacial ice for transportation, shelter, or proximity to an abundant food resource. Sea ice, moreover, is more widespread and diverse than glacial ice, which tends to occur only near tidewater glaciers. The relatively short lifespan of glacial ice in comparison with that of sea ice (its life is measured in hours rather than months) inhibits seals from using it for passive transportation. In addition, harbor seals must leave the glacial ice and move into ice-free waters in the outer portion of the fjords or in the open sea to forage. Glacial ice does not provide as much diversity of habitat as sea ice. However, the size of glacial floes varies to a certain extent, and Hoover (1983) found that harbor seal females giving birth and raising a pup prefer larger floes than other harbor seals. Although the terrain of large bergs offers little shelter, the glacial ice in deep fjords provides protection from high swells and storms. Rafted glacial ice further reduces the effect of heavy seas. In addition, it is easier for harbor seals, especially pups, to haul out on glacial ice than on land (Fig. 3.10).

ADAPTATIONS FOR LIFE ON ICE

Pinnipeds have adapted to the harsh and demanding polar environment in their behavior as well as body morphology. Most of their adaptations are associated with either maintaining breathing holes in unbroken ice or migrating annually in relation to the moving ice pack. Southern elephant seals, walruses, and bearded seals often use their heavy skulls to break ice. Fay (1974b) points out, however, that since the skulls of some pinnipeds living in ice-free waters are of comparable size, selective pressures other than ice may have influenced their development. Certainly, though, all the seals that use their heads to break holes in the ice possess large skulls. Southern elephant seals create breathing holes about 60 centimeters in diameter simply by ramming the ice from below with their heads until it breaks (Laws 1953a; Ling and Bryden 1981). Bearded seals use their heads to bash holes in the ice as well, although the ice they prefer to inhabit is relatively thin (generally no more than 10 centimeters thick). Walruses use the same strategy to break ice up to 22 centimeters thick. They keep ice holes open by abrading the ice with their massive tusks, either with chopping motions or by swinging their heads from side to side while they press the shafts of their tusks against the edges of the ice (Burns and Fay, unpubl. data). Captive walruses even "chop" in their concrete pools, causing tremendous wear to tusks, often to the point of exposing the pulp cavity to infection (King 1983). Walruses also use their tusks to help them haul out on the ice, to anchor themselves to the edge of the ice while they sleep in the water (Fay 1982), to fight with other walruses (Zabel et al. 1987), and occasionally to impale seals for food (Lowry and Fay 1984).

The claws of all pagophilic Arctic phocids are substantially larger than those of phocids living in ice-free waters (Fig. 3.11) (Fay 1974b). Ringed seals often use their heavy foreflipper claws to make and maintain breathing holes in the ice (Vibe 1950; McLaren 1958; Stirling 1977). The closely related Caspian seal and Baikal seal (these seals, along with the ringed seal, make up the subgenus *Pusa*) also use their strong claws to maintain ice holes (Ognev 1935; Kozhov 1963). The claws of each species vary in size with the thickness of the ice each inhabits. The Baikal seal, which lives on the hardest freshwater ice, has the strongest claws, followed by the ringed seal, which inhabits areas of softer and less dense sea ice. The Caspian seal tends to live in areas of the thinnest ice and accordingly has the weakest claws. Burns (unpubl. data) has found a similar, though less pronounced, difference in claw size and structure between harbor seals that inhabit ice and those that live in the ice-free waters of the Bering

FIG. 3.11. Ringed seal pup; note strong claws. *Photo by Kathy Frost.*

Sea; those that live on ice have stronger claws. Bearded seals are also known to use their claws to maintain breathing holes in the thinner ice. The ribbon seal and largha seal seem unable to make and maintain breathing holes in the ice; these species live primarily along the ice front where there are large areas of open water and thin ice. Yet possibly these seals could make breathing holes in thinner ice; J. A. Estes (pers. comm.) has observed largha seals near what appeared to be ice holes.

The Arctic Eskimos use the breathing holes made by seals, which they call *ullas*, to help them capture bearded seals and ringed seals. Each seal maintains numerous *ullas* because it must surface for air every quarter hour or so. In the dead of winter the sea ice may be several feet thick, with a covering of snow that often conceals the opening of the *ulla*. A hunter may fortuitously stumble upon a breathing hole, but more often his sled dogs help him locate one. After finding an *ulla*, the hunter removes the snow and opens the ice dome. With a special tool, often made from the horn of a musk ox, he scoops out the floating ice chunks and with a caribou antler explores the shape of the breathing hole to determine how the seal will position itself when it next surfaces. These preliminaries completed, the hunter replaces the snow, leaving a tiny hole at the top fitted with an intricate device containing the downy feather of a swan or

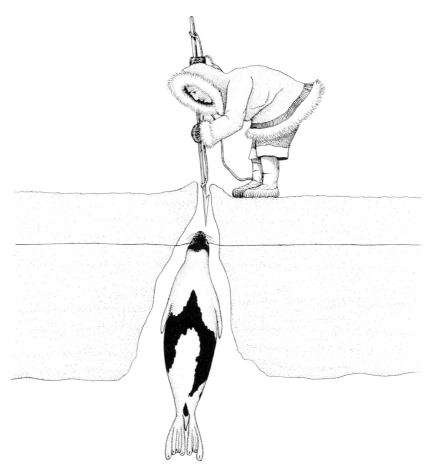

FIG. 3.12. Eskimo hunter at an ice hole, waiting for a seal to surface. *Drawing by Pieter Folkens.*

other water bird; this device serves as a telltale sign that the seal is surfacing into the air space some distance below the hunter. Standing on a piece of skin or fur, the hunter may remain motionless for hours, bent at the waist and holding his harpoon poised to thrust it into the hole (Fig. 3.12). The harpoon has a toggle point at one end that is attached to a hide rope so that the pointed tip disengages when the harpoon is thrown. K. Rasmussen (1941) tells of one twelve-hour vigil and relates stories of some hunters who spent more than forty-eight hours at a hole in this uncomfortable position. The hunter may build a small wind break of drifted snow for shelter but is otherwise unprotected from the subfreezing wind and cold. Several hunters may spread themselves over the ice at various breathing holes in hopes that a seal may visit at least one of them. Once a seal is struck, all the other hunters rush to help haul the animal onto the ice.

In the Antarctic fast ice, Weddell seals maintain breathing holes by gnawing on the ice with their strong teeth, using them in a manner similar to that of walruses, who ream the ice (see, e.g., Bertram 1940; Sapin-Jaloustre 1952; Kooyman 1969). The outer upper incisors and upper canines of Weddell seals project more horizontally than in other seals, presumably to enhance the process of abrading the ice. The Weddell seal's teeth become very worn, often to the pulp cavity, from the constant abrasion. In fact, the serious wear on the teeth caused by chewing on ice contributes to mortality in Weddell seals, since a seal with poor teeth has no means of maintaining breathing holes (Stirling 1969b). For this reason, the use of claws rather than teeth to abrade ice would seem preferable, since claws continue to grow, while regrowth is not possible in teeth (Stirling 1977).

Another adaptation for living on ice is the specialized woolly natal fur, or lanugo, which helps seal pups born on ice to retain body heat. If the pup lost heat, it would melt quickly into the ice. Moreover, if the fur were to become wet, its thermoregulatory properties would be substantially reduced (Davydov and Makarova 1965; Burns 1970). In contrast to the seal pups, walrus calves, like adult walruses, have very sparse coats. The calves are able to maintain body heat because of their relatively large size and because they huddle with other walruses (Fay and Ray 1968). Burns (1970) suggests that young walruses as well as bearded seals (which also have sparse coats and large bodies) do not require the extra insulation provided by the lanugo of smaller seals because these larger species have a higher ratio of body volume to surface area. In addition, young walrus calves might have trouble swimming with the herd in a bulky lanugo coat and would be unable to dive well because dense lanugo traps air. Moreover, the long fur would probably freeze when the calves hauled out of the water.

The white natal, or birth, coat of seals such as the ribbon seal, harp seal, ringed seal, and largha seal is often considered an adaptation for growing up in snow or ice because of the protective coloration it gives to pups (Fig. 3.13). As Burns (1970) points out, however, coloration of the natal pelage should make little difference to a ringed seal pup, which is one of the most vulnerable species to predation, when it is hidden in a birth lair. In addition, other pinnipeds breeding in ice-free areas, such as grey seals, also have pups with white coats, while a number of ice-

FIG. 3.13. Harp seal pup, showing white lanugo fur. *Photo by Graham Worthy.*

inhabiting seals, such as walruses, bear pups with dark coats. Walrus calves, however, would appear to have no need for a white coat for cryptic coloration since newborns are always within a walrus herd and accompanied by their mothers and are even carried on their mothers' backs in the water.

Ringed seals and Baikal seals give birth and raise their pups in concealed caves and crevices in the ice and snow, called subnivean birth lairs. The seals' practice of raising their young in ice caves apparently enhances the pups' chances of surviving by providing them protection from the wind and cold and from land predators such as polar bears and foxes. Mothers and pups are restricted in their movements, and they are especially vulnerable to predators when they are visible on the ice. Young harp seals and largha seals may also take shelter from predators and inclement weather in pressure ridges (Burns 1970; Fay 1974b).

Pinnipeds inhabiting Arctic pack ice undergo seasonal migrations that are related to the seasonal advance and retreat of the ice. Arctic seals that migrate to maintain a year-round association with ice include species such as bearded seals, harp seals, and walruses (see Table 14). In addition, the cycles of reproduction and molting in Arctic pinnipeds have evolved to coincide with the most favorable conditions for hauling out, breeding, and raising young on the ice. Many of the Arctic ice-inhabiting seals give birth within a 6-week period between April and mid-May, when ice conditions are optimal for breeding (Burns 1970). Hooded seals and harp seals pup earlier in the year, however, with hooded seals giving birth during the last two weeks of March and harp seals pupping between late February and April, depending on their location (Lavigne and Kovacs 1988). Ringed seals, ribbon seals (Fig. 3.14), and largha seals tend to give birth during the time of slowest ice movements, weakest ocean currents, and maximum snow cover (Fedorova and Yankina 1963). The pups are generally weaned at an optimal time in the spring when the ice pack is receding northward. At this time, food resources begin to increase, air temperatures are much warmer, the direction of the prevailing winds changes, the wind-chill factor is substantially reduced, and there are extensive areas of open water. Bearded seals and walruses give birth during mid-to-late April or early May, when weather conditions are much milder and many areas of open water are present; accordingly, they do not depend as much on relatively stable ice during the pupping season, and the young of both species are able to swim soon after birth (Burns 1970).

Fay (1974b) points out that relatively few evolutionary changes in ice-inhabiting seals appear to be directly associated with living on ice. This

FIG. 3.14. Ribbon seal in slush ice. *Photo by Kathy Frost.*

would seem to suggest that the ice itself has not posed strong selective pressures or that the time needed for adaptation to the Arctic environment has been fairly short. Indeed, most Arctic seals probably did not inhabit areas of ice until three to four million years ago, at the beginning of the Pleistocene. In the Arctic, extensive areas of sea ice did not develop until at least this time, and perhaps even more recently.

4

Predation on Pinnipeds

Because they live on both land and water, pinnipeds are subject to the double burden of terrestrial and aquatic predators. The impact each predator has on seal populations varies with the species of pinniped and its habitat. For instance, while predation on land is generally not a problem for seals that haul out and breed on isolated islands in nonpolar waters, terrestrial predators—especially polar bears—are an important cause of mortality among some ice-inhabiting seals in the Arctic. In contrast, for other seals inhabiting islands in tropical or temperate regions, predation by sharks may be a significant cause of mortality. Shark predation on the small and dwindling population of Hawaiian monk seals or on Galápagos otariids, for example, may well be a limiting factor in population growth. Heavy shark predation may even influence maternal care strategies among Galápagos fur seals (this issue is discussed in chapter 8). Unfortunately, it is difficult to study predation as a potential variable regulating the population dynamics and biogeography of pinnipeds and other marine mammals (McLaren and Smith 1985), and there is much yet to learn about the relationships between pinnipeds and their predators (see Table 6).

Exploitation by Man

Historically, human predators have had by far the most dramatic impact on seal populations throughout the world. Numerous species of pinnipeds valued for their blubber and fur had been hunted to the brink of extinction by the late nineteenth century. The rich oil derived from a seal's blubber stores was in great demand because of its many uses: it was burned to provide light, and it was used to lubricate engines as well as to produce soap, paint, and clothing. An average-sized elephant seal male might yield around 100 gallons of oil, although an extremely large bull could produce twice that amount.

Pinnipeds such as harp seals, Caspian seals, hooded seals, elephant seals, walruses, and all the fur seals were especially heavily exploited. Sealing, in fact, was as important a commercial undertaking as whaling in its day. Most pinnipeds, however, either have made or seem to be making a successful comeback from near extinction. Some species, such as Antarctic fur seals, have made spectacular recoveries and may be as abundant today as before they were commercially hunted. Yet many of the heavily exploited pinnipeds recolonized their ranges slowly at first, so that there was a long lag between exploitation and recovery. Today, the monk seals are the only species in danger of extinction.

Researchers studying recovered seal populations are interested in the genetic impact on a species that has been reduced to near extinction. The northern elephant seal is one of these species, having made a remarkable comeback from near extinction in the late 1800s. The current northern elephant seal population descended from a remnant population of between twenty and one hundred animals that managed to survive on the rugged and remote island of Guadalupe, off Mexico. In a species that undergoes a population "bottleneck" as severe as that of northern elephant seals, one would expect a considerable loss of genetic variability. Close inbreeding and the highly polygynous mating system of elephant seals could be expected to intensify this genetic fixation. In fact, Bonnell and Selander (1974) found that northern elephant seals do indeed lack genetic variability in comparison with other seals and land mammals. They analyzed 21 blood proteins in 125 seals from various rookeries and found no polymorphisms; in other words, all proteins were monomorphic, as in genetic twins. Despite this loss of genetic variability, the northern elephant seal population has successfully repopulated much of its former range. It remains to be seen whether the reduced gene pool will cause problems for this species in the future as environmental conditions change.

Among the species still subject to seasonal commercial harvesting are the northern fur seal, South American fur seal, crabeater seal, leopard seal, Weddell seal, hooded seal, and harp seal (Bonner 1982; Gentry and Kooyman 1986). South American fur seals are still hunted for their oil

FIG. 4.1. Blueback hooded seal pup. *Photo by Graham Worthy.*

and pelts. In addition, the fur seal teeth are used to make handicrafts for small local markets, and the genitals of adult males, believed to possess aphrodisiac properties, are exported to Asia. The commercial killing of harp seal pups, called whitecoats, and hooded seal pups, called bluebacks (Fig. 4.1), has been prohibited within the past year by the Canadian government, but several hundred whitecoat pups are still taken each year in Norway. Public outcry over the clubbing of whitecoats began to discourage commercial sealing as early as 1984, but it is still legal to harvest harp seal beaters (weaned pups) and adults (D. Bowen, pers. comm.).

The 1972 convention for the conservation of Antarctic seals permits the hunting of crabeater seals, leopard seals, and Weddell seals within certain catch limits. But it prohibits the hunting of Weddell seals over one year of age between September 1 and January 31 each year to help protect the breeding stock. Southern elephant seals, Antarctic fur seals, and Ross seals are also protected by this agreement (Bonner 1982).

While modern commercial sealing in the eighteenth and nineteenth centuries yielded some economically valuable commodities, for the Arctic Eskimos sealing underlay an entire traditional culture. The Arctic Eskimos once populated a large region around the Arctic circle, from Siberia to Alaska and eastward across Canada to Greenland. Hunting seals on the ice not only allowed these people to survive the severe winters but also provided the basis for their society and customs. An individual hunter's skill determined whether his family and village survived or starved. Much of this traditional lifestyle is documented in a masterful ecological study by R. K. Nelson (1969), *Hunters of the Northern Ice.*

Extensive rules governed the disposal of the seal meat, and even more complex customs determined the handling and eating of the carcass. Once a seal was killed, the hunting party quickly shared a feast of hot raw liver. The butchering and disposal of the actual carcass, however, was tightly controlled by kinship and allegiance. Sharing the seal was based in part on partnership, which typically lasted throughout the partners' lifetimes and was often passed on from father to son. K. Rasmussen (1941) notes that hunters often addressed one another by the name of the part they exchanged. If two men always gave each other the shoulders of the seal carcass, they would call the other "my shoulder" instead of their regular names. All parts of the seal were enjoyed: brains made an especially good pâté, blood made the best soup, and tongue was considered an aphrodisiac.

Netsilik Eskimos believed that since seals lived in salt water, they were always thirsty and in fact would even let themselves be killed for a drink of fresh water (K. Rasmussen 1941). Therefore, a careful hunter always gave a seal a drink when it was killed. Although the seal's body was perishable, its soul was considered immortal. If a hunter treated the soul respectfully, the same seal could return to be caught again and again. A seal's bones were carefully tossed back into the sea so that they would again become seals, in a variation on the Nuliajuk creation story (see pp. 55–56). When a new winter camp was to be established, the skulls of the seals already killed were set out on the ice with their noses turned toward the new hunting ground so that the seals could follow the hunters from place to place.

Aboriginal or subsistence hunting still takes place on some seal species, primarily in the Arctic. Subsistence sealing is generally thought of as differing from commercial sealing in that the subsistence hunter uses the seal products himself or perhaps trades for products directly related to survival or livelihood. According to D. L. Taylor et al. (1985), the native harvest of walruses in Alaska varies from approximately two to four thousand annually and is increasing each year. Fay (1982) summarizes the harvest of walruses by Eskimos and other natives, estimating that between four and seven thousand have been killed each year since the mid-1960s. Hooded seals, harp seals, and ringed seals are currently also hunted for subsistence purposes by Greenlanders and Canadian Eskimos (Reeves and Ling 1981).

Regional or international regulations currently govern most of the commercial and aboriginal sealing operations. I do not discuss in detail the human exploitation of pinnipeds, but excellent reviews of both exploitation and conservation efforts are available in Bonner (1982),

Hofman and Bonner (1985), Busch (1985), and Lavigne and Kovacs (1988).

In addition to commercial and aboriginal sealing, human activity affects seals in incidental ways. Seals may drown if they become entangled in nets either during fishing operations or afterward, when fishing gear is lost or discarded; or they may swallow fishing hooks. Moreover, whale-watching vessels and other boats disturb the seals; commercial fisheries compete with them for food resources; and humans degrade the seals' habitat as they develop heavily populated or industrialized regions along the coast and exploit oil and gas resources in coastal areas (Bonner 1982). Human-related disturbances to Pacific harbor seals (*Phoca vitulina rich-ardsi*), for example, include offshore oil development, boating activities, logging, and construction of homesites. All these activities encroach on the habitat of the sensitive harbor seal (Calambokides et al. 1978; Allen et al. 1979; Murphy and Hoover 1981).

The use of large gill and seine nets by growing commercial fisheries has resulted in considerable incidental mortality among pinnipeds as well as other marine mammals, such as sea otters (Wendell et al. 1985) and cetaceans, especially porpoises (Hofman and Bonner 1985). Many South African fur seals are caught and killed in trawl nets. According to Shaughnessy and Payne (1979), 2.3 percent of the South African fur seal population become entangled in nets each year, and 1.3 percent of the population drown in gill nets. The use of synthetic, nonbiodegradable fibers in fishing nets compounds the problem of pinnipeds' becoming entangled in fishing gear after it is lost or discarded. Antarctic fur seals, California sea lions, Hawaiian monk seals, and northern fur seals have become enmeshed in such gear (Bonner and McCann 1982; Gilmartin 1983); thousands of northern fur seals are caught and killed each year in this way (Fowler 1982). Discarded plastic packing straps have also created a major problem for northern fur seals on the Pribilof Islands. The heavy plastic tape can slip over a seal's head and tightly encircle the body, cutting off circulation and often leading to a slow death.

Aquatic Predators

GREAT WHITE SHARKS AND OTHER LARGE SHARKS

Great white sharks naturally arouse curiosity and fear among most people, probably in part because of the recent movie series *Jaws* (Fig. 4.2). Occasionally white sharks do attack people, but probably only because

FIG. 4.2. Great white shark at Año Nuevo mainland beach. This 15½-foot shark had recently eaten a four-year-old male elephant seal. *Photo by author.*

they mistake people for pinnipeds. Great white sharks eat or attack other marine animals besides pinnipeds—porpoises, whales, and sea otters—as well as sea turtles and an assortment of fishes (see, e.g., Ames and Morejohn 1980; Tricas and McCosker 1984). Along the eastern North Pacific coast, pinnipeds appear to be a major source of food for great white sharks, which are known to prey on harbor seals, northern elephant seals, California sea lions, and Steller sea lions (Le Boeuf et al. 1982; Ainley et al. 1985). In the Southern Hemisphere, white sharks prey on the eight species of southern fur seals (*Arctocephalus* spp.) near South Africa, Australia, New Zealand, Mexico, South America, and the Galápagos (Bonner 1981a; Warneke 1982; Trillmich 1984; Eibl-Eibesfeldt 1984b). White sharks also prey on New Zealand sea lions (Gentry, pers. comm.) and on Australian sea lions, especially near Dangerous Reef in South Australia (Walker and Ling 1981a). White shark attacks (Table 6) on large Australian fur seal bulls have been observed at Dangerous Reef as well (McCosker 1985).

In eastern Canada, the white shark may be an important predator of harbor seals (Boulva and McLaren 1979) and grey seals (Brodie and Beck 1983). A number of other shark species also occur off eastern Canada (e.g., porbeagles, blue sharks, smooth hammerheads, short-fin makos,

TABLE 6

Nonhuman Aquatic and Terrestrial Predators of Pinnipeds

SPECIES	PREDATORS	REFERENCES
Hawaiian monk seal	Tiger sharks Grey reef sharks Reef white-tip sharks	Kenyon 1973; DeLong & Brownell 1977; Taylor & Naftel 1978; Balazs & Whittow 1979; Alcorn & Kam 1986; Johanos & Kam 1986
Northern elephant seal	Great white sharks Killer whales	Ainley et al. 1981, 1985; Le Boeuf et al. 1982
Southern elephant seal	Killer whales* Leopard seals*	Carrick & Ingham 1962; Condy 1977a; Condy et al. 1978; Laws 1984
Weddell seal	Leopard seals* Killer whales	Øritsland 1977; Kooyman 1981d; Bonner 1982; Laws 1984
Ross seal	Leopard seals?	King 1983
Crabeater seal	Leopard seals* Killer whales	Condy 1977b; Øritsland 1977; Siniff & Bengtson 1977; Laws 1977b, 1984; Bonner 1982; T. G. Smith et al. 1981
Leopard seal	Killer whales Sharks?	Øritsland 1977; Bonner 1982
Hooded seal	Polar bears Greenland sharks? Killer whales?	Sergeant 1976, 1977; McLaren 1977; Reeves & Ling 1981; Kovacs & Lavigne 1986a
Bearded seal	Killer whales Walruses Polar bears	Zenkovich 1938; Smith 1980; Fay 1982; Lowry & Fay 1984
Harp seal	Polar bears? Walruses	Stirling 1983; Lowry & Fay 1984
Ribbon seal	Polar bears†	King 1983
Caspian seal	Wolves Eagles	Naumov 1933; Ognev 1935; Rumyantsev & Khiras'kin 1978
Baikal seal	Wolves Eagles	King 1983
Ringed seal	Walruses Steller sea lions Polar bears Arctic foxes* Walruses* Red foxes*† Wolves† Wolverines† Dogs† Ravens†	Burns 1970; Stirling 1974a; Stirling & Archibald 1977; Smith 1980; Gentry & Johnson 1981; Fay 1982; King 1983; Lowry & Fay 1984; Hammill & Smith 1987; Kelly et al. 1987

Continued on next page

NOTE: Question mark indicates probable predator (no evidence conclusively proves predation).

* Preys mainly on juveniles or pups.

† Probably an insignificant predator.

SPECIES	PREDATORS	REFERENCES
Largha seal	Walruses	Fay 1982; Lowry & Fay 1984

Harbor seal	Great white sharks
	Other shark spp. (off
	   eastern Canada)
	Killer whales
	Steller sea lions†
	Polar bears
	Coyotes*
	Eagles*
		Ainley et al. 1981, 1985; Le Boeuf et
		   al. 1982; Boulva & McLaren 1979;
		   King 1983; Steiger et al. 1985; J. A.
		   Estes, pers. comm.

Grey seal	Great white sharks Other shark spp. Killer whales†	Bonner 1981a; Brodie & Beck 1983

California sea
  lion
	Great white sharks
	Bull sharks?
	Killer whales
		Ainley et al. 1981, 1985; Le Boeuf et
		   al. 1982; Estes, pers. comm.

Galápagos sea
  lion
	Tiger sharks
	White-tip sharks
		B. Nelson 1968; Barlow 1972; Eibl-
		   Eibesfeldt 1984b

Steller sea lion	Great white sharks
	Killer whales
		Rice 1968; Le Boeuf et al. 1982;
		   Ainley et al. 1985; Loughlin &
		   Livingston 1986

Southern sea
  lion
	Sharks (various spp.)
	Killer whales
	Leopard seals*
	Pumas*
		Hamilton 1934; J. Wilson 1975;
		   Bartlett & Bartlett 1976

Australian sea lion	Great white sharks	Walker & Ling 1981a; King 1983
New Zealand sea lion	Sharks†	King 1983

Northern fur
  seal
	Great white sharks
	Killer whales
	Steller sea lions*
		Gentry 1981a; Gentry & Johnson
		   1981; Scheffer et al. 1984; Hansen
		   1987

Guadalupe fur seal	Great white sharks Killer whales?	Fleischer 1978b, 1987; Bonner 1981b
Juan Fernán- dez fur seal	Great white sharks? Blue sharks? Killer whales? Leopard seals?	Bonner 1981b; Torres 1987

Galápagos fur
  seal
	Tiger sharks
	Other shark spp.
	Killer whales
	Feral dogs
		Bonner 1981b; Trillmich 1984,
		   1987b

South Ameri-
  can fur seal
	Sharks
	Killer whales
	Southern sea lions*
	  (preys on adult
	  females and
	  juveniles)
		Vaz-Ferreira 1950; Gentry & Johnson
		   1981; Bonner 1981b

TABLE 6, *continued*

SPECIES	PREDATORS	REFERENCES
New Zealand fur seal	Great white sharks Killer whales New Zealand sea lion*†	Mattlin 1978, 1987; Bonner 1981b
Antarctic fur seal	Leopard seals* Killer whales	Bonner 1981b; Kooyman 1981b; McCann & Doidge 1987
Subantarctic fur seal	Leopard seals*? Killer whales? Sharks	Bonner 1981b; Bester 1987
South African fur seal	Great white sharks Killer whales? Black-backed jackals* Brown hyenas*	Shaughnessy 1979, 1982; Bonner 1981b; David 1987b
Australian fur seal	Great white sharks Killer whales?	Bonner 1981b; Warneke 1982; McCosker 1985
Walrus	Killer whales Polar bears*	Belopol'skii 1939; Zenkovich 1938; Nikulin 1941; Brooks 1954; Loughrey 1959; Lono 1970

Atlantic sharpnose sharks, ocean white-tip sharks, and dusky sharks), but Brodie and Beck (1983) do not specify which species are responsible for most of the predation on seals; they suggest that declining shark stocks off eastern Canada (resulting from changes in fishery practices) may have promoted an increase in the grey seal population over the past twenty years in this area.

In eastern Pacific waters, evidence suggests that sharks prefer to feed on elephant seals and harbor seals rather than other pinnipeds (Le Boeuf et al. 1982; Ainley et al. 1985). In particular, white sharks are found near the rookeries of elephant seals. These large phocids may provide a more substantial meal, or perhaps sea lions can escape from attacking sharks more often than elephant seals. According to Ainley et al. (1985) great white sharks tend to take their victims unaware, attacking them from the rear. They found that most surviving shark-bitten sea lions had lower-body and hindflipper injuries whereas the majority of elephant seals and harbor seals sustained upper body injuries. The nature of their injuries suggests that when attacked from the rear, sea lions and other otariids, which swim using their foreflippers, are more likely to escape than are phocids, which propel themselves with their lower bodies.

One might wonder how biologists know what the white shark eats, given its elusive behavior, which makes it difficult to observe in the field. A conclusive method is to examine the contents of its stomach. The remains of elephant seals and harbor seals are often found in the stomach contents of white sharks caught or found dead off the coast of California (Le Boeuf et al. 1982). A large white shark is able to kill and swallow sizable seals. White sharks have enormous flexible jaws that can open extremely wide to accommodate a large piece of flesh or even an entire animal (Tricas 1985). For instance, one 15½-foot female white shark that washed ashore dead near the Año Nuevo elephant seal breeding rookery had recently devoured an entire four-year-old male elephant seal, large chunks of which were found in the shark's stomach; the head had been completely severed from the body. Male seals of this age measure about 10 or 11 feet and weigh perhaps half a ton. After eating one large seal, a white shark may be able to wait as long as forty-five days before feeding again (Carey et al. 1982).

Other evidence, as presented in Le Boeuf et al. (1982) and Ainley et al. (1985), indicates that white sharks like to eat seals. For example, these sharks are known to frequent waters near elephant seal rookeries such as those of the Farallon Islands and Año Nuevo Island in California and Guadalupe Island in Mexico. There may be other ecological reasons why sharks are found near seal rookeries, since the islands where seals breed are usually characterized by rich upwelling or current confluence, a high level of primary productivity, and large fish populations. Yet because seals use the islands for breeding and molting on a predictable yearly schedule, the sharks presumably always know where and when they they will find a desirable food source. In fact, Ainley et al. (1985) suggest that the same sharks (perhaps not more than six different individuals) may have been responsible for much of the predation on Farallon pinnipeds during each breeding season from the mid-1970s to early 1980s. Many researchers have directly observed sharks attacking seals near seal rookeries, es-pecially on the Farallon Islands off San Francisco, where most of the victims appear to be elephant seals (Ainley et al. 1981; 1985).

Pinnipeds bitten by white sharks are often seen hauled out on islands along the coast of California and Mexico (although California sea lions in the Gulf of California may also be preyed on by bull sharks, according to J. A. Estes, pers. comm.). Especially near large elephant seal rookeries, one can spot a number of seals of all ages with recent shark injuries or large scars from past attacks. Even though the actual shark attack was not observed, the injuries are known to have been inflicted by great white sharks because the lacerations bear telltale signs of the white shark bite (Figs. 4.3, 4.4). Remarkably, most injured seals that make it to land manage to survive, although pregnant elephant seal females that survive their

FIG. 4.3. Northern elephant seal with injury from great white shark. *Photo by Burney Le Boeuf.*

FIG. 4.4. Sea lion bitten by great white shark. *Photo by Burney Le Boeuf.*

injuries usually lose their pups. Most of these females either give birth to a stillborn pup, abandon the pup soon after giving birth, or fail to wean the pup successfully (Le Boeuf et al. 1982).

In central and northern California, the frequency of shark attacks on elephant seals seems to decrease in the springtime and increase during the fall when juvenile elephant seals are most abundant. It is unclear why this is so. White sharks may fast while breeding in the spring (Springer 1967), or they may migrate elsewhere (Squire 1967). Such seasonal trends in the frequency of shark predation might have influenced the timing of reproductive events among elephant seals and possibly harbor seals. Newly weaned elephant seal pups, for example, which may be especially vulnerable to shark attack, leave the breeding rookery for sea in the spring at a time when they are less likely to encounter white sharks (Le Boeuf et al. 1982). But the seasonal upwelling of nutrient-rich waters in central California may be the primary influence on the timing of the breeding season, benefiting the pups that venture out to sea when food resources are more plentiful. Harbor seals in central California wean their pups in late summer, when white sharks are less abundant in inshore waters, at least near the Farallon Islands (Ainley et al. 1985).

White shark predation on seals and attacks on humans as well as the number of shark sightings in California have increased over the past several years. Apparently these increases relate to the sharks' attraction to areas where pinnipeds live and to the substantial growth of seal populations along the California coast over the past twenty or thirty years. The populations of elephant seals especially have expanded dramatically in recent times (see, e.g., Le Boeuf 1977; Le Boeuf and Bonnell 1980).

Although white shark attacks on humans are still rare, they have been increasing in central and northern California since 1955. A total of fifty-nine attacks on humans occurred from 1926 to 1984 along the coasts of California and Oregon. Most of the incidents (fifty-four) took place in California, and nearly all the attacks in which the species of the predator could be identified involved white sharks (D. G. Miller and Collier 1981; Lea and Miller 1985). Twelve of the attacks occurring between 1980 and 1984 involved one scuba diver, one paddle boarder (on a long surf board), one swimmer, three skin divers, and six surfers on short boards. Most of the incidents took place at the surface, as do most successful attacks on seals, since an underwater approach seems to allow the shark to take its prey by surprise more easily (McCosker 1985). As Lea and Miller (1985) point out, the increased popularity of ocean sports such as surfing, diving, and wind surfing in recent years makes it difficult to compare past and recent data on the number of humans attacked by sharks.

Why are people swimming, diving, or surfing in the ocean occasion-

FIG. 4.5. Great white shark underwater, looking at a harbor seal (right) and human surfer on a short board (left) at the surface. From the shark's perspective, the seal and surfer silhouettes are similar. *Drawing by Brad Decaussin.*

ally attacked by great white sharks? Wet-suited humans probably resemble the large marine prey of white sharks, which apparently attack the humans by "mistake." The shark often "spits out" the unfortunate victim after the first bite, allowing the person to escape, although even one bite inflicted by a large white shark can cause serious injury or death. Mc-Cosker (1985) believes that this "bite-and-spit" approach may lessen the chance of injury to the shark from struggling prey. An intriguing, if disturbing, theory concerning the increase in white shark attacks on surfers is presented by Tricas and McCosker (1984). They believe that the recent popularity of the short tri-fin surf board may be related to the shark attacks: from below, the silhouette of a surfer paddling on a short board looks very much like that of a harbor seal (Fig. 4.5).

REQUIEM SHARKS AND OTHER SHARK SPECIES

Requiem sharks of the family Carcharhinidae include the reef white-tip shark, grey reef shark, and tiger shark. These three species are common in the warmer waters near Hawaiian monk seal rookeries in the Northwestern (Leeward) Hawaiian Islands where at least tiger sharks are known to prey on monk seals. Tiger shark predation appears to cause significant injury and mortality among monk seals of all ages and may even be partially responsible for the decline of the monk seal population in some areas (A. M. Johnson et al. 1982; Alcorn and Kam 1986). Remains of monk seals have been found inside the stomachs of several tiger sharks at French Frigate Shoals (L. R. Taylor and Naftel 1978). In addition, tiger sharks have been observed inflicting injuries on monk seals (R. Johanos, unpubl. data) as well as feeding on dead monk seals that they had probably killed (Balazs and Whittow 1979; R. Johanos and Kam 1986). Alcorn and Kam (1986) actually saw a fatal attack on a live monk seal by a tiger shark on Laysan Island in May 1982. When the shark attacked an adult male swimming with what appeared to be a subadult female, the male behaved aggressively toward the shark, swimming rapidly between it and the female. Aggressive behavior of adult male monk seals toward large tiger sharks during the breeding season has also been observed on Lisianski Island (Loughlin and Kooyman, unpubl. data; R. Johanos and Kam 1986). Only after the male had hauled out on shore did the shark fatally attack the subadult female seal. After the shark rapidly approached her, there was a sudden violent splashing and the female was forcibly jerked underwater. As the water turned a bloody red, at least ten small sharks (believed to be grey reef sharks) appeared in the vicinity, circling within 15 meters. The attacked monk seal was not observed again.

Many monk seals also bear massive scars on their bodies and are missing flippers because of shark attacks. Although some of the more severely injured seals die of their wounds, many appear to recover remarkably well, considering their substantial injuries (Fig. 4.6). As many as 12 percent of 235 monk seals observed during one census of the French Frigate Shoals had scars due to shark attack (Kenyon 1973). During a 1977 census of Pearl and Hermes reefs, De Long and Brownell (1977) noted that 35 percent of the 34 adult, subadult, and juvenile seals had sustained injuries inflicted by sharks. In comparison, 13 percent of the seals at Lisianski Island bore the scars of shark-inflicted wounds, suggesting that rates of predation vary among islands, as do rates of population decline in each area. Since the late 1950s many observers have reported seeing monk seals that had sustained shark injuries (Kenyon and Rice 1959; Wirtz 1968; B. W. Johnson and P. A. Johnson 1978; Alcorn 1984).

L. R. Taylor and Naftel (1978) made some fascinating observations of monk seals while scuba diving at La Perouse Pinnacles in the French Frigate Shoals. They saw monk seals exhaling large amounts of air in coral caverns well below the surface. The exhaled air made up a large bubble of breathable air in the roof of the cave. Taylor and Naftel (1978) suggest that this supply of stored air may allow the monk seals to take refuge from

FIG. 4.6. Hawaiian monk seal with partially healed shark injury. *Photo by David Olsen.*

FIG. 4.7. Galápagos sea lion bull. *Photo by Phil Thorson.*

sharks inside the caves when they are unable to escape onto land since the pinnacles are located well offshore of the islands.

Shark predation on the sea lions and fur seals of the Galápagos Islands appears to be a regular and common occurrence. A large proportion of Galápagos sea lions and fur seals hauled out on land bear fresh lacerations as well as scars probably due to attacks by tiger sharks and white-tip sharks (Trillmich 1984; Eibl-Eibesfeldt 1984b). Such predation may have a significant impact on the population dynamics, life history, and behavior of these pinnipeds. For instance, Trillmich (1984) suggests that the extended period of maternal care (from two to three years) in Galápagos fur seals may in part be related to shark predation, which is a significant cause of mortality in young seals. Because this long period of maternal care reduces the time the pup needs to spend foraging in the water, it should considerably enhance the pup's chances of survival.

In addition, Trillmich (1984) speculates that Galápagos fur seals feed primarily on moonless nights not only to take advantage of food resources that become more abundant at these times but also to avoid sharks. The seals tend not to feed during nights when the moon is full. Bright moonlight would appear to help sharks in detecting and approaching fur seals, which would be silhouetted against the surface, easy targets for sharks approaching from the darker waters below (Trillmich and Mohren 1981). Although white sharks are presumably attracted to the pinniped silhouette on the surface (McCosker 1985), they are also known to use olfactory, electrical, and magnetic field stimuli to locate prey (Tricas and McCosker 1984).

Territorial Galápagos sea lion bulls (Fig. 4.7) actually herd sea lion pups from deep water toward shore when sharks are present (B. Nelson 1968; Barlow 1972; Eibl-Eibesfeldt 1984b). Either singly or in groups of two or three, bulls approach a shark that swims close to shore, sometimes actually swimming in a cooperative manner to chase it away. Once a bull was even observed to bite at a shark's tail during such a "mobbing" incident. The behavior of the young sea lions in the presence of a shark is quite different; they usually rush toward the shore to escape danger. Although Barlow (1972) has called the protective behavior of the bulls paternal (and even though such behavior may protect a bull's offspring in a general sense), bulls do not actively direct paternal care to their own offspring.

KILLER WHALES

Killer whales consume a variety of marine vertebrates, including pinnipeds. They are known to prey on southern sea lions, California sea lions, Steller sea lions, South American fur seals, walruses, bearded seals, southern elephant seals, northern elephant seals, harbor seals, Weddell seals, and leopard seals. They probably include many more species of pinnipeds in their diet as well (see, e.g., Zenkovich 1938; Siniff and Bengtson 1977; T. G. Smith et al. 1981; J. A. Estes, pers. comm.). For instance, the eight species of *Arctocephalus* fur seals appear to be subject to killer whale predation throughout their range (Bonner 1981b). Like large white sharks, killer whales are capable of consuming large amounts of food in a short time. As a perhaps extreme example, in 1861 the Danish biologist Eschrict reportedly found the remains of fourteen seals and thirteen porpoises inside the fore-stomach of a 21-foot killer whale. If this report is true, the remains probably accumulated over time, so it is highly unlikely that the whale ate all these mammals at once.

Killer whales often forage cooperatively, employing sophisticated and highly coordinated maneuvers to capture their warm-blooded prey. For instance, T. G. Smith et al. (1981) observed a group of killer whales cooperatively hunting a crabeater seal in the Antarctic; they suggest that such predation on crabeater seals may be relatively common. Seven whales swimming in a group frequently rose vertically out of the water, a movement called spyhopping, apparently searching the ice floes for prey. The whales swam in echelon formation past an ice floe with a crabeater seal resting on it, creating a large wave that tipped the floe at a steep angle and slid the seal into the water. Whether the seal was eaten was not determined, but the whales' intent was clear. Apparently, killer whales also tilt ice floes so that basking Weddell seals slip into the water where the whales can seize them (Kooyman 1981d).

At sea, killer whale pods often herd sea lions into a tightly packed group while individual whales take turns plowing through the almost solidly bunched school, feeding on the trapped sea lions (Zenkovich 1938; Norris and Prescott 1961; Leatherwood and Samaris 1974). J. Wilson (1975) observed a group of killer whales feeding on a school of southern sea lions. The whales fed mainly on the two- to three-month-old pups, collectively killing over twenty in an hour. Foraging killer whales sometimes catapult adult southern sea lions as high as 30 feet into the air with their powerful forebodies, possibly to stun the animals (Bartlett 1976).

Killer whales also cooperatively encircle walruses and feed on them in a similar manner. In fact, besides polar bears, killer whales appear to be the major predators of walruses (Belopol'skii 1939; Fay 1982). Zenkovich (1938) described his observations of attacking killer whales:

> They acted like wolves on land: they surrounded the group of walruses on all sides; then, 6 or 7 on each flank formed straight lines, each whale just behind the head of the next; 5 approached the walruses from the front and 10 came in from behind. Then one of the whales which had come in from the rear burst into the herd and divided it, whereupon the others moved into that location, and the water there boiled as in a cauldron.

During another incident, Zenkovich (1938) observed fifteen killer whales surround sixty to seventy walruses, separating out and killing ten to twelve of them.

Walruses of both sexes and all ages are eaten by killer whales, although some believe that predation is directed particularly at calves and juveniles (Zenkovich 1938; Nikulin 1941; Brooks 1954). It may be that single killer whales prey mainly on calves or sick walruses, while groups of whales are able to capture adults easily. Scammon (1874) even reports that killer whales will ram a mother from below to dislodge the calf she is carrying on her back. Frequently killer whales seem to use their heavy blunt snouts as "battering rams" to stun and kill walruses as well as other marine

FIG. 4.8. Killer whale seizing southern elephant seal weaned pup on beach in Patagonia, Argentina. *Photo by Jeff Foott.*

mammals (see, e.g., Norris and Prescott 1961; Rice 1968; Fay 1982). Walruses attacked by killer whales often sustain internal injuries.

Like white sharks that aggregate near elephant seal colonies, killer whales often group around the rookeries of pinnipeds such as southern elephant seals and southern sea lions, especially during the breeding season. In subantarctic waters, the whales are major predators of southern elephant seals, especially the juveniles (Carrick and Ingham 1962; Condy et al. 1978). Condy et al. (1978) found that killer whales are observed most often at Marion Island during the southern elephant seal breeding season, although the seasonal occurrence of penguins may also attract killer whales. Killer whale predation appears to be an important cause of elephant seal mortality, especially among juvenile seals. Small groups of killer whales patrol the shoreline, scanning the beaches and sometimes venturing into the surf zone to grab a seal (Fig. 4.8).

Along the Valdés Peninsula of Argentina, killer whales capture many southern sea lion pups. The whales wait in shallow waters near the pinniped colonies for unsuspecting pups and young juveniles to venture into the surf. At the Valdés Peninsula sea lion rookeries, killer whales sometimes lunge toward the beach at full speed, and in the ensuing panic among the sea lions, a few whales are able to catch up to twenty pups in

FIG. 4.9. Killer whale seizing southern sea lion pup in Patagonia, Argentina. *Photo by Jeff Foott.*

one hour. Most of the whales, however, capture only a few sea lion pups in a single feeding bout (J. Wilson 1975; Bartlett and Bartlett 1976). Sometimes killer whales even attempt to capture seals by breaching onto the beach (Fig. 4.9). So at times it seems that seals are not safe from aquatic predators even on land.

PINNIPEDS

Under certain circumstances, one species of pinniped preys on another. Walruses, Steller sea lions, southern sea lions, New Zealand sea lions, and leopard seals are known to eat other pinnipeds. Interesting as this type of predation is, it does not generally represent a significant cause of mortality, with the possible exception of leopard seal predation on crabeater seals (Fig. 4.10). Pinniped predation on other pinnipeds is discussed in chapter 5. See Table 8 for the pinniped species preyed on by other seals.

Terrestrial Predators

POLAR BEARS

Polar bears (*Ursus maritimus*), which some consider to be marine mammals themselves, are versatile predators well adapted for catching Arctic pinnipeds. Because the bear is able to spend long hours in the freezing Arctic waters, it can venture onto the unstable offshore ice in search of seals and walruses. Polar bears prey on many Arctic pinnipeds, including bearded seals (T. G. Smith 1980), hooded seals (Reeves and Ling 1981), ribbon seals (King 1983), ringed seals (Stirling 1977; T. G. Smith 1980), walruses (Fay 1982; Calvert and Stirling 1987) and occasionally harbor seals (King 1983).

Ringed seals are preyed on especially heavily by polar bears (Fig. 4.11). Stirling (1974a) has found that ringed seals basking on ice in the summer have a difficult time relaxing and constantly lift up their heads every few seconds to scan for polar bears. Bears trying to sneak up on basking seals consequently have a poor rate of success. In contrast, ribbon seals, which are probably not preyed on as heavily by polar bears or other land predators, rest for long periods without scanning for predators and are easily approached by people. Eskimos as well as the Arctic explorers Freuchen (1935) and Nelson (1969) say that a polar bear on ice or snow stalks seals on three legs, covering its telltale black nose with a paw. Whether or not this is true, anyone watching a polar bear creep up on a seal can see that the bear moves as stealthily as possible, cautiously and stiffly, with its neck stretched out low to the ground.

Polar bear predation on ringed seals takes place primarily in the offshore ice areas, especially near the unstable ice flaw zone. In Alaska most of the ringed seals killed by polar bears are over six years of age (Eley 1978a). Yet in the Canadian Arctic, it is mainly the younger seals (the one-

FIG. 4.10. Leopard seal. *Photo by Gerry Kooyman.*

FIG. 4.11. Polar bear eating ringed seal. *Photo by Steve Amstrup.*

to two-year-olds) that are taken, at least during the period from March to the breakup of the ice (Smith 1976). Polar bears also prey heavily on pups in areas such as the Barrow Strait (Hammill and Smith 1987) and the western Beaufort Sea (Kelly et al. 1987), where subnivean birth lairs (those constructed beneath the snow) offer some degree of protection for the pups. It has been estimated that predation rates of polar bears are equivalent to a minimum of one ringed seal every week per bear (R. C. Best 1977).

At hooded seal breeding grounds, polar bears may also cause considerable pup mortality (McLaren 1977). Bearded seals in the Chukchi and Bering seas are sometimes subject to polar bear predation as well. Although the extent of such predation is not known, it probably occurs most often during late summer and fall when both the bears and bearded seals are concentrated in the same areas along the margin of multiyear ice (Figs. 4.12, 4.13) (Burns 1981a).

The extent of polar bear predation on walruses is not well known and may vary from area to area. Calvert and Stirling (1987) believe that the polar bear may be a significant predator of walruses in the Canadian High Arctic, where bears are often seen stalking and chasing walruses. Atlantic walruses appear more vulnerable to polar bear predation in the Canadian Arctic, where the ice is heavily frozen over and breathing holes

Fig. 4.12. Polar bear mother and pups. *Photo by Steve Amstrup.*

are scarce, than in the Bering Sea, where walruses inhabit areas of
consolidated pack ice. Walruses may even become "frozen out" in the
Canadian areas and be forced to travel long distances across the fast ice,
where they are easier targets for bears. In the Bering Sea, the rate of bear
predation on walruses appears to be relatively low, and predation occurs
mainly during the summer (Fay 1982).

FIG. 4.13. Polar bear mother and pup swimming. *Photo by Brian Fadely.*

In the Bering Sea area, walrus calves, young juveniles, and sick walruses seem to be more vulnerable to polar bear predation (Lono 1970; Fay 1982) than healthy adults. In fact, on occasion, it is the bears that have been threatened by or have fallen victim to walruses (Freuchen 1935; Pedersen 1962). But sometimes polar bears and walruses in the same area seem to ignore one another entirely (Mohr 1952; Brooks 1954; Nyholm 1975). Polar bears observed consuming walrus flesh or bears found with walrus remains inside their stomachs appear to have been feeding on seal carrion in some instances (Mohr 1952; Manning 1961; Lono 1970; Fay 1982).

Soviet biologists and other scientists have actually observed polar bears stalking or attacking walruses (Nikulin 1941; Popov 1958, 1960b, translated in Fay 1982; Loughrey 1959; Shults 1978, in Fay 1982; Calvert and Stirling 1987). During many of these hunting episodes, the bear typically succeeds only in frightening the entire walrus herd into the water without capturing a single animal. The females, calves, and young walruses appear to be most wary of an approaching polar bear and are the first to escape into the sea, while the adult males are the last to leave and often threaten the bear when it comes close. Popov (1958) speculates that although most of the walruses can easily escape an approaching bear, occasionally some of the calves are crushed in the stampede into the water. The bear's strategy of dashing toward the herd is therefore a good one; even though it may not catch a walrus, it may profit by causing calves to be crushed or abandoned.

Sometimes polar bears try to hide from walruses by digging pits in the sand in which to conceal themselves or by stacking driftwood to hide behind. Bears apparently observe the herd for a time in their "blinds" before leaping out and rushing toward the walruses (Popov 1958, 1960b). Polar bears have even been reported to throw chunks of ice at resting walruses or other seals. Standing erect on hindlegs, the bear hurls the projectile at the herd, sometimes killing or at least injuring one of the animals and increasing its chances of capturing its prey (R. K. Nelson 1969; Beck 1980).

OTHER LAND CARNIVORES

Other terrestrial carnivores—smaller and perhaps less threatening than polar bears—also prey on seals, particularly the vulnerable pups. Most other land predators, with the exception of the Arctic foxes that frequently prey on ringed seal pups, do not appear to be a significant cause of mortality (Fig. 4.14). Yet we know very little about the importance of terrestrial predation on pinnipeds. Young ringed seals (Fig. 4.15) are also

FIG. 4.14. Arctic fox. *Photo by Susan Dudenhoffer.*

FIG. 4.15. The movements of ringed seals are now being tracked using satellite packages, which are attached to the seal's back. The three photographs show (*top of page*) the capture, handling, and tagging; (*above*) release; and (*opposite*) acoustic recording of a ringed seal at Wolstenholme Fjord, Greenland, on June 5, 1988. This is the first successful use of remote sensing using satellite signals to track the movements of a wild seal. The research is being conducted by Steve Leatherwood, Brent Stewart, Pam Yochem, and Mads-Peter Heide-Jorgensen. *Photos by Steve Leatherwood and Brent Stewart, Hubbs Marine Research Center.*

attacked by red foxes, wolverines, wolves, dogs, and ravens (Burns 1970). Wolves and eagles prey on the closely related Caspian seals, especially pups, which inhabit the freshwater Caspian Sea (Naumov 1933; Ognev 1935; Rumyantsev and Khiras'kin 1978).

Ringed seal predation by Arctic foxes and polar bears may have partly promoted the female's habit of giving birth and raising her pup in small ice lairs or caves, a practice that in turn appears to have had an important structuring effect on the breeding system (see chapter 6). The foxes appear to kill only pups. In fact, Arctic fox predation in the Amundsen Gulf represents the most significant cause of mortality in newborn pups (T. G. Smith 1976). In a study of predation on ringed seals in the western Beaufort Sea, Kelly et al. (1987) found that Arctic foxes entered about 15 percent of the birth lairs studied and killed pups in 20 percent of these lairs.

Along the Pacific coast of North America, grizzly bears were probably an important predator of pinnipeds in the past and may have determined where pinnipeds such as the northern elephant seal could breed in California. It is unlikely that elephant seals were ever mainland breeders prior to the extinction of grizzly bears in California during the late 1800s, since breeding seals on land are extremely vulnerable to terrestrial predators. Coyotes also prey on seals along the Pacific Northwest coast. In Puget

Sound, Washington, coyotes are responsible for 16 percent of the mortality of harbor seal pups (Steiger et al. 1985).

In the Southern Hemisphere, pumas prey on southern sea lion pups (Cabrera and Yepes 1940). The South African fur seals that breed along the mainland coast of South Africa are preyed on by hyenas and silverbacked jackals. Again, it is primarily the pups that are attacked (Shaughnessy 1979; Bonner 1981b).

The effect of terrestrial and aquatic predation on the behavior of Arctic versus Antarctic pinnipeds has been reviewed by Stirling (1977, 1983). Arctic pinnipeds that are subject to predation by land predators—specifically polar bears—tend to take refuge in the water. Most Arctic species, for example, mate aquatically. In contrast, Antarctic seals, which evolved with aquatic predators such as leopard seals and killer whales, take refuge on ice. Stirling (1977) has compared the behavior of the Arctic ringed seal in relation to predation with that of the Antarctic Weddell seal, concluding that these species differ significantly in distribution patterns, morphology, choice of pupping sites, means of underwater communication, and various other behaviors.

5

Diet and Food Resources

The ocean provides an abundant and diverse array of food resources, such as zooplankton, fish, squid, crustaceans, molluscs, and even birds and mammals. The carnivorous pinnipeds have taken advantage of each type of food resource, consuming a wide variety of organisms found on or beneath the surface of the ocean. The various species of pinnipeds are characterized by one or more of the following diet types: marine zooplankton or krill; marine fish and squid; marine molluscs and crustaceans; marine mammals and birds; and freshwater fish (Table 7). There is often considerable overlap in the food resources used by a particular species. The krill-eating crabeater seal, for instance, may also consume fish at times. The leopard seal has a diverse diet that includes marine birds and mammals, squid, fish, krill, and various invertebrates.

The diet of many pinnipeds varies seasonally and/or geographically in a relatively predictable way, especially for seals living in polar and sub-polar environments: the Antarctic phocids, northern fur seals, Antarctic fur seals, ringed seals, bearded seals, harp seals, and Caspian seals. Caspian seals like to eat the seasonally abundant sculpin fish so much that they will actually gorge themselves to death when they have access to a seemingly unlimited supply (King 1983). But this behavior is rare, and seals in general are not gluttonous.

Food resources change predictably throughout the year for many pinnipeds. As Gentry and Kooyman (1986) point out, the predictability of a seal's environment increases as one moves from the equator (the least predictable environment) to the polar regions, where food resources are most abundant and predictable from year to year. In contrast, the diet of

TABLE 7
Diet of Pinnipeds

SPECIES	FOOD RESOURCE	EXEMPLARY FOOD TYPES	REFERENCES
Hawaiian monk seal	Fish, cephalopods, other invertebrates	Flatfish, scorpaenids, larval fish, moray eels, conger eels, octopuses, spiny lobsters	Kenyon 1981
Mediterranean monk seal	Fish, cephalopods	Barbouni, gopa, pargo, rays, octopuses	Kenyon 1981; Marchessaux & Muller 1985
Northern elephant seal	Fish, cephalopods, tunicates*	Pacific whiting, squid, octopuses, pelagic red crabs	Condit & Le Boeuf 1984; Hacker 1986; Antonelis et al. 1987
Southern elephant seal†	Fish, cephalopods, other invertebrates	Antarctic blennies, squid, octopuses, amphipods	Laws 1956a; Yaldwyn 1958; Csordas 1965; Ling & Bryden 1981; Clarke & Macleod 1982b; Laws 1984
Weddell seal‡	Fish, cephalopods, krill, other invertebrates	Atlantic cod, squid, octopuses, *Euphausia superba*, amphipods	Dearborn 1965; Kooyman 1968; Øritsland 1977; Clarke & Macleod 1982a; Laws 1984
Ross seal‡	Cephalopods, fish, other invertebrates, krill	Squid, octopuses, cuttlefish, *E. superba*, amphipods, lantern fish, dragon fish	King 1969; Øritsland 1977; Laws 1984
Crabeater seal	Krill, cephalopods,* fish*	*E. superba, E. crystallorophias*, octopuses	E. A. Wilson 1907; Marr 1962; Øritsland 1977; Laws 1984
Caspian seal§	Fish, crustaceans	Sculpin, gobies, herring	Vorozhtsov et al. 1972; King 1983

NOTE: See Table 8 for the warm-blooded prey of pinnipeds.

* Known to be a relatively minor food resource.

† Diet known to vary with age.

‡ Diet shows distinctive geographic variation.

§ Diet known to vary seasonally.

SPECIES	FOOD RESOURCE	EXEMPLARY FOOD TYPES	REFERENCES
Leopard seal‡	Birds, pinnipeds, krill, fish, cephalopods, other invertebrates	Penguins, crabeater seals, *E. superba,* squid, octopuses	E. A. Wilson 1907; Marr 1962; Penney & Lowry 1967; Hofman et al. 1977; Laws 1977a,b; Siniff & Bengtson 1977; Øritsland 1977; Bonner 1982
Hooded seal	Fish, cephalopods, other invertebrates	Redfish, cod, capelin, herring, squid, octopuses, mussels, shrimps	Sergeant 1976; Reeves & Ling 1981
Bearded seal	Bivalve molluscs, crustaceans, cephalopods, fish, other invertebrates	Clams, gastropods, shrimps, crabs, worms, cod, octopuses	Chapskii 1938; Fedoseev & Bukhtiyarov 1972; Lowry et al. 1977, 1978, 1980a; Burns 1967, 1981a
Harp seal†‡§	Fish, crustaceans, krill	Capelin, cod, herring, crabs, shrimps, euphausiids, amphipods	Mansfield 1967; Sergeant 1973b; Ronald & Healey 1981
Ribbon seal	Crustaceans, fish, krill, cephalopods	Crabs, shrimps, mysids, cod, pollock, eelpouts, squid, amphipods	Arseniev 1941; Wilke 1954; Shustov 1965a,b; Lowry et al. 1977, 1978; Frost & Lowry 1980; Burns 1981b
Baikal seal	Fish, crustaceans	Gobies, *Comephorus*	Pastukhov 1969; Vorozhtsov et al. 1972; King 1983
Ringed seal‡§	Fish, crustaceans, krill	Cod, shrimps, amphipods, euphausiids	McLaren 1958; Fedoseev 1965; M. L. Johnson et al. 1966; Lowry et al. 1980b; Frost & Lowry 1981

Continued on next page

SPECIES	FOOD RESOURCE	EXEMPLARY FOOD TYPES	REFERENCES
Largha seal†	Fish, cephalopods, crustaceans	Cod, rockfish, sculpin, flounder, octopuses, amphipods	Tikhomirov 1966; Goltsev 1971; Bigg 1981; Bukhtiyarov et al. 1981
Harbor seal†‡§	Fish, cephalopods, krill, other invertebrates	Herring, cod, flounder, sculpin, gadoids, salmon, octopuses, whelks, shrimps, amphipods	Sergeant 1951; Spalding 1964; Bigg 1973; Rae 1973; Pitcher & Calkins 1979; Bigg 1981
Grey seal‡§	Fish, cephalopods, crustaceans	Salmon, cod, herring, mackerel, squid	Rae 1968, 1973; Mansfield & Beck 1977
California sea lion	Fish, cephalopods	Anchovies, herring, Pacific whiting, rockfish, hake, salmon, squid, octopuses	Fiscus & Baines 1966; Keyes 1968; Morejohn 1977; Bailey & Ainley 1981/1982; Ainley et al. 1982; Antonelis et al. 1984
Galápagos sea lion	Fish, cephalopods	Octopuses	Eibl-Eibesfeldt 1984b
Steller sea lion‡	Fish, cephalopods, crustaceans, bivalve molluscs, pinnipeds	Rockfish, sculpin, capelin, flatfish, squid, octopuses, shrimps, crabs, northern fur seals	Mathisen et al. 1962; Fiscus & Baines 1966; Keyes 1968; Gentry & Johnson 1981
Southern sea lion‡	Fish, crustaceans, cephalopods, other invertebrates, birds, pinnipeds	Anchovies, rockfish, *Munida* (crustacean), squid, octopuses, penguins, South American fur seals	Vaz-Ferreira 1950, 1981; Boswall 1972; Aguayo & Maturana 1973; Gentry & Johnson 1981; Majluf 1987; F. Lanting, pers. comm.
Australian sea lion‡	Fish, cephalopods, crustaceans, birds	Whiting, salmon, squid, rock lobsters, crayfish, penguins	Wood Jones 1925; King 1964a; Marlow 1975; Walker & Ling 1981a

SPECIES	FOOD RESOURCE	EXEMPLARY FOOD TYPES	REFERENCES
New Zealand sea lion‡	Fish, cephalopods, crustaceans, bivalve molluscs, birds	Flounder, squid, octopuses, crabs, mussels, penguins	Gwynn 1953; King 1964a; Gaskin 1972; Marlow 1975
Northern fur seal‡§	Fish, cephalopods, birds*	Pollock, herring, lantern fish, cod, rockfish, squid, loons, petrels	F. H. C. Taylor et al. 1955; Niggol et al. 1959; Lander & Kajimura 1976; Gentry 1981a; Kajimura 1984
Guadalupe fur seal	Fish, cephalopods	Rockfish?	Fleischer 1978b; Perez & Bigg 1986; M. Pierson, pers. comm.
Juan Fernández fur seal	Fish, cephalopods, crustaceans	Squid, lobsters	Castilla 1981; Torres 1987
Galápagos fur seal	Fish, cephalopods	Anchovies, mackerel, small squid, lantern fish, deep sea stints	Clarke & Trillmich 1980; Trillmich 1984
South American fur seal	Fish, cephalopods, crustaceans, other invertebrates	Anchovies, carangids, sea trout, squid, sea snails	Vaz-Ferreira 1979, 1982a; Majluf 1987; Trillmich 1987c
New Zealand fur seal	Cephalopods, fish, birds, crustaceans	Octopuses, squid, barracouta, penguins, crabs, rock lobsters	Bailey & Sorensen 1962; Street 1964; Marlow & King 1974; Crawley & Wilson 1976
Antarctic fur seal‡	Krill, fish,* cephalopods,* birds*	*E. superba,* squid, notothenid fish, penguins	Bonner 1968, 1981b; Bonner & Hunter 1982; North et al. 1983; Croxall & Pilcher 1984; Doidge & Croxall 1985
Subantarctic fur seal‡	Cephalopods, fish, krill, birds	Notothenid fish, squid, euphausiids, penguins	Paulian 1964; Rand 1956; Condy 1981; Bester & Laycock 1985; Kerley 1987

Continued on next page

TABLE 7, *continued*

SPECIES	FOOD RESOURCE	EXEMPLARY FOOD TYPES	REFERENCES
South African fur seal†	Fish, cephalo-pods, crusta-ceans, birds	Mackerel, pilchard, maasbanker, ancho-vies, squid, octo-puses, cuttlefish, shrimps, rock lob-sters, penguins	Rand 1959; Cooper 1974; Bonner 1981b; David 1987a
Australian fur seal	Cephalopods, fish, crustaceans	Squid, octopuses, snook, rock lob-sters, mullet, parrot fish, whit-ing, barracouta	Warneke 1982; Warneke & Shaughnessy 1985; Shaugh-nessy & Warneke 1987
Walrus	Benthic inverte-brates, fish, cephalopods, pinnipeds	Bivalve molluscs (soft-shelled clams, cockles, Arctic rock borers), octopuses, polar cod, annelid & sipunculid worms, crabs, shrimps, amphi-pods, mysids, sea cucumbers, ringed seals	Chapskii 1936; Nikulin 1941; Vibe 1950; Brooks 1954; Mansfield 1958a; Lough-rey 1959; Krylov 1971; Fay et al. 1977; Fay 1982; Oliver et al. 1983, 1985; Lowry & Fay 1984; Fuku-yama & Oliver 1985

many pinnipeds, especially otariids that live in temperate and tropical climates, varies more from year to year. This variability is caused pri-marily by the ENSO — El Niño Southern Oscillation — events, discussed in chapter 3, which occur every two to ten years and affect the abundance of food resources in an unpredictable way. The occurrence of El Niño also affects the diet of temperate-water phocids such as harbor seals and probably affects other species of pinnipeds as well.

A number of generalizations can be made about the relation of food resources to the various taxonomic and geographic groups of pinnipeds and to pinnipeds as a whole. For most species of pinnipeds, fish and squid are the principal food resource. The next most common food seems to be other invertebrates (especially crustaceans and bivalve molluscs), fol-lowed by zooplankton and warm-blooded prey (birds and other pinnipeds). Although most seals do not specialize in zooplankton or krill,

krill is the most heavily exploited food resource because the extremely abundant crabeater seals eat it almost exclusively.

Variation in Diet and Foraging Behavior

Diet varies with age in a number of pinnipeds. The diets of newly weaned or juvenile seals often differ from those of adults of the same species. Harp seal adults, for example, feed mainly on loose groups of fish (especially pelagic schooling fish) and some crustaceans (Fig. 5.1). The weaned pups, in contrast, which are solitary for their first year, eat mainly zooplankton (pelagic crustaceans such as the euphausiid *Thysanoessa* and the amphipod *Anonyx*) in surface waters. Juvenile harp seals more than one year old feed at intermediate depths on capelin fish, whereas adults dive much deeper, to 150–200 meters to feed on bottom fish such as herring and cod (Sergeant 1973b). Although largha seal adults eat mainly fish along with some cephalopods and crustaceans, the newly weaned pups specialize in zooplankton in the form of small amphipods near ice floes (Goltsev 1971). The diet of adult harbor seals (*Phoca vitulina richardsi*) in the Pacific Northwest is similar to that of largha seals, but the weaned pups feed mainly on benthic crustaceans, especially on the shrimp *Crangon*, during the first few months of life (Bigg 1973). In at least two other subspecies of harbor seal (*P.v. concolor* and *P.v. vitulina*), weaned pups also eat mainly small invertebrates like shrimp, while adults consume primarily fish along with some cephalopods, crustaceans, and molluscs (King 1983). In southern elephant seals, weaned pups tend to eat small crustaceans, especially in intertidal areas, whereas adults feed mainly on fish and squid (Ling and Bryden 1981). The stomach contents of yearling South African fur seals show that they consume large amounts of shrimp (*Micropetasma africana*) (Rand 1959).

Why do juveniles and adults eat different foods? Perhaps the foods juveniles consume are easier to capture. It is known that juvenile sea otters, for instance, tend to obtain food that is less calorically rewarding but easier to capture than foods eaten by adults (J. A. Estes et al. 1981; Riedman et al. 1988). Possibly juvenile seals are also relegated to areas with different food resources, or they may not be able to dive deep enough to forage on the same prey as adults.

The diet of female harp seals and perhaps the females of other species as well vary according to reproductive condition. During and immediately following lactation, female harp seals consume decapods (Ronald and Healey 1981). The diet of some female sea otters also varies with

FIG. 5.1. Harp seals under the ice. *Photo by W. Curtsinger, International Fund for Animal Welfare.*

reproductive status, that is, whether or not they have a pup or are in estrus (Lyons and Estes 1985; Riedman et al. 1988). Although similar changes may occur in other pinniped females, such dietary variation is difficult to document in free-ranging seals.

In species that supplement their diets with warm-blooded prey, it seems to be the adult or subadult males that take penguins or other seals. This is true of Steller sea lions, southern sea lions, leopard seals, Antarctic fur seals, walruses, and possibly other species as well. For instance, it is usually the older subadult and adult male walruses that are seen eating seals (Chapskii 1936; Mansfield 1958a; Rausch 1970; Fay 1960; Fay et al. 1977; Lowry and Fay 1984). Only rarely are females observed consuming seal flesh. While all of the carnivorous walruses in the Bering Strait region observed by Lowry and Fay (1984) were males, two seal eaters in the Chukchi Sea were females, indicating that female walruses may be carnivorous under certain circumstances.

Although male seals, rather than females, tend to be responsible for the predation on mammals and birds, the reason for this tendency is unclear. But it reflects a similar possible trend among other mammals. For instance, although the diverse diet of chimpanzees consists largely of fruit and vegetables, adult male chimpanzees capture and eat other mammals and birds (such as red colobus monkeys, bush pigs, and baboons) far more frequently than do female chimpanzees (Goodall 1986). In addition, the hunting of mammals among chimpanzees is more common in some areas than others, which seems to be the case among pinnipeds as well. Among California sea otters there may be a similar tendency for adult males in a certain area to eat warm-blooded prey in the form of various seabirds, although the diet of California otters typically consists of various marine invertebrates (Riedman and Estes 1988).

Recent research on the foraging ecology of various animals is yielding an important and intriguing finding: that individuals feeding under similar conditions may specialize in particular foods or foraging strategies, irrespective of age, sex, and body morphology. The existence of such variation in feeding behavior indicates the importance of studying animal foraging systems, along with other aspects of their biology, at the level of the individual.

Although individual variation in diet has rarely been documented conclusively in pinnipeds (perhaps because of logistic difficulties in observing the foraging behavior of free-ranging seals), it is possible that specialization in certain types of prey by individuals foraging in the same area does occur; it has been documented in species as diverse as sea otters (Lyons and Estes 1985; Riedman et al. 1988), tropical marine snails (West

1986a,b), chimpanzees (Goodall 1986) and Cocos finches of Costa Rica (Werner and Sherry 1987). The variation in depth and diving patterns of pinnipeds bearing Time-Depth Recorders at least suggests that individual differences in the type of fish and cephalopods eaten may occur in species such as northern elephant seals (Le Boeuf et al. 1988), northern fur seals (Gentry, Kooyman, and Goebel 1986), Antarctic fur seals (Kooyman, Davis, and Croxall 1986), and Galápagos fur seals (Kooyman and Trillmich 1986a).

California sea lions have learned to aggregate near the mouths of freshwater rivers in Washington State, where they prey on large numbers of spawning steelhead. State Fish and Game biologists found that the same individuals repeatedly used this foraging strategy, preying on fish in the Ballard Lock area not only throughout the four- to five-month spawning season, but also returning over a period of three to four years to capture steelhead in the same area. In any given year, only a small number of individuals (between three and twenty) were responsible for the substantial impact on spawning fish populations. Biologists identified the sea lions, all subadult and adult males, by scars and other natural markings. A distinctive hump on the back of one seal earned him the name Humpback. Another seal used a unique foraging tactic while eating the fish. While most of the sea lions thrashed the 18- to 20-pound fish on the surface to break them into bite-sized chunks, this particular male tossed fish high up in the air to "tenderize" them for consumption (B. Byrne, pers. comm.).

Especially among pinnipeds that include warm-blooded prey in their diets, certain individuals consistently seem to prey on birds and mammals. This appears to be true of leopard seals, walruses, southern sea lions, and perhaps other species as well. It also seems to be true of other mammals that occasionally supplement their diets with warm-blooded prey. The same individual California sea otters, for example, appeared to be responsible for several cases of predation on seabirds in the Monterey Bay area (Riedman and Estes 1988). Similarly, certain male southern sea lions appear to specialize in penguins as a food resource. According to F. Lanting (pers. comm.), the same individual bulls—identifiable by distinctive scars on their faces—consistently and predictably attacked penguins. A single bull might consume six penguins each day. Some male walruses prey on seals for food more persistently than other walruses. Biologists and Eskimos identify these seal-eating males in the field by their tusks, which are distinctively stained with seal blubber and scratched from tussles with seals (Fay 1982; J. A. Estes, pers. comm.).

Finally, there is some evidence to suggest that only a few adult leopard seals, apparently males, are responsible for most of the predation on penguins at Cape Crozier, Ross Island, where seals have captured an estimated 15,000 penguins in a fifteen-week period (Penney and Lowry 1967) at a rookery that contains 300,000 penguins. Both Penney and Lowry (1967) and Kooyman (1965a) saw only four seals at one time at the penguin rookery, although one would surely think that more than four seals were responsible for the capture of 15,000 penguins each season. Kooyman (1965a) did observe one especially successful seal that ate 6 penguins in seventy minutes. And according to Lipps (1980), one leopard seal was found with the remains of 16 adult Adélie penguins inside its stomach. If so few seals were really involved in "penguin hunts," then perhaps some adult male seals have learned to specialize in penguins as their principal food resource in this area. In addition, it is possible that certain individual leopard seal males at South Georgia Island specialize in fur seals and other pinnipeds.

Pinniped Predation on Warm-blooded Prey

PREDATION ON BIRDS BY LEOPARD SEALS

The interesting pinniped practice of eating birds and mammals has not been thoroughly reviewed before. That it is primarily the males, and possibly the same individuals, who seem to be responsible for most of this predation makes this foraging tactic even more interesting, and I therefore review it in detail, beginning with the leopard seal, which is probably responsible for more of the predation on warm-blooded prey than any other pinniped. In fact, penguins and pinnipeds appear to form a staple of the diet among some populations of leopard seals (Fig. 5.2), although these seals also consume a wide variety of other foods, including fish, squid, crustaceans, and krill (J. E. Hamilton 1939; Gwynn 1953; K. G. Brown 1957). Leopard seals also feed on the carrion of seals and whales on occasion. One male leopard seal was even found near Sydney, Australia, with an adult duckbill platypus in its stomach (Troughton 1951). Other seabirds besides penguins, such as giant petrels (*Macronectes* spp.), are also eaten by leopard seals (Table 8) (T. Øritsland 1977).

Leopard seals are known to eat emperor, gentoo, chinstrap, crested, Magellanic, king, and especially Adélie penguins (see, e.g., Markham 1971; T. Øritsland 1977; Horning and Fenwick 1978). Adélie penguins, the most abundant penguin throughout the range of the leopard seal, fall prey to leopard seals more frequently than other species.

The composition of a leopard seal's diet tends to vary geographically.

Table 8
Mammalian and Avian Prey of Pinnipeds

SPECIES	AGE & SEX	PENGUINS/ OTHER SEABIRDS	PINNIPEDS/ OTHER MAMMALS
New Zealand fur seal		Rock hopper penguins	
Antarctic fur seal	Subadult males	Macaroni & gentoo penguins	
Subantarctic fur seal		Rock hopper & crested penguins	
Northern fur seal		Beal petrels, red-throated loons, rhinoceros auklets	
South African fur seal		Jackass penguins, gannets	
New Zealand sea lion		Rock hopper, gentoo, & yellow-eyed penguins	New Zealand fur seals*
Australian sea lion		Little penguins	
Southern sea lion	Adult males	Rock hopper, Magellanic, jackass, & gentoo penguins	South American fur seals* (pups & adult females also eaten)
Steller sea lion	Adult & juvenile males		Northern fur seals,* ringed seals, harbor seals, sea otters?
Walrus	Adult males		Ringed seals,† bearded seals,† largha seals,† harp seals, narwhals
Leopard seal	Adult males	Adélie, emperor, gentoo, chinstrap, crested, Magellanic, & king penguins; giant petrels, diving petrels, blue petrels, prions, Kerguelen diving petrels, blue-eyed cormorants, Cape pigeons, Southern black-backed gulls, terns	Crabeater seals,* Ross seals?, Weddell seals,* southern elephant seals,* southern sea lions,* subantarctic fur seals,* Antarctic fur seals, possibly other *Arctocephalus* spp.

NOTES: The importance of each food resource in the diet of a particular species varies widely with geography. See the text and Table 6 for references and the Index for scientific names.

* Primarily juvenile pinnipeds taken.

† May have been carrion in some cases, although most of the seal eating appears to be predation on live seals.

FIG. 5.2. Leopard seal eating penguin, its primary prey in some areas of the Antarctic. *Photo by Gerry Kooyman.*

Leopard seals may forage heavily on pinnipeds in one region; in another, especially near pack ice, they may eat mainly krill. Interestingly, near Palmer Station, in the lower latitudes of the Antarctic Peninsula where many leopard seals and penguin rookeries are located together, it is rare to observe seals preying on penguins. The seals' diet in this area appears to consist mainly of krill (Hofman et al. 1977; Kooyman 1981b). Similarly, leopard seals near the South Shetland Islands feed heavily on krill, which makes up over half of the diet, with the remainder consisting of fish, cephalopods, and penguins, at least during August and September (T. Øritsland 1977).

Like their pinniped predators, penguins also feed on fish, cephalopods, and krill — with some species able to dive to depths of 100 meters or more. Because penguins must leave their rookeries to feed at sea, they are constantly vulnerable to predation by pinnipeds, especially as they swim to and from land in shallow waters close to shore.

The largest of the four Antarctic phocids, the leopard seal, with its incredibly powerful jaws and long sinuous neck, is well equipped to capture large birds and seals. The leopard seal can actually coil its neck back and then strike its prey like a snake. When hunting penguins, leopard seals swim back and forth on the surface close to shore, where most of the attacks on penguins take place. The seals have also learned to detect penguins walking on thin ice, however, and are even able to break through the ice to capture one. How adept penguins are at escaping attacking leopard seals is uncertain, but Lipps (1980) tells of a wounded penguin that suddenly leapt aboard his wooden whaleboat with a large leopard seal in hot pursuit. The penguin huddled in the middle of the boat while the seal circled and followed the vessel to shore. Eventually the leopard seal gave up its pursuit, and, happily, the penguin recovered within a few days.

Adult penguins are usually captured by leopard seals in an underwater chase. Fledgling penguins are so helpless in the water that the leopard seal doesn't even have to pursue them but merely swims up and catches them as they splash about on the surface. Leopard seals often congregate around penguin-laden ice floes and wait for a bird to slip or leap into the water. Groups of penguins may be seen clustered tightly on an ice floe, seeming on the brink of diving into the water but hesitating to be the first to "splash down" because of the danger of leopard seals. Sometimes the birds even seem to try to push one another off the ice floe; if one of them falls, the other penguins soon know whether a leopard seal is lurking nearby (Fig. 5.3). As soon as one penguin dives in safely, the rest

FIG. 5.3. Group of penguins on an ice floe, with a leopard seal resting nearby. *Photo by Frank Todd.*

FIG. 5.4. Penguins and Antarctic fur seals resting together on land. This fur seal sometimes preys on penguins in the water. *Photo by Dan Costa.*

usually follow. The leopard seal that catches a penguin proceeds to eat it by chewing its skin away and swinging it back and forth on the surface, sometimes actually turning the penguin inside out. Most leopard seals do this to decapitate the bird and then chew off its strong feet. They do not appear to like the feathers, skin, feet, or beaks. Anyone viewing a scene as gruesome as the one I have just described and considering the high rate of leopard seal attacks on penguins in some areas understands why penguins are so careful to "look before they leap."

PREDATION ON BIRDS BY OTHER PINNIPEDS

No Antarctic phocid other than the leopard seal has been seen eating penguins or other seabirds. A number of otariids (four fur seals, three sea lions) in the Southern Hemisphere, however, include penguins in their diets to varying degrees. A wide variety of penguin species are preyed on by otariids: crested penguins, emperor penguins, gentoo penguins, jackass penguins, king penguins, little penguins, macaroni penguins, Magellanic penguins, rock hopper penguins, and yellow-eyed penguins.

Six otariids eat penguins: New Zealand fur seals (Street 1964; Csordas and Ingham 1965), Antarctic fur seals (Fig. 5.4) (Bonner and Hunter

1982), subantarctic fur seals (Paulian 1964), South African fur seals (J. Cooper 1974), New Zealand sea lions (Gwynn 1953; A. M. Bailey and Sorensen 1962; Gaskin 1972; R. Gentry, pers. comm.), Australian sea lions (Wood Jones 1925), and southern sea lions (Boswall 1972). In addition, the northern fur seal sometimes eats birds such as the red-throated loon, rhinoceros auklet and beal petrel. These seabirds, however, appear to represent an insignificant part of the northern fur seal's diet (Niggol et al. 1959). South African fur seals may also take gannets occasionally (J. Cooper 1974).

Although harbor seals are not generally known to prey on birds, isolated observations of harbor seal (*Phoca vitulina richardsi*) attacks on birds have been reported in Monterey Bay, California. In one case, a rehabilitated harbor seal juvenile male raised as an orphaned pup was observed to kill and eat two pigeons near the outdoor tidal basin of the Monterey Bay Aquarium (J. Hymer, pers. comm.). This particular seal was fed regularly by aquarium staff in this area. Less than 300 meters away, at Hopkins Marine Station, M. Staedler (pers. comm.) observed a harbor seal attacking and struggling with a brown pelican, which eventually escaped.

As is true of leopard seals, predation on penguins by sea lions and fur seals in the Southern Hemisphere appears to be more common in some areas than in others. For instance, subantarctic fur seals eat mainly crested penguins and squid at Amsterdam Island, but at Marion Island, cephalopods, fish, and krill form the mainstay of the diet. Southern sea lions eat penguins (and possibly other seabirds) only in some areas, such as the Falkland Islands (J. E. Hamilton 1946; Boswall 1972). Three- to six-year-old male Antarctic fur seals prey heavily on macaroni penguins (and possibly on gentoo and king penguins as well) at Bird Island, South Georgia. For some reason, the fur seals do not always eat the penguins after killing them, but giant petrels (*Macronectes* spp.) immediately devour any available penguin carcasses. Sometimes the aggressive petrels even try to steal a penguin from a fur seal. This additional food resource for petrels has resulted in a dramatic increase in the breeding population of giant petrels at Bird Island (Bonner and Hunter 1982).

Adult male southern sea lions frequently prey on penguins (rock hopper, Magellanic, and especially gentoo) at New Island and Bonshane Island in the Falklands. Frans Lanting (pers. comm.) observed some fascinating interactions between the penguins and sea lions on these islands during February 1987. The behavior of gentoo penguins indicated that the birds were well aware of the ever-present danger of the sea lion bulls. In the morning before their daily feeding trip to sea, gentoo penguins congregated in groups of up to five hundred birds on the beach, moving in synchrony back and forth just in front of the surf. Hesitantly, the birds entered the water in groups of perhaps a dozen, only to turn and scramble back up the beach if there was a bull offshore. This was a daily morning ritual. The birds' reluctance to enter the water is understandable, since large sea lion bulls often waited for them in the shallow water near the beach. Large flocks of giant petrels, or "stinkers," hovering offshore may have warned the penguins that there were sea lion bulls in the water, for the petrels knew within seconds when a penguin had been killed and waited aloft to scavenge the remains. If no sea lions were evident, groups of perhaps a dozen penguins would hesitantly enter the sea at intervals until all the birds had gone into the water for their daily feeding trip. None of the penguins seemed to want to be first or last in the water. Once the penguins reached deeper water, they seemed to be able to outswim the sea lions.

The gentoo penguins went to great lengths to avoid an encounter with foraging sea lion bulls. One morning, a persistent sea lion bull remained in the surf for two hours, blocking the penguins' access to the ocean. After waiting two hours for the bull to move on, the penguins finally must have decided that a little hike to a safer entry point would be preferable to risking an attack by the bull. The group of birds walked a kilometer along the coast over rough terrain to another beach where they could safely enter the water, at which point one would imagine they had worked up quite an appetite.

The penguins were not necessarily safe on land, however. Periodically, a large sea lion bull bodysurfed onto the beach, singling out and seizing a penguin separated from the confused mass of birds. The presence of a sea lion bull on the beach generally sent most of the birds running into the surf or further inland. After emerging from the water, rock hopper and gentoo penguins would rapidly hop or run from the surf zone, scrambling swiftly up the beach and rocks until they were safely away from the shoreline. The birds were then able to preen in an area relatively secure from surprise attacks by sea lions. A sea lion successful in capturing a penguin would return to the water, swinging the bird until the skin, head, and feet were removed. Many of the "penguinophilic" bulls bore recent flesh wounds on their heads, apparently inflicted by the penguins' bills. On one of the Falkland Islands, southern sea lion bulls have even learned to prey on jackass penguins on land by raiding the burrows where the young penguins are raised (Tony Chater, unpubl. data).

PINNIPEDS AS A FOOD RESOURCE

A number of pinnipeds, including leopard seals, New Zealand sea lions, southern sea lions, Steller sea lions, and walruses, prey on other species of pinnipeds. Again, the males appear to be responsible for such predation. Southern sea lion males prey on South American fur seal pups and adult females in the Falkland Islands (Gentry and Johnson 1981; Majluf 1987). In at least one case, remains of a New Zealand fur seal pup were found in the stomach contents of a male New Zealand sea lion at Snares Islands (Mattlin 1987). A similar situation is found in the Northern Hemisphere, where Steller sea lion males prey on northern fur seals as well as ringed seals, harbor seals, and possibly sea otters (Tikhomirov 1959, 1964a; Gentry and Johnson 1981; Pitcher and Fay 1982). Interestingly, in all three species it is either the sea lion adult or juvenile males that capture the comparatively small-sized juvenile or female fur seals, rather than the fur seal males that eat the sea lion juveniles.

Steller sea lion predation on northern fur seals has been especially well documented since 1967 at St. George Island in the Pribilofs, although such predation has never been seen at St. Paul Island, which is only 65 kilometers away. It may be that some of the St. George sea lion males, like leopard seals, have learned to specialize in fur seals, while the St. Paul sea lions have not acquired a taste for other pinnipeds. Gentry and Johnson (1981) observed such predatory behavior in detail and found that juvenile males hunting alone were responsible for nearly all the predation on sea lion pups. A sea lion hunted by swimming along the shore past a group of pups, submerging within 20 meters of the intended victim, and grabbing the pup from beneath the surface. Sometimes a male chased pups or rode waves into the middle of pup groups in the surf zone, seizing a pup in the process. The frequency of predation was highest from August to November, when many of the fur seal pups depart for sea. An estimated 3.4–6.8 percent of the pups born in 1975 (between 2,708 and 5,416 pups) were killed by young male sea lions. The fur seal pups are thought to supplement rather than to constitute the principal diet of sea lions (Gentry and Johnson 1981).

Walruses also consume pinnipeds, apparently much more frequently than previously believed. Lowry and Fay (1984) found that seal eating was much more common in the 1970s and early 1980s in the Bering Strait region than it had been during the previous thirty years. They trace the increased predation on seals to a larger walrus population as well as unusually restrictive ice conditions (especially in 1979), which resulted in a greater overlap in the distribution of walruses and other seal species. In addition, they speculate that walrus predation on seals could have a substantial impact on seal populations in some areas.

Although in the past it was not always clear whether walruses actually preyed on the seals or just ate carrion they happened to encounter (Fay 1960; Krylov 1971), more recent observations indicate that most seal eating is predation rather than scavenging (Lowry and Fay 1984). The remains of bearded seals (Brooks 1954; Fay et al. 1977; Lowry and Fay 1984), ringed seals (Nikulin 1941; Brooks 1954; Shustov 1969; Krylov 1971; Lowry and Fay 1984), and largha seals (Shustov 1969; Lowry and Fay 1984) have been found in the stomach contents and fecal material of walruses. Lowry and Frost (unpubl. data) observed a walrus eating a freshly killed largha seal pup and found the carcass of another largha seal pup (perhaps similarly dispatched) nearby. Attacks on a harp seal pup and bearded seal were also observed. In each case, the walrus stabbed the seal with its tusks (Lowry and Fay 1984). Occasionally walruses have been observed feeding on narwhals; besides leopard seals, which occasionally eat whale carrion, walruses are the only pinnipeds known to eat cetaceans. Whether walruses actually kill narwhals is unknown, but in at least two cases the whales had died very recently (King 1983).

FIG. 5.5. Crabeater seal with scars inflicted by a leopard seal. *Photo by Roger Gentry.*

Leopard seals capture and eat pinnipeds, particularly juveniles, of the following species: crabeater seals, Weddell seals, southern elephant seals, southern sea lions, subantarctic fur seals, Antarctic fur seals, and possibly other species of southern fur seals as well as Ross seals (Hamilton 1934; Erickson and Hofman 1974; Condy 1977a; Hofman et al. 1977; T. Ørits-land 1977; Siniff and Bengtson 1977; Laws 1977a; Ling and Bryden 1981; Kooyman 1981b; Stone and Siniff 1981; King 1983; McCann and Doidge 1987). In certain areas of their range, pinnipeds appear to form an important part of the leopard seal's diet. At South Georgia, for example, leopard seals often feed on various species of southern fur seals (*Arctocephalus* spp.) (Kooyman 1981b). Most of the fur seals taken are juveniles or females.

Crabeater seals seem to be an especially important food resource for leopard seals. The characteristic parallel tooth scars of leopard seal attacks are often visible on crabeater seal survivors (Fig. 5.5); almost all crabeater seals over one year of age bear the distinctive scars inflicted by leopard seals (Gilbert and Erickson 1977; Hofman et al. 1977). As many as 156 (78 percent) of 200 crabeater seals observed by Siniff and Bengtson (1977) bore scars from leopard seal attacks. Crabeater seals less than eighteen months old appear to be most vulnerable to leopard seal attack. Laws (1977b) noted that 83 percent of fresh wounds on seals were found on those less than eighteen months old. He suggests that pups younger than six months probably do not survive encounters with leopard seals.

Named for the pattern of dark spots and splotches on their light-colored undersides, leopard seals have acquired a reputation for being aggressive and dangerous to man; in truth, however, they are usually wary and shy around humans. Although there are numerous reports of leopard seals attacking or chasing humans both above and below water (Penney 1969; Lipps 1980; Kooyman, 1981d), it is likely that leopard seal attacks on humans on land are simply cases of mistaken identity; that is, the seals mistook humans for penguins. In his book *Endurance* (1959), which is about Shackleton's Antarctic expedition, Lansing reports one frightening incident:

> Returning from a hunting trip, Orde-Lees, traveling on skis across the rotting surface of the ice, had just about reached camp when an evil, knoblike head burst out of the water just in front of him. He turned and fled, pushing as hard as he could with his ski poles and shouting for Wild to bring his rifle. The animal—a sea leopard— sprang out of the water and came after him, bounding across the ice with the peculiar rocking-horse gait of a seal on land. The beast looked like a small dinosaur, with a long, serpentine neck. After a

half-dozen leaps, the sea leopard had almost caught up with Orde-Lees when it unaccountably wheeled and plunged again into the water. By then Orde-Lees had nearly reached the opposite side of the floe; he was about to cross to safe ice when the sea leopard's head exploded out of the water directly ahead of him. The animal had tracked his shadow across the ice. It made a savage lunge for Orde-Lees with its mouth open, revealing an enormous array of sawlike teeth. Orde-Lees' shouts for help rose to screams and he turned and raced away from his attacker. The animal leaped out of the water again in pursuit just as Wild arrived with his rifle. The sea leopard spotted Wild, and turned to attack him. Wild dropped to one knee and fired again and again at the onrushing beast. It was less than 30 feet away when it finally dropped.

Lipps (1980) recounts a terrifying underwater encounter. Three leopard seals approached two Antarctic scientists who were scuba diving at a depth of 22–32 meters. One of the seals began to behave aggressively toward the men. Retreating up the underwater cliff with the seals following close behind, the divers found two meter-length pieces of angle iron, which they used to shield themselves. The leopard seals repeatedly dove from the surface to bite at the irons, coiling up their necks to strike in much the same way they probably attack penguins. Because the seals stayed above the divers, blocking their escape route, the confrontation lasted forty-five minutes. Finally, the seals were distracted by the dive tender, who slapped the surface noisily with the boat oars, and the divers were able to escape. Only a few days later, the same scientists had a similar encounter with a smaller leopard seal.

Krill

Krill — the tiny planktonic food of seals and whales — is especially abundant in the cold, nutrient-rich polar seas, especially in the Antarctic. Plankton (a Greek word *planktos*, meaning "drifting") is composed of millions of minuscule algae, tiny copepods, amphipods, jellyfish, molluscs, fish eggs, and other crustaceans. Small shrimplike crustaceans of the family Euphausiidae, which constitute a major portion of the zooplankton, are collectively termed krill (Fig. 5.6).

The pinniped that specializes in zooplankton is the extremely abundant Antarctic crabeater seal, which eats krill almost exclusively. Crabeater seals appear to eat mainly the two-inch-long euphausiid, *Euphausia superba,* but in the southern portion of their range they may consume

FIG. 5.6. Krill (*Euphausia superba*). Drawing by Monterey Bay Aquarium Graphics.

FIG. 5.7. Ribbon seal mother and pup. *Photo by Kathy Frost.*

E. crystallorophias (Marr 1962). Although crabeater seals also prey on fish and squid occasionally, krill makes up 94 percent of their diet (Perkins 1945; T. Øritsland 1977). The three other Antarctic phocids do not eat krill exclusively, but the Weddell seal, Ross seal, and especially the leopard seal also include some krill in their diets. Krill makes up over one-third of the leopard seal's diet on average, although the proportion varies with geographic location (Bonner 1982).

A number of Arctic phocids also consume zooplankton (mostly pelagic crustaceans and amphipods): ringed seals, ribbon seals (Fig. 5.7), harp seals, largha seals, and some harbor seals. Primarily the juveniles

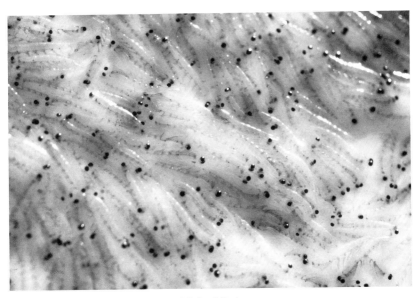

FIG. 5.8. Swarm of krill. *Photo by Michael Farley.*

and newly weaned pups of the last three species eat krill. With the exception of the Antarctic fur seal and subantarctic fur seal, no otariids or odobenids are known to eat zooplankton. As might be expected, the fur seals that do eat krill live in southern oceans where it is abundant. The Antarctic fur seal eats large-sized krill (*E. superba*) almost exclusively while the subantarctic fur seal consumes mainly krill (Fig. 5.8) and penguins.

Gastroliths

Many pinnipeds engage in the seemingly odd practice of ingesting pebbles and rocks. These ingested stones are called gastroliths (from the Greek *gastēr*, "stomach," and *lithos*, "stone"). Species found with gastroliths or observed swallowing rocks include South African fur seals, Australian sea lions, New Zealand sea lions, southern sea lions, Steller sea lions, southern elephant seals, northern elephant seals, and walruses. Some of the ingested stones are the size of a small orange. Up to 11 kilograms of pebbles and stones (some with sharp sides) have been found in the stomach of a single southern sea lion (King 1983). One large southern elephant seal male dragged about a load of 35 kilograms of ingested pebbles (Ling and Bryden 1981). In some cases, the seals may swallow stones incidentally while feeding. Walruses ingest sand and small stones, probably because they forage by using suction to dislodge molluscs from sand and mud. Gastroliths found in the stomachs of walrus pups that are still suckling were probably ingested by mistake (King 1983).

Most of the time, however, seals appear to ingest stones intentionally. No one knows why, although a number of theories have been offered to explain this unusual behavior. Some biologists believe that the stones have a stabilizing effect, providing ballast to help the seal regulate buoyancy and balance while swimming and diving. The Nile crocodile, for example, routinely swallows stones, which appear to stabilize it. The extra weight allows young crocodiles to achieve a slight negative buoyancy while on the surface and helps crocodiles of all ages stay on the bottom when there is a strong current (Cott 1961). Gastroliths were also common in aquatic dinosaurs.

Other pinniped biologists believe that seals eat stones primarily during times of fasting during the breeding and molting seasons, to fill the stomach when no food is ingested. For instance, southern elephant seals have been observed swallowing small pebbles and sand while on land. Although many gastroliths are found in southern elephant seal stomachs during periods of fasting (Laws 1956a), there is no evidence that swallowing stones would benefit the seal during this time.

Still another theory is that gastroliths may help break apart large pieces of food in the stomach or "grind up" parasitic worms that commonly infest seals. No evidence, however, has yet been found to support this idea. Finally, King (1983) suggests that the gastroliths, by providing additional weight, may aid the stomach muscles in ejecting fish bones and squid remains from the mouth, as sea lions often do.

"Feeding Equipment": Teeth and Jaws

The teeth of pinnipeds differ from those of land carnivores. Land carnivores have a greater number of teeth than pinnipeds, with a more complex structure, designed for grinding and shearing flesh. Most pinnipeds, in contrast, have pointed, cone-shaped teeth used mainly for catching and holding slippery fish and squid (Fig. 5.9). Small prey are usually swallowed whole; larger fish are brought to the surface to be consumed in bite-sized chunks, so little chewing or grinding is needed. Pinnipeds also lack the carnassial teeth (specialized molars or premolars for shearing flesh) characteristic of most terrestrial carnivores. A seal's molars and premolars have small pointed cusps to help grip slick food.

The three-cusped cheek teeth of the crabeater seal and leopard seal

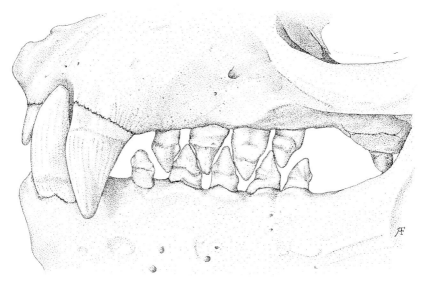

FIG. 5.9. Steller sea lion dentition, illustrating the typical pointed conical teeth of pinnipeds. *Drawing by Pieter Folkens.*

FIG. 5.10. Specialized cheek teeth of a crabeater seal, designed for straining krill. *Drawing by Pieter Folkens.*

are specialized for straining krill, which is a major food resource for these species (Fig. 5.10). In fact, crabeater seals possess some of the most strongly modified teeth of any mammal. There are five postcanine, or cheek, teeth in either jaw in the crabeater seal. When the seal closes its mouth after taking in a mouthful of krill, the tiny animals are trapped inside as the water filters back through the small spaces between the tubercles (knoblike protuberances) on the teeth (King 1961). The leopard seal's cheek teeth are similar to those of crabeater seals, although the occlusion is not quite as tight and there are fewer tubercles (T. Øritsland 1977). Unlike the crabeater seal, the leopard seal has elongated canines — necessary for preying on warm-blooded animals. Bonner (1982) believes that after locating a dense swarm of krill, a leopard seal or crabeater seal sucks individual krill into its mouth by depressing its tongue, then expelling water through the sieve created by the cheek teeth. This method of feeding is unlike that of the baleen whale, which simply swims through the water with its enormous mouth open, filtering large amounts of krill as it moves. The Antarctic fur seal, which feeds heavily on zooplankton as well, has the smallest postcanine teeth of any fur seal, possibly modified for straining krill (Bonner 1968; Repenning et al. 1971).

The Ross seal has specialized throat muscles and teeth — sharp, recurved canines — that appear to be an adaptation for catching and holding slippery cephalopods, which make up most of this seal's diet (Fig. 5.11). The throat muscles used for securing and swallowing food are exceptionally well developed. These modifications are reviewed in detail by King (1964b; 1965; 1969) and Bryden and Felts (1974).

The teeth of walruses are unusual in that the upper canines, in both males and females, are elongated into a huge pair of tusks. Although male

FIG. 5.11. Ross seal. *Photo by Gerry Kooyman.*

FIG. 5.12. Group of walruses. *Photo by Kathy Frost.*

tusks are heavier than those of females, female tusks tend to be longer. The average tusk length is about 35 centimeters (a little over 1 foot), but they can grow to 100 centimeters (over 3 feet) (Fay 1982). The heavy ivory tusks of walruses — allegedly harder and less liable to yellow than elephant ivory — have been highly valued for centuries (Fig. 5.12).

Walrus tusks help the animal to pull itself onto ice floes, to chop apart ice and enlarge breathing holes, and to defend its young. Among adult walruses, tusks also appear to serve a social function. Males may use them to advertise their dominance rank to females and other males while females wield them in aggressive encounters. At times mothers with calves are more aggressive than males and use their tusks to protect themselves as well as their young, especially when they are on land in dense groups (Zabel et al. 1987).

Contrary to what was once believed, a walrus does not appear to use its tusks to dig up clams (although it sometimes tries to stab seal prey with its tusks). Instead, a walrus forages by swimming along the bottom "upside down," stirring up mud and sediment while "rooting" for molluscs with its sensitive whiskers and snout and probably looking for conspicuous prey. Walruses may also excavate prey by squirting a powerful jet of water into the soft sediment. Foraging walruses produce long, continuous pit furrows in the mud. Remarkably, they may be able to excavate and eat more than six clams per minute, efficiently sucking the meat out of the shell or possibly squirting strong jets of water into the shell (J. S. Oliver et al. 1983a; C. H. Nelson and Johnson 1987).

Foraging Behavior

Although much remains to be learned about underwater foraging behavior, pinnipeds can be thought of as employing two basic foraging strategies, depending on their food resource and its distribution: individual foraging or cooperative foraging. An individual forager may be alone or in a loose group that does not feed cooperatively. Seals that pursue an individual feeding strategy do not necessarily feed alone; concentrated food resources may attract loosely spaced groups that can better find patchily distributed foods. Nonetheless, the individuals foraging in this way do not engage in intraspecific cooperation. A cooperatively foraging seal, in contrast, coordinates its feeding efforts with conspecifics or animals of another species to increase its foraging success.

INDIVIDUAL FORAGING STRATEGIES

Food resources such as nonschooling fish, slow-moving or sessile invertebrates, or relatively small-sized warm-blooded prey such as penguins and other pinnipeds are most efficiently exploited by an individual pinniped. Solitary foraging strategies therefore tend to characterize pinnipeds that use these food resources, which are often found in coastal waters, bays, and rivers, although northern elephant seals prey on fish that live at great depths in the open ocean. Food resources exploited by solitary foragers are often, but not always, distributed uniformly, at least in relation to the distribution of schooling fish and squid, which are characterized by a more "clumped," or patchy, distribution. The benthic molluscs eaten by walruses are also located in clumps or patches. Although walruses often feed on their own, they do so in small to large groups, a practice that helps individuals to locate the patchily distributed invertebrates. Foraging walruses even surface and dive in unison (Tomilin and Kibal'chich 1975; Fay 1982). Solitary foraging strategies characterize many of the phocids, including elephant seals and harbor seals. Of course, since many phocids and other pinnipeds are difficult to observe as they feed in the ocean depths, it is unclear whether they feed alone or in groups.

COOPERATIVE FORAGING STRATEGIES

Cooperative foraging appears to improve the efficiency of each individual's search for food. It occurs most frequently among the pinnipeds that forage in a coordinated unit to exploit large, patchily distributed schools of fish and squid in pelagic waters. Many sea lions and fur seals cooperate in a loose group to locate and exploit such schools, helping to herd them. Similar selective benefits of foraging in a group for clumped or unpredictable food resources have been documented in a number of birds (Krebs 1974), African ungulates (Jarman 1974), and cetaceans (Herman and Tavolga 1980).

There is a great deal of plasticity in otariid foraging strategies, however, and otariids may forage singly or in cooperative groups, depending on the type and distribution of food resources. The foraging strategies of California sea lions, South African fur seals, southern sea lions, and Steller sea lions often vary with the availability of schooling fish. These otariids tend to feed alone or in small groups unless large schools of fish or squid are present, in which case they may feed in large cooperative groups.

The foraging tactics of Steller sea lions, for example, vary with time of day, availability of food resources, and geographic location. Steller sea lions forage together in large groups when feeding on schooling fish or squid, but when such schools are not available, the sea lions feed alone or in groups of two to five animals (Fiscus and Baines 1966). Most of the massing together of sea lions to feed on large fish schools occurs during the day. At night, sea lions tend to feed singly or in small groups. Although Sandegren (1970) found that Alaskan Steller sea lions often leave the rookery to feed at night, Fiscus and Baines (1966) found that Steller sea lions in Alaska usually feed during the day, leaving for sea in tight groups of up to several thousand animals that break up into smaller groups of fifty or so, which regroup in large aggregations as they return to land. In California, female Steller sea lions frequently feed singly at night during the breeding season (May–July). At the end of this season females leave for sea in large groups, but because most females return alone (Gentry 1970), it is unclear whether they remain in these large aggregations for feeding.

Just before leaving for sea, female Steller sea lions in California group together on land, milling about near the water, vocalizing, and engaging in stereotyped behaviors with other females (Gisiner 1985). Such behavior may help to synchronize cooperative group foraging and, according to Le Boeuf (1978), could serve a function like that of the prehunt greeting ceremony of some canids (see also Estes and Goddard 1967).

Southern sea lions and California sea lions even forage symbiotically with feeding cetaceans and seabirds. Large groups of southern sea lions are frequently observed feeding along with cetaceans or seabirds on schools of fish, especially near Lobos Islands, Uruguay, where anchovies are abundant (Vaz-Ferreira 1981). Sizable feeding aggregations of southern sea lions, dusky dolphins, and seabirds have also been seen off Argentina (Würsig and Würsig 1980).

California sea lions sometimes feed cooperatively with harbor porpoises on schooling fish in Monterey Bay, California (Fink 1959), and in Mexico. Sea lions have also been seen among large feeding aggregations of seabirds, northern right whale dolphins, Risso's dolphins, and especially white-sided dolphins off the coast of central California (A. Baldridge, unpubl. data). Similar observations in Baja California (R. Luckenbach, pers. comm.) indicate that the appearance of flocks of birds (comprising several different species) over large schools of fish may attract a number of marine mammals such as Bryde's whales, fin whales, and several species of dolphins or porpoises as well as California sea lions. Sometimes these "feeding frenzies" persist for many hours. The precise symbiotic relationship among the various species in such feeding aggregations is obscure but could be similar to the "social parasitism" of various species of dolphins that associate with other species with superior food-locating abilities (Norris and Prescott 1961). When one species locates the anchovy schools, other species of birds, dolphins, and seals arrive to participate in the feast and, as late arrivals, are parasitic on the original feeders.

Ryder (1957) has documented various avian-pinniped feeding associations in Pacific walruses, ringed seals, Steller sea lions, and northern fur seals. In most of these cases, the birds appeared to be the parasites, feeding on fish scraps, small live fish, and possibly the seals' feces. Ryder, however, believes that the pinnipeds may sometimes locate fish by swimming to flocks of feeding birds. In addition, gulls may benefit walruses by removing ectoparasites from animals sleeping on the ice.

Würsig and Würsig (1980) have suggested that dusky dolphins are able to locate seabird aggregations visually from a distance of several kilometers by leaping high into the air to gain an even better viewing perspective (although such leaping may serve other functions as well). Since California sea lions often porpoise out of the water, they too may locate these feeding aggregations visually.

PLATE 1. Ringed seal. *Photo by William Stortz.*

PLATE 2. Harbor seal. *Photo by W. E. Townsend, Jr.*

PLATE 3. Southern sea lion adult male. *Photo by John Francis.*

PLATE 4. Hawaiian monk seal. *Photo by David Olsen.*

PLATE 5. Northern elephant seal bulls fighting on Guadalupe Island. *Photo by W. B. Tyler.*

PLATE 6. Adult male Guadalupe fur seal. *Photo by Mark Pierson.*

PLATE 7. Northern elephant seal.
Photo by Doc White.

PLATE 8. Ribbon seal pup. *Photo by Kathy Frost.*

PLATE 9. Hooded seal adult male with red nasal sac inflated. *Photo by Kit Kovacs.*

PLATE 10. California sea lion. *Photo by Doc White.*

Plate 11. Northern fur seal breeding colony. *Photo by Dan Costa.*

Plate 12. Bearded seal pup. *Photo by Kathy Frost.*

PLATE 13. Walruses. *Photo by Kathy Frost.*

PLATE 14. Female New Zealand fur
seal. *Photo by Rob Mattlin.*

PLATE 15. Harp seal mother nursing pup. *Photo by Kit Kovacs.*

OTHER FORAGING STRATEGIES

A number of pinniped feeding tactics cause problems for fisheries. Good general discussions of pinniped-fishery interactions are available in Mate (1980), Strombom (1981), Bonner (1982), Harwood (1983), and Beddington et al. (1985). To the dismay of those who fish in the Pacific Northwest, California sea lions have learned to wait at mouths of freshwater streams and fish ladders, particularly during steelhead trout runs. In fact, heavy predation on steelhead in the Puget Sound area appears to have had a major impact on spawning fish populations in a number of river systems, including the Green River system, Snohomish River system, and Ballard Lock. According to Bob Byrne of the Washington State Department of Wildlife, the problem has been especially well documented at Ballard Lock, where it has escalated over the past four years. In the past, perhaps sixteen hundred steelhead escaped upstream to spawn in this area, protected from all forms of hunting by humans but not, of course, by seals. During the 1985–86 spawning season, which lasts from late November to April, state Fish and Game biologists found that only four hundred of the sixteen hundred steelhead spawned. The rest were apparently eaten by as few as three to five sea lions working the river over a four- to five-month period. (The sea lions had also eaten considerable numbers of fish during the previous spawning season.)

The number of sea lions involved in the steelhead predation increased each year—from seven to twelve individuals in 1986–87 to twelve to twenty in 1987–88. Still, relatively few animals were responsible for extensive predation. All the sea lions were adult and subadult males (no females inhabit Washington waters). Some of the same individuals even returned year after year.

To deter the seals from eating steelhead, Fish and Game biologists exploded firecrackers and used other forms of acoustic harassment; they also tried "taste aversion" in the form of lithium chloride placed inside fish. None of these methods worked for long. Eight thousand firecrackers were set off in 1986–87, yet the seals still consumed 53 percent of the spawning fish. A net barrier of nylon cord (allowing fish but not sea lions to pass through) was set up during the 1987–88 season, with some degree of success. According to Bob Byrne, a total of $150,000 was spent in an attempt to safeguard the steelhead during this season, but so far the sea lions seem to be "winning."

Many pinnipeds have learned to steal fish from commercial fishing nets at sea, including grey seals and harbor seals (see, e.g., Bonner 1982), southern sea lions (Vaz-Ferreira 1981), California sea lions, Steller sea lions (Loughlin and Nelson 1986), northern fur seals (Fowler 1982), and South African fur seals (P. B. Best 1973; Shaughnessy and Payne 1979; Shaughnessy et al. 1981). The African fur seals actually wait until a purse seiner has encircled a school of fish and then jump over the corkline into the net to feed. Southern sea lions have learned to tag along after fishing vessels to fishing grounds, where they remain with the boat for one or more days, stealing fish from the nets (Vaz-Ferreira 1981).

Grey seals often damage fishing nets in their attempts to feed on captured fish, such as salmon, mackerel, and herring, or to escape when entrapped inside the net (Rae 1960; Rae and Shearer 1965; Mansfield and Beck 1977). In one case, which seems humorous in retrospect though it must have been infuriating at the time, a grey seal inside a trap net burst out of the netting when frightened by the fisherman, creating a huge hole and followed by most of the salmon catch (Bonner 1982).

Harbor seals steal salmon from the gill nets of commercial salmon fishing vessels in the Pacific Northwest (Imler and Sarber 1947; Fisher 1952) and the western Atlantic (Bonner 1982). The startling or uncomfortable sounds produced by acoustic harassment devices designed to deter seals from stealing salmon have been ineffective for several reasons: seals habituate to the sounds, or individual seals have different tolerances to them, or their intensity is not great enough to cause "unconditioned aural pain" (see, e.g., Geiger 1985). Anderson and Hawkins (1978) reached similar conclusions about the effects of these devices on harbor seals.

In another attempt to scare pirating South African fur seals (Fig. 5.13), Shaughnessy et al. (1981) weighted firecrackers that exploded underwater. Although most of the fur seals near the fishing nets scattered in response, they returned within a few minutes and soon habituated to the firecrackers. The researchers also played back the underwater vocalizations of killer whales. Many of the fur seals were aware of and curious about the sounds but did not leave the area around the fishing vessel. Other seals ignored the killer whale sounds and continued to feed on the netted fish. Models of killer whale dorsal fins displayed in the water along with the playback did not cause any reaction or changes in behavior. As Bonner (1982) puts it, "It seems almost certain that in the absence of some regular reinforcement, namely the killing or injury of seals, a sound meant to deter would eventually come to be regarded merely as a dinner gong." Other, more destructive, methods to deter hungry seals have been employed by angry fishermen, who shoot the seals that are looting nets, poison them with tainted bait, or kill them with explosives.

FIG. 5.13. Young male South African fur seal. *Photo by Roger Gentry.*

FORAGING STRATEGIES OF FUR SEALS

In their excellent review, Gentry and Kooyman (1986) have summarized some basic foraging strategies among the fur seals, primarily in relation to maternal care strategies. They offer the following conclusions on the foraging behavior of fur seals. All the fur seals dive primarily at night. The majority of them (with the exception of the Galápagos fur seal) dive deeper during the day than at night. Most fur seals dive to relatively shallow depths for their prey — usually to 50 meters or less — except for the northern fur seal, which dives most frequently to two mean depths: 60 meters and 175 meters.

Adult female fur seals show two different foraging patterns in relation to maternal care strategies. The tropical and temperate fur seal females feed at sea for short periods and have a small number of feeding bouts per trip to sea. These species often have a prolonged pup dependency period — several months to three years. The two subpolar fur seal females (northern fur seals and Antarctic fur seals) go out to sea for longer feeding trips and have many feeding bouts on each trip. Subpolar fur seal females care for their pup for only about four months. Apparently, the subpolar species must travel greater distances to feeding grounds than other fur seals. Subpolar seals forage in patterns that suggest that their food supply is more difficult to locate, their food resources more distant,

FIG. 5.14. Weddell seal outfitted with Time-Depth Recorder. *Photo by Dan Costa.*

or their energy requirements greater for each trip than is the case for tropical species. The food resources of subpolar species, however, are typically more abundant and predictable than those of tropical otariids.

FORAGING DIVE PATTERNS

The recent use of Time-Depth Recorders that can be attached to free-ranging seals has made possible some remarkable discoveries about their diving and foraging behavior (see Table 1). The Time-Depth Recorder (Fig. 5.14) was originally developed by James Billups and Gerry Kooyman, and is described in detail in Kooyman (1965b) and Gentry and Kooyman (1986). The recorders have shown that phocids tend to make deeper and longer dives than otariids. In particular, some phocids are able to dive to incredible depths of thousands of feet to capture their food. As I mentioned in chapter 1, the Weddell seal and northern elephant seal are perhaps the most prolific deep divers of all the pinnipeds. Weddell seals may remain submerged for up to seventy-three minutes, and dive to a maximum depth of 600 meters (nearly 2,000 feet). However, these seals usually dive for less than fifteen minutes to a depth of 400 meters or less, where they feed on fish such as Antarctic cod (Kooyman 1981d).

Kooyman (1981c) found that Weddell seals actually have two dive patterns, one for exploratory and one for deep dives. Exploratory, or search, dives are uncommon (Kooyman et al. 1980; Kooyman et al. 1983). They are long (twenty to seventy-three minutes) and relatively shallow (the surface of the water is always in view) and require anaerobic metabolism. Deep-feeding dives, in contrast, are brief (five to twenty-five minutes) and usually within the aerobic dive limit (before oxygen is depleted and muscles switch to anaerobic functioning). Seals making these dives may descend to depths of 400 meters but travel only a short distance from the ice hole.

One of Kooyman's (1981c) most significant discoveries is that seals (and diving vertebrates in general) seldom resort to anaerobic metabolism on foraging dives. Of the 4,600 dives by free-ranging Weddell seals that he measured, only 3 percent lasted longer than twenty-five minutes, the aerobic dive limit, beyond which the seal metabolizes anaerobically, producing lactic acid that builds up in the blood, incurring an oxygen debt. The amount of time a seal can remain submerged and still function aerobically depends on body size, with the larger seal able to dive for a longer period before switching to anaerobic metabolism. A seal making repeated dives of short duration could spend more time feeding, since it would not have to recycle lactate by means of gluconeogenesis (producing new glucose, or sugar, by releasing glycogen from the pancreas; that is, by converting a substance [e.g., fat] other than carbohydrate).

Northern elephant seals are able to dive deeper than any other pinniped (Le Boeuf et al. 1985, 1986, 1988, in press, pers. comm.). Time-Depth Recorders were attached to adult female northern elephant seals at the Año Nuevo rookery just before they left for sea after their thirty-four-day breeding season fast. The seal's modal, or most common, dive depth was between 350 and 650 meters. Although most of the females dove to a depth of between 400 and 450 meters, the seal identified as Dot preferred a depth of 350–400 meters, whereas Snl and Td were deeper divers, preferring depths of 500–550 and 600–650 meters, respectively. The deepest dive recorded was 1,250 meters, or 4,100 feet—the deepest dive known for any pinniped. Most dives lasted about twenty minutes, with the longest period of submersion sixty-two minutes. Older and larger females dove for longer periods than younger females. Because adult males are much larger than females, they may be able to remain submerged for still longer periods, but males have not yet been followed. Yearlings (20 months of age) had dives similar in mean depth and duration to those of adult females, although their maximum depth was not as great (Le Boeuf et al., in press).

Perhaps even more astonishing than the depth of some dives was the females' "machinelike" pattern of diving. After leaving the breeding grounds, females dove continuously throughout their entire period at sea, unlike other pinnipeds that tend to engage in distinct foraging bouts alternating with periods of rest. One female dove nonstop during her first eleven days at sea, for a total of 653 dives. Each hour the females dove two to three times (the mean dive rate being 2.7 times per hour). Elephant seals spent very little time resting at the surface—only about 2.8 minutes or so after each dive. Consequently, only about 15 percent of their time at sea (or 3.5 hours per day) was spent above water. The amount of time spent on the surface did not change no matter how long the preceding dive had been. With so much continuous feeding, one would expect the seals to gain back much of the weight they had lost during their lactation fast. The females that were weighed upon returning to the rookery after a mean of 72.6 days at sea were found to have gained a mean of 76.5 kilograms—a rate of slightly more than 1 kilogram per day.

At night the elephant seals tended to dive more often, less deep, and for shorter periods than during the day. As Le Boeuf et al. (1988) point out, this diving pattern seems to be closely associated with the migratory movements of their prey. Northern elephant seals feed on a variety of deep-water squid and fish, such as rays, sharks, ratfish, Pacific hake, and rockfish, which are found either midwater or on the bottom at great depth (Condit and Le Boeuf 1984). At night, when the seals' dives are shallower, many of these prey migrate toward surface waters.

As I have mentioned, Le Boeuf et al. (1986, 1988) suggest that the seals may even sleep at great depths, both to conserve energy and to avoid predators. In deep waters elephant seals may be safe from the attacks of the great white sharks that appear to prey heavily on them. Most of the shark attacks take place near the surface. By spending long periods of time far beneath the surface, elephant seals may lessen the chance of their being attacked by white sharks. But this idea, intriguing as it may be, remains highly speculative.

Diving behavior in the fur seals and the Galápagos sea lion is reviewed in Gentry and Kooyman (1986). The mean and maximum duration of dives is very similar for all species. The mean dive duration ranges from 2 to 2.5 minutes, and the maximum dive duration from 5 to 7.5 minutes. Fur seals and Galápagos sea lions therefore remain submerged for a much shorter time than Weddell and northern elephant seals. With the exception of northern fur seals, the mean depth of dives is also similar for all species, ranging from 30 to 45 meters. (The mean depth of dives for northern fur seals is 68 meters.) Maximum depth of dive varies from 101 meters in the Antarctic fur seal to 207 meters in the northern fur seal. In addition, the mean number of dives per hour is similar for all species, ranging from 14.3 for the South American fur seal to 19.3 for the Antarctic fur seal. The depth and timing of Antarctic fur seal dives, like those of northern elephant seals, correspond to the daily fluctuation in the distribution of their main prey in summer—krill. Foraging dives peaked between 2000 and 2100 hours and between 0200 and 0300 hours, times closely associated with the nightly rise of krill (Kooyman et al. 1986).

Recent diving studies of New Zealand sea lions have shown that their diving behavior does not fit the pattern typical for fur seals (Gentry et al. 1987). The timing and depth of the sea lions' dives did not change with time of day as they do among other otariids. Instead of feeding in discrete three-hour periods as fur seals do, the sea lions dove for two to three days at a time with few breaks. In addition, both the sea lions' deepest dive of 460 meters and their maximum dive duration of twelve minutes substantially exceeded those recorded for fur seals.

6

Mating Systems, Breeding Behavior, and Social Organization

The Evolution of Pinniped Mating Systems

Pinnipeds have diverse mating systems. Many species are highly polygynous and sexually dimorphic; that is, males are much larger than females. Others are monogamous, or at least serially monogamous, in which case males and females are nearly the same size and look very much alike. What accounts for this diversity in breeding systems?

First, the inherently very different reproductive strategies of each sex set the stage for the evolution of mating systems. Female mammals are a valuable as well as limiting resource in the sense that they produce only one or a few eggs each year and therefore a limited number of eggs in a lifetime. Males, in contrast, can produce millions of sperm every day during the breeding season. (As reproductive biologists say, "Sperm are cheap, eggs are costly.") Males therefore compete for females, but females can afford to be choosy about their mates.

Perhaps more important, females not only provide the "expensive" egg, but also carry and nourish the young within them and produce milk after birth. Among most species of mammals, including all pinnipeds, females provide all parental care, nursing and protecting the young after they are born. That male mammals, in contrast, cannot contribute much as parents promotes the polygynous mating systems so prevalent in mammalian social organization. When males are freed from the burden of parenting, they can best maximize their reproductive success by fertiliz-

FIG. 6.1. Juan Fernández fur seals, showing extreme sexual dimorphism. The male and female are mating. *Photo by John Francis.*

ing as many females as possible. Milk and mothering are therefore important structuring elements of mammalian societies, whether terrestrial or marine.

The environment, too, plays a critical role in structuring the mating systems of pinnipeds as well as those of other mammals and birds (Emlen and Oring 1977). The evolution of pinniped mating systems, a fascinating and complex subject, has been explored in a number of thought-provoking articles (Bartholomew 1970; Stirling 1975b, 1983; Jouventin and Cornet 1980; Le Boeuf 1985, 1986). Much of the following discussion is derived from the work of I. Stirling (1983) and B. J. Le Boeuf (1986). The determining factor in the evolution of mating systems is the distribution of the female seals, which in turns depends on breeding habitat. The distribution of females, along with a number of other environmental variables, affects the way that males compete for mates and consequently influences the entire breeding system.

Pinnipeds can be divided into three categories according to breeding habitat: (1) those that breed on land, (2) those that breed on pack ice (floating ice), and (3) those that breed on fast ice (ice attached to land). Of the 33 species of pinnipeds, 20 breed on land, while the remaining 13 breed on ice (see Table 5). Eighteen of the 20 species of land-breeding pinnipeds are highly polygynous and strikingly sexually dimorphic, and they breed in moderate-sized to extremely large colonies (Fig. 6.1). Highly or moderately polygynous land-breeding pinnipeds include all sea lions and fur seals, northern and southern elephant seals, and the grey seals. The remaining land breeders, which are probably polygynous to some degree, are the harbor seals and two species of monk seal. In contrast, 11 of the 13 species of ice-breeding seals appear to be either monogamous, serially monogamous, or slightly polygynous, as well as monomorphic (that is, the sexes are similar in size and appearance). Except for the walruses, all of the ice breeders are phocid seals. The two ice breeders that do not fit the pattern, showing moderate polygyny, are the Weddell seal and the walrus. A moderate to high degree of polygyny is therefore found in all otariids and in the walrus, but in only four phocids (Table 9). The term "extreme polygyny" refers to males that have the opportunity to monopolize and mate with a large number of females (at least 15–20 and often more) during a breeding season.

Most of the land breeders congregate on islands, although some breed on isolated mainland beaches and sandbars. Islands offer protection from mainland terrestrial predators, such as wolves, grizzly bears (now extinct), mountain lions, and humans. Not many islands favorable to seals exist, however, and on the few available islands, breeding sites are restricted. Often seals can haul out and breed only on certain areas of an island, such as sandy beaches. So females are forced to clump together, a tendency sometimes called gregariousness. The harassment of isolated females by low-ranking males further encourages females and pups to group together where the territorial male or alpha bull offers them protection from other males.

These large aggregations of females provide the opportunity for some males to mate with a great many females. This results in intense competition among the males for mates, which leads to sexually dimorphic traits and polygyny; that is, one male monopolizes and mates with many females. Among highly polygynous pinnipeds, males are much larger than females (Fig. 6.2) and have developed secondary sexual characteristics, such as the pendulous nose and frontal chest shield of the elephant seal or the hooded seal's balloonlike inflatable nasal sac.

Because islands are permanently fixed land masses, unlike ice, and because most seals return to the same spot year after year to give birth, a tendency called philopatry or site fidelity, males can predict when and where land-breeding females will arrive each year to breed. Males arrive early in the breeding season in anticipation of the females soon to follow.

Fig. 6.2. Hooded seal male, female, and pup. *Photo by Kit Kovacs.*

Sea lion and fur seal bulls generally set up territories, while elephant seals and grey seals establish dominance hierarchies. But the nature of territoriality and the defense of females in dominance hierarchies may change with such variables as climate, topography, and breeding substrate (these are discussed later in this chapter).

Among many species of polygynous land-breeding otariids and phocids, most high-ranking males remain on land throughout the breeding season, trying to mate with as many females as possible. In some species, males fast for as long as two to three months during the breeding season; a top bull that left the breeding beach to feed would no doubt lose many mating opportunities, since it would be difficult for him to regain his high-ranking position in the dominance hierarchy. At the end of the breeding season, for example, the same elephant seal bulls that arrived padded in rolls of fat look like virtual skeletons of their former selves. The larger the bull and the greater its reserve of blubber, the more likely it is to survive its fast during the breeding season—another reason that large size is promoted in polygynous land breeders.

Seals that breed on polar ice are subject to selective pressures different from those that affect land breeders. Female ice-breeding seals are not forced to group together because they have vast expanses of available breeding space, but females of some species clump loosely. Arctic

TABLE 9

Breeding Environments, Mating Systems, and Social Organization during the Breeding Season

SPECIES	BREEDING ENVIRONMENT	SOCIAL ORGANIZATION*	MATING SYSTEM†	SEXUAL DIMORPHISM‡
Hawaiian monk seal	Islands	Small to moderate-sized groups; females well spaced	Slight polygyny or "promiscuity"	Slight reverse
Mediterranean monk seal	Islands; remote mainland beaches; sea caves	Solitary, mother-pup, or small groups; females widely spaced	Slight polygyny or "promiscuity"?	Monomorphic
Northern elephant seal	Islands; remote mainland beaches	Colonial; moderate-sized to extremely large groups; females densely spaced	Extreme polygyny	Pronounced§
Southern elephant seal	Islands; remote mainland beaches (occasionally fast ice in southern part of range)	Colonial; moderate-sized to extremely large groups; females densely spaced	Extreme polygyny	Pronounced§
Weddell seal	Fast ice; Antarctic	Colonial; moderate-sized groups; females well spaced	Moderate polygyny	Slight reverse or monomorphic
Ross seal	Pack ice; Antarctic	Solitary, mother-pup pairs, or small groups; females widely separated	Unknown	Slight reverse or monomorphic
Crabeater seal	Pack ice; Antarctic	Solitary, male-female pairs, triads, occasionally small groups; females widely separated	Serial monogamy	Slight reverse or monomorphic
Leopard seal	Pack ice; Antarctic	Solitary, mother-pup pairs; females widely separated	Unknown	Slight reverse
Hooded seal	Pack ice; Arctic	Solitary, triads, mother-pup & several males; females widely spaced	Serial monogamy or slight polygyny	Moderate§

Species	Habitat	Grouping	Mating system	Dimorphism
Bearded seal	Pack ice; Arctic	Solitary, mother-pup pairs; small groups; females widely separated	Serial monogamy?	Slight reverse or monomorphic
Harp seal	Pack ice; Arctic	Moderate-sized to large groups; females well spaced	Slight polygyny or "promiscuity"	Monomorphic
Ribbon seal	Pack ice; Arctic	Solitary to small groups; females well spaced	Slight polygyny	Monomorphic
Caspian seal	Fast ice; Arctic (no birth lairs)	Solitary, mother-pup pairs, triads; females widely separated?	Serial monogamy	Slight or monomorphic
Baikal seal	Fast ice; Arctic (birth lairs used)	Solitary, mother-pup pairs, triads; females widely separated?	Serial monogamy	Slight or monomorphic
Ringed seal	Fast ice; occasionally pack ice; Arctic (birth lairs used)	Solitary, mother-pup pairs, triads; females widely separated	Serial monogamy	Slight or monomorphic
Largha seal	Pack ice; Arctic	Mother-pup pairs, triads; females widely separated	Monogamy or serial monogamy	Slight or monomorphic
Harbor seal‖	Islands; remote mainland coasts; Arctic pack ice	Small to large groups; females well spaced to widely separated	Slight polygyny or "promiscuity"	Slight
Grey seal‖	Islands; remote mainland beaches; pack ice; fast ice	Colonial; small to extremely large groups; females densely to well spaced	Moderate to extreme polygyny	Pronounced§
California sea lion	Islands; remote mainland beaches	Colonial; moderate-sized to large groups; females densely spaced	Moderate to extreme polygyny	Pronounced§
Galápagos sea lion	Islands	Colonial; small to moderate-sized groups; females well spaced	Polygyny	Pronounced§
Steller sea lion	Islands	Colonial; moderate-sized to extremely large groups; females densely spaced	Extreme polygyny	Pronounced§

Continued on next page

TABLE 9, *continued*

SPECIES	BREEDING ENVIRONMENT	SOCIAL ORGANIZATION*	MATING SYSTEM†	SEXUAL DIMORPHISM‡
Southern sea lion	Islands; remote mainland coasts	Colonial; moderate-sized to large groups; females densely spaced	Moderate polygyny	Pronounced§
Australian sea lion	Islands; remote mainland coasts	Colonial; small to large groups; females well spaced	Slight to moderate polygyny	Pronounced§
New Zealand sea lion	Islands; remote mainland beaches	Colonial; large groups; females densely spaced	Extreme polygyny	Pronounced§
Northern fur seal	Islands	Colonial; extremely large groups; females densely spaced	Extreme polygyny	Pronounced
Guadalupe fur seal	Islands; sea caves on islands	Colonial; small groups; females well spaced	Moderate polygyny	Pronounced
Juan Fernández fur seal	Islands	Colonial; small to moderate-sized groups; females well spaced	Moderate polygyny	Pronounced
Galápagos fur seal	Islands	Colonial; small groups; females well spaced	Polygyny	Pronounced
South American fur seal	Islands; remote mainland coasts	Colonial; small groups; females well spaced	Polygyny	Pronounced
New Zealand fur seal	Islands; remote mainland coasts	Colonial; small to large groups; females well spaced	Polygyny	Pronounced

Antarctic fur seal	Islands	Colonial; small to large groups; female spacing variable	Moderate to extreme polygyny	Pronounced
Subantarctic fur seal	Islands	Colonial; small to large groups; females well spaced	Moderate to extreme polygyny	Pronounced§
South African fur seal	Islands	Colonial; moderate-sized to extremely large groups; females densely spaced	Moderate to extreme polygyny	Pronounced
Australian fur seal	Islands; remote mainland coasts	Colonial; small to large groups; females densely spaced	Moderate to extreme polygyny	Pronounced
Walrus	Pack ice; Arctic	Colonial; extremely large groups; females densely spaced	Polygyny	Pronounced§

NOTE: References are given in the text.

* Triads consist of a mother-pup pair and an adult male. In densely spaced groups, females are positively thigmotactic and gregarious; in well-spaced groups, females are negatively thigmotactic and gregarious; in widely spaced groups, females are separated by great distances and are nongregarious.

† Serial monogamy is actually a form of polygyny since a male may mate with more than one female during a breeding season, although with only one female at a time. I use the term "serial monogamy" to avoid confusion, since other writers have established its use. In mating systems characterized by extreme polygyny, males have the opportunity to mate with some 15–20 females during the breeding season, often more; in mating systems characterized by moderate polygyny, males mate with some 5–15 females; in mating systems characterized by slight polygyny, males usually mate with ≤5 females.

‡ Sexual dimorphism pertains to size; secondary sexual characteristics may also be present in males (e.g., elongated nose, chest shield, sagittal crest). The categories of sexual dimorphism are monomorphic (sexes are approximately equal in size); slight (one sex is about 10% larger or smaller than the other); moderate (the male is >10% and <1.5 times larger than the female); pronounced (male is 1.5–5 times larger than the female); reverse (female is larger than the male).

§ Secondary sexual characteristics are present in males.

‖ Mating system may vary with breeding habitat.

pinnipeds such as ribbon seals, harp seals, hooded seals, bearded seals, and walruses breed on pack ice, much of it like little islands that offer isolation and protection from most predators. In addition, pack ice often floats directly over rich and abundant sources of food for seals. Unlike islands, however, the abundant pack ice provides not only a great deal of breeding space but also unrestricted access to the water (Fig. 6.3).

Because female seals breeding on pack ice are widely spaced, males are unable to mate with and monopolize many females. Moreover, the location of pack-ice breeding sites changes from year to year, and males cannot anticipate where females will spend the breeding season. The breakup of the unstable floes of moving pack ice, which can happen at any time, is hazardous to young pups. Because of the danger to them, females tend to give birth at the same time during the short period when ice conditions are most stable. The breeding season is therefore quite short, sometimes lasting only a few days, and all of the females come into estrus synchronously when their pups are weaned. Consequently, a male has enough time to mate with only one or a few sexually receptive females during the brief breeding season. Aquatic copulation, common in many ice-breeding seals (Table 10), makes the control of females even more difficult, if not impossible, for males. All of these selective pressures have

FIG. 6.3. Ribbon seal male on pack ice. *Photo by Kathy Frost.*

TABLE 10

Aquatic versus Terrestrial Copulation in Pinnipeds

AQUATIC	TERRESTRIAL	BOTH AQUATIC & TERRESTRIAL
Hawaiian monk seal	Northern elephant seal‡	Grey seal#**
Mediterranean monk seal*	Southern elephant seal‡	Harp seal§††
Weddell seal	Crabeater seal§	Hooded seal§††
Ringed seal	Leopard seal*§	
Harbor seal†	California sea lion‡‖	
Walrus	Galápagos sea lion‡‖	
Bearded seal	Steller sea lion	
	Southern sea lion‡	
	Australian sea lion	
	New Zealand sea lion‡	
	Northern fur seal	
	Guadalupe fur seal‖	
	South American fur seal	
	Galápagos fur seal‖	
	New Zealand fur seal	
	Antarctic fur seal	
	Subantarctic fur seal	
	South African fur seal	
	Australian fur seal	

* Small sample size.
† Copulations occasionally observed on land.
‡ Copulations occasionally observed in water.
§ Mating occurs on ice.
‖ Otariids in warmer climates also copulate in shallow-water tidepool areas.
Mating occurs on both ice and land.
** Higher proportion of mating occurs on land.
†† Higher proportion of mating occurs in water.

inhibited the evolution of extreme polygyny in seals that breed on the pack ice. Instead, a very low level of rather unstructured polygyny or monogamy persists.

The remaining ice-breeding seals inhabit fast ice, which consists of vast continuous expanses of ice attached to land. It offers what seems like an enormous area, relatively stable in time and space, to breeding seals. But seals can breed only in certain areas of the fast ice along shore or where cracks in the ice, open breathing holes, and polynyas are located. Because breeding space is abundant, females are widely dispersed and breed in small- to moderate-sized well-spaced colonies. A male has the

FIG. 6.4. Weddell seal colony. *Photo by Dan Costa.*

opportunity to mate with only as many females as are gathered around a breathing hole. Antarctic Weddell seals, for example (Fig. 6.4), breed in well-spaced colonies in open areas free of land predators. A hole in the ice may attract females, but the number of females that can share a hole at one time is limited. A Weddell seal male is able to monopolize a group of females, but he is limited both by the well-spaced distribution of the small female groups and by the practice of copulating in the water. Male Weddell seals actually maintain aquatic territories (called maritories) beneath the ice near breathing holes. Although a moderate degree of polygyny has therefore evolved among Weddell seals, this species shows reverse sexual dimorphism, females being slightly larger than males. Small size in males is thought to promote agility underwater, where males defend territories and mate with females.

Unlike the situation in the Antarctic, where there are no land predators, polar bears and Arctic foxes inhabit the Arctic. As I mentioned in chapter 4, polar bears prey on several Arctic pinnipeds, including hooded seals, bearded seals, ringed seals, and walruses (Stirling 1975b). Such predation is another variable that influences the distribution of females and therefore the mating systems among Arctic seals breeding on fast ice. A female ringed seal, for instance, may try to avoid dangerous predators by giving birth and raising her pup alone in a small protected cave under an ice ridge (although cave breeding may also provide a warmer, more sheltered environment for the young pup). Each female maintains a concealed breathing hole in her shelter, which she defends against intruding females. As a result, ringed seal females do not clump together at all, and one male usually mates with only one female. The breeding system is either monogamous or slightly polygynous—if the male can find more than one female to mate with (Fig. 6.5). An interesting side note on ringed seals is that males give off a strong and, at least to humans, nasty odor during the breeding season. According to King (1983), the early specific epithet *foetida* was derived from the foul-smelling breeding males.

When the breeding system is not polygynous, there is no need for the male to develop a large size or other secondary sexual traits, since he is not competing for females by trying to dominate other males. Pronounced sexual dimorphism is especially useless for seals that copulate in the water. Monomorphism (equal size) or even a slight reverse sexual dimorphism (the female larger than the male) has come about in these seals. Large size in a pagophilic, or ice-loving, female seal helps her to provide large quantities of fat-rich milk for her pup and protects her from the frigid polar temperatures. With the exception of the hooded seal and

FIG. 6.5. Ringed seal. *Photo by Brendan Kelly.*

FIG. 6.6. Harp seal. *Photo by Graham Worthy*

walrus, ice-breeding seals of both sexes are of similar size, as are harbor seals and monk seals (Fig. 6.6).

As Stirling (1983) and Le Boeuf (1986) point out, among ice breeders, walruses are the exception in that females do form dense aggregations on the pack ice. Fay (1982) suggests that the clumping of female walruses on pack ice may help them use a patchy food resource. Walruses show marked sexual dimorphism and appear to be polygynous. Their mating is unusual among the pinnipeds, however, in that they seem to gather in a lek, an area used exclusively for courtship displays. Animals such as North American grouse and African antelope similarly gather to mate in lek-king grounds. Among walruses, one or more males remain near each female-laden ice floe and produce strange bell-like sounds and visual signals in ritualized displays designed to attract females. Female walruses appear to choose a displaying male and mate with him in the water.

Among land-breeding phocids, the two species of monk seal and the harbor seal are the only species not markedly sexually dimorphic and not characterized by a highly polygynous breeding system. Hawaiian monk seals breed in the subtropical waters near the Northwestern Hawaiian Islands. Their mating system seems to be somewhat polygynous. Adult females, slightly larger than the adult males, are neither solitary nor gregarious but instead form well-spaced small groups on the expansive beaches where they breed. The monk seal population is currently at a very low level. Their sparse population, as well as the presence of roomy beaches, does not promote clumping as in other land-breeders. In addition, monk seals copulate in the water.

In monk seals, climate has also influenced the structure of the mating system. Seals inhabiting areas characterized by moderate, aseasonal climatic conditions tend to have an extended pupping season. Among Hawaiian monk seals the pupping season lasts about eight months, from late December to mid-August (Kenyon 1981). But the high degree of asynchrony in female receptivity has, in effect, the same consequence as a brief breeding season coupled with a high degree of female synchrony: males are unable to monopolize many mates at once. It is apparently too energetically costly for Hawaiian monk seal males either to defend restricted geographic areas or to control the females themselves for the length of the breeding season. Because few females become sexually available at any time, a mildly polygynous opportunistic system has developed in which males continuously search for receptive females (Fig. 6.7).

Among otariids, the Australian sea lion is less polygynous than other species. These sea lions have an extended year-round pupping season as

FIG. 6.7. Hawaiian monk seal breeding colony, sharing the beach with sea turtles. *Photo by Chip Deutsch.*

FIG. 6.8. Grey seals on Sable Island. *Photo by Sheila Anderson.*

well as an unusual eighteen-month breeding cycle. Because of this unique, prolonged pupping season, larger numbers of males have opportunities to mate with females. About 57 percent of breeding males observed from December to May (1986–87) at Seal Bay, South Australia, mated with only one female. Territorial behavior was flexible in space and time, depending on the number of estrous or pregnant females available. Some males were present on a territory only one day each week; others occupied a territory continuously (Higgins 1987).

Although the widespread harbor seal and the largha seal are largely monomorphic, their breeding systems appear to vary throughout their extremely wide range in the Northern Hemisphere. Largha seals that breed in pack ice form triads (mother-pup pair and adult male), as might be expected (Fay 1974a). Harbor seals (*Phoca vitulina richardsi*) that breed on glacial ice form similar triads (Hoover 1983). Other harbor seals that breed on land (such as *P.v. richardsi* and *P.v. stejnegeri*) congregate in sizable groups. Sullivan (1981) found that male harbor seals (*P.v. richardsi*) in these groups repel other males and attract females by engaging in frequent noisy displays of "lobtailing" and splashing during the breeding season. This land-breeding subspecies of harbor seal is probably polygynous to some degree; the higher-ranking dominant males are able to mate with more females. Copulation, however, appears to take place in

FIG. 6.9. Grey seal breeding colony on a cobble beach. *Photo by Sheila Anderson.*

the water. Land-breeding harbor seals are also more sexually dimorphic than harbor seals that breed on ice (Bigg 1981).

The grey seal ranges widely throughout the North Atlantic Ocean and breeds in a variety of different habitats: islands (Fig. 6.8), sandy mainland beaches, rocky inlets, and fast and drift ice. As one might expect, the mating system of grey seals varies with the breeding substrate as well as the density of each breeding colony. For instance, the Canadian grey seal population (studied by Mansfield [1966]) breeds in two different habitats: on ice in the Gulf of St. Lawrence and on ice-free islands east of Nova Scotia. The land-breeding grey seals at the Basque Islands aggregate since space is limited, allowing the dominant bulls to control harems of six or more females. In contrast, at Sable Island there is a great deal of space for females to spread out on long sandy beaches. Here the system is virtually monogamous: few males are able to monopolize more than two females, and most males have only one. Most of the ice-breeding seals in the Gulf of St. Lawrence pup on pack ice in the Northumberland Strait. The grey seals in this area are either monogamous or slightly polygynous. Since mothers and pups are well spaced on ice floes, males are unable to monopolize groups of females. Instead, adult males are usually seen with only one female and pup. The eastern Canadian grey seals therefore

FIG. 6.10. Grey seal pup and mother, with icicles hanging off her snout, on ice in eastern Canada. *Photo by Graham Worthy.*

demonstrate that mating systems not only are strongly influenced by the environment but also may show a great deal of flexibility among the same population of seals. Boness (1979) and Anderson and Harwood (1985) have also shown that the degree of polygyny and the mating behavior in grey seals vary with breeding habitat (Figs. 6.9, 6.10).

The Diversity of Mating Strategies in the Male

HOW MALES COMPETE FOR MATES

Whether the mating system is highly polygynous or largely monogamous, males generally monopolize females and acquire mates either by territorial defense ("resource defense polygyny") or by dominating other males in agonistic encounters ("female defense polygyny"). Most pinniped species set up a conspicuous and often rigid system of either territoriality or dominance hierarchies in competing for females. Nearly all adult male sea lions and fur seals and a few male phocid seals are highly territorial during the breeding season, defending a particular area, often characterized by fixed boundaries, rather than a group of females. In the dominance system characteristic of many phocid seals, males fight one another to possess females, not a particular area; the goal in this system is

FIG. 6.11. Largha seal mother and pup on ice floe. *Photo by Kathy Frost.*

to achieve a high social rank. Yet among walruses, southern sea lions, and a number of phocid seals, competition among males is less well defined or not as conspicuous, involving such variations on the main themes of territoriality or dominance hierarchies as lekking and unstructured polygamy, or promiscuity.

Even when the mating system is relatively unstructured, as it appears to be for the Caspian seal and the Hawaiian monk seal, one male must often dominate another to secure a mate. Yet there appears to be no strict hierarchical ordering among the competing males of these species. Among walruses, which are characterized by lek competition, females choose among displaying males in the water. But adult males also repel other males from their "display station" and the water surrounding the ice floes on which females aggregate. In the case of the monogamous or slightly polygynous largha and hooded seals, a male defends primarily the female, but he also protects the area around "his" ice floe (Fig. 6.11). The resident male hooded seal constantly chases challenging males away from the ice floe by fighting and using noisy displays and impressive expansions of the snout (Øritsland and Benjaminsen 1975). The male hooded seal is even said to "play" with its hood while resting quietly on the ice—moving the air back and forth between the front and back of the ballooning sac.

Competition for females between Hawaiian monk seal males would probably involve a dominance hierarchy if females aggregated and came into estrus close to the same time. But because the pupping season is prolonged, very few females become sexually available at any one time. This being the case, males compete for access to one estrous female at a time. Deutsch (1985) has found that male monk seals engage in two types of competition (as defined by Barash [1978]): "scramble" competition, in which adult males opportunistically cruise beaches in search of estrous females, and "contest" competition, in which a male, having found a female, defends her against intruding males. As Barash (1978) puts it, scramble competition is like "an Easter egg hunt in which every partici-pant ignores the other and simply concentrates upon finding as many eggs as possible." Contest competition occurs when the participants argue and fight among themselves for access to eggs. In fact, much of the interaction between pinniped males during the breeding season seems to involve contest competition. But monk seals must also "scramble" to find the rare estrous female. Since males outnumber females by a ratio of three to one, each basking female is disturbed frequently. Male monk seals do not really defend a territory but instead remain close to the female wherever she goes. An attending male follows "his" female into the water to copulate but may face intense competition among other males for the estrous female; several other males may attempt to displace the "resident" male. Despite this "free-for-all" or opportunistic system, some males, presumably the older and larger bulls, appear to be socially dominant over others (Fig. 6.12).

Breeding behavior among the widespread and generally polygynous grey seals is difficult to categorize neatly. In most grey seal populations, male breeding behavior involves an unstable and nonlinear dominance ranking as well as what some authors have called a loose type of ter-ritoriality, unlike that seen in otariids. Grey seals do not defend a spatially defined territory, and the area controlled by a bull may change from day to day (Anderson et al. 1975). There are no boundary displays and fighting is infrequent. In crowded colonies, bulls may lie only 3–4 meters from one another (Bonner 1981a). The reproductive strategy of bulls involves some degree of scramble competition, in which males spend their time attempting to mate with as many females as possible, rather than fighting with other males and trying to exclude competitors.

In fact, Boness (1979) and Boness and James (1979) believe that grey seals on Sable Island do not really defend territories or form dominance hierarchies but instead compete among themselves for tenure, or the privilege of remaining with the shifting population of females. The

Irene Campagna © 1985

FIG. 6.12. Hawaiian monk seals. A female on the beach is threatening the male in the surf. *Drawing by Irene Campagna.*

turnover of tenured males occupying a particular area is high, with the average tenure being about ten days. Such an unstable breeding system among the males can lead to considerable confusion at the breeding colony. With the dominance-ranking system constantly in flux, the scene at the breeding beach at peak season is anything but orderly. According to Boness and James (1979), the breeding beach "looks more like a rowdy house than the site of an organized society."

As I have mentioned, breeding behavior among the grey seal populations may also vary with the population density and breeding habitat or substrate type. The breeding substrate can range from rocky shores in Scotland and sandy beaches at Sable Island to ice in the Baltic Sea and Gulf of St. Lawrence. Among the ice-breeding seals, males engage in agonistic encounters less frequently as the breeding systems tend toward monogamy or slight polygyny. In the British Isles, breeding bulls station themselves at strategic positions in the water at the approach to an isolated cove containing a small number of females (Hewer 1957; Fogden 1971). In the more crowded colonies where females aggregate far inland, successful bulls position themselves ashore near the females and away from the water (Anderson et al. 1975). Older, more experienced bulls tend to position themselves in the best inland areas, while subordinate males hang about the fringes of the female area. The largest males may sire about ten times as many offspring as the smallest breeding males (Anderson and Fedak 1985). Male grey seals fast throughout the breeding season. The costs of reproduction place a substantial drain on the male's resources, although the daily weight loss of a male is proportionately about two-thirds less than that of a female, who also must contend with the energetic demands of lactation (Anderson and Fedak 1985).

Pinnipeds that are monogamous are often serially monogamous, since the male, given the opportunity, will search for a second female or mother-pup pair after mating. Actually, serial monogamy may be thought of as a form of sequential polygyny, since one male mates with more than one female during a breeding season. Monogamy or serial monogamy characterizes such phocids as ringed seals, crabeater seals, and, in some cases, hooded seals. (Sometimes, however, male hooded seals may be able to monopolize and defend small groups of females at the same time [Boness et al. 1987].) In serially monogamous pinnipeds, the male may have to wait before the female comes into estrus and is receptive to mating. After copulation, the male may leave to search for another mother-pup pair or receptive female. Whether he finds one depends on the timing of the breeding season and his own ability to monopolize females.

TERRITORIALITY AND BREEDING BEHAVIOR IN OTARIIDS

An otariid male that secures a territory will probably mate with most of the females who haul out on his "turf" to give birth and raise a pup. Without a territory inhabited by breeding females, a male has difficulty obtaining access to receptive mates. A male's territorial tenure—the time he remains on his territory during the breeding season—depends on a number of related variables: a male's size and age, his ability to compete with other males, his ability to fast, the location of his territory, the distribution of females, the population density, the climate, and the accessibility of water for thermoregulation (McCann 1987b). Previous experience in holding a territory and mating with females can also be important for a male. Gisiner (1987) conducted a long-term study of Steller sea lions on Año Nuevo Island and found that the breeding success of males depended heavily on past experience and that most successful males began their reproductive histories by holding inferior territories. Steller sea lion males (Fig. 6.13) could even remember their relationships with territorial neighbors from year to year, which helped them employ the most successful reproductive strategy in maintaining and expanding their territory.

FIG. 6.13. Steller sea lion territorial male threatening another male. *Photo by Bob Gisiner.*

Some otariids breed on sandy beaches, while others prefer rocky shores and cobble coves that are close to the water and exposed to surf and sea spray. These areas are cooler and cleaner as well, since waste materials and decomposing carcasses are regularly washed away. Other male otariids, especially fur seals, may hold landlocked territories well inland from the water. When otariids breed in warm areas, however, access to water and shady areas is important to breeding males as well as females. Territorial boundaries often extend to the edge of the water and even beyond, to deeper water. On the desolate lava-rock island of Guadalupe, for example, shady areas are at a premium, and some males even hold territories inside damp sea caves. In warm-climate breeding areas, such as the Gulf of California and the Galápagos Islands, California sea lions (Fig. 6.14) remain closer to the shoreline and cool off in the water more often than they do in the temperate regions of southern California (Heath 1985).

Since males have a more difficult time controlling females in the water, Stirling (1983) suggests that the nature of territorial behavior may change in warm areas. In fact, among a number of otariids, the animal's thermo-regulatory needs as well as the topography of the rookery are important in structuring the male's breeding behavior and in determining his mat-

FIG. 6.14. California sea lions. *Photo by Frans Lanting.*

ing success. One of the most dramatic examples of the significant relationships between climate, rookery topography, and thermoregulatory
behavior has been documented in the southern sea lion by Campagna
(1987) and Campagna and Le Boeuf (1988). They compared two breeding
rookeries on the Valdés Peninsula in Argentina: Puerto Piramide, characterized by rocky shelves containing tidepools and boulders; and Punta
Norte, which had wide, uniform beaches containing no tidepools. At
Punta Norte the sea lions had no way to relieve heat stress, whereas at
Puerto Piramide they did. Effective thermoregulation is extremely important to the sea lions during the hot, dry breeding season. At Punta Norte,
all territories were similar, with none superior to the others. As a result,
males employed a strategy of herding or sequestering females and defending territories with more flexible boundaries along the high-tide
mark. At the Puerto Piramide rookery, competition for and defense of
preferred territories (those with tidepools and boulders for shade) was
more intense. The males' success in mating varied more in this area: those
whose territories contained the largest tidepools copulated more than
those in poorer territories.

The topography of the rookery and thermoregulatory requirements
similarly influence breeding behavior in a number of other sea lions and
fur seals. On the Channel Islands off southern California, the location of

sea lion territories on open, sandy areas may shift with the tides, air temperature, time of day, and location of females (Peterson and Bartholomew 1967), whereas territory boundaries on rocky beaches are generally stable throughout the breeding season (Odell 1972). Among Australian fur seals, Gentry (1973) found that 84 of 109 copulations (82 percent) during one breeding season were by males that had water on their territories. In the hot, dry Galápagos Islands, most Galápagos fur seal territories extend to the water, a strategy that may also be associated with the small low-density breeding colonies characterizing this species (Trillmich 1984). Male California sea lions may be extremely territorial in the water as well as on land. Skin divers have been taken by surprise by sea lion bulls charging and barking at them underwater in Mexico (J. A. Estes, pers. comm.) and the Galápagos (R. Luckenbach, pers. comm.).

Female otariids choose which territory to haul out in and often, but not always, move freely in and out of a male's territory. Among most sea lions and fur seals, a male's attempts to control females usually do not work well. Although a resident bull may try to herd females to prevent them from leaving his territory, a female determined to leave can usually do so. The bulls may attempt to restrain a departing female by blocking her way or leaning on her, often while uttering a warbling whimper. In fact, Guadalupe fur seal bulls seem so concerned that a female might leave their territory that often the subtlest shift of position by the female is enough to evoke a herding response from the territorial male.

The males of several species, however, are able to herd females effectively, preventing them from leaving their territories. Among these species are the southern sea lions (Campagna 1987; Campagna and Le Boeuf 1988), Australian sea lions (Marlow 1975), and northern fur seals when female groups are small and widespread (Francis 1987). When northern fur seal females are numerous and densely grouped, males give up even attempting to herd them (R. Gentry, pers. comm.). Herding bulls sometimes treat females very roughly, going so far as to throw a departing female back into their territory.

Otariid bulls generally arrive at the rookery earlier in the breeding season than the females to establish territories. Most territories are relatively small, and territorial boundaries usually fall on a distinctive natural feature such as a rock edge or crevice. Nonbreeding bulls and subadult males congregate in "bachelor" groups near the breeding areas. In many species, territorial bulls show pronounced "site fidelity," returning to the same territory year after year. Among Steller sea lions, some bulls have held the same territory for up to seven consecutive breeding seasons, although the average lifetime territorial tenure for a bull is two to three years (Gisiner 1985). The territory of a successful Steller sea lion male may contain up to thirty females. Successful northern fur seal males may hold the same territory for as long as four years, mating with perhaps fifty females during the breeding season. Top bulls may mate with as many as a hundred females in one season (Gentry 1980). Australian fur seal bulls are able to hold a territory for six years, although the average male's lifetime tenure is 1.8 years (Warneke and Shaughnessy 1985). The longest occupation of the same territory recorded for an otariid male was nine breeding seasons. The record-setting seal, named Lefty, was identifiable because he was missing his left hindflipper (Pierson 1987).

Much of the fighting between otariid males takes place during the first few days of the breeding season when territorial boundaries are being established. Fights involve chest-to-chest pushing, lunging, and slashing at an opponent's flippers, chest, and vulnerable hindquarters. Once boundaries are established, fighting and agonistic displays between neighboring males become much less frequent; evidently neighboring males become habituated to one another (Gentry 1975; McCann 1980b; Roux and Jouventin 1987). A bull established on a territory rarely fights, although he may advertise his ownership by constant vocalizing and aggressive displays (Figs. 6.15–6.16). Territorial bulls often engage in a ritualized boundary ceremony. During this display, two barking bulls rush at each other openmouthed, weave their heads threateningly from side to side, then puff out their chests to project the largest possible image, while performing the "oblique stare" at one another. Because the displays are carefully choreographed, there are no surprise attacks or physical contact—which is probably just as well for the two opponents since neither gets hurt yet each is reminded of the other's "territorial rights" and boundary limits (Figs. 6.17–6.19).

Most otariid bulls fast during the breeding season while maintaining their territories. A male that left his territory to feed not only would lose mating opportunities but would likely have to expend considerable energy chasing away intruding males and reestablishing his territorial boundaries. A male's territorial tenure varies with each species. Among California sea lions, breeding males do not appear to fast for long periods, and few bulls manage to hold one territory for more than two weeks. The territorial California sea lion bulls continuously displace one another throughout the breeding season. Male Steller sea lions and northern fur seals, in contrast, generally remain on their territories for the entire breeding season, fasting for two to three months until the last females to leave have been inseminated. The large size of many otariid bulls helps them not only to compete with other males but also to stay on

FIG. 6.15. Juan Fernández fur seal bulls displaying during the breeding season. *Photo by John Francis.*

FIG. 6.16. New Zealand sea lion bulls threatening one another. *Photo by Roger Gentry.*

Fɪɢ. 6.17.

Fɪɢ. 6.18.

Fɪɢs. 6.17–6.19. Guadalupe fur seal territorial bulls performing oblique stare (*top*) and boundary display (*bottom*) and engaging in open-mouthed threat and retreat (*opposite page*). *Photos by Mark Pierson.*

FIG. 6.19.

their territories without feeding throughout the breeding season. After all, a male can feed during most of the year, but he can mate for only a few weeks during the breeding season — if he is lucky enough to acquire a territory in the first place.

Campagna (1987) and Campagna and Le Boeuf (1988) found that the breeding behavior of southern sea lion males differs from that of other sea lions in a number of respects. Unlike most other species of sea lions, the southern sea lion male sequesters females and, as mentioned previously, prevents them from leaving his territory by blocking their path, or grasping them in his jaws and hurling them back. The only other sea lion bull known to herd females successfully is the Australian sea lion (Marlow 1975), also characterized by a small harem and male group raids, like the southern sea lion. The practice of sequestering pre-estrous females leads to several other differences in territorial and breeding behavior between the southern sea lion and other sea lion species: the southern sea lion adds female defense polygyny to the strategy of resource defense polygyny (or territoriality); the male remains close to females at all times to prevent them from leaving; fewer females inhabit each male's territory since the male is limited by the number he can successfully herd; and the variation in mating success among males is reduced (Campagna 1987).

DOMINANCE HIERARCHIES IN PHOCID SEALS

A dominance hierarchy is like a pecking order, in which one male is dominant over another, who is in turn dominant over the male below him in the hierarchy, and so on down the line. A dominance-ranking competition among males is characteristic of many of the phocids that breed on land in large to moderate-sized colonies, including northern elephant seals, southern elephant seals, and some populations of grey seals. Monogamous or slightly polygynous phocids that breed on pack ice — harp seals and largha seals, for example — usually defend females from other males (although they may also engage in some territorial defense), but there is no elaborate hierarchical ranking system in phocids, such as elephant seals, that breed in colonies.

Among elephant seals, the dominance hierarchy is established at the beginning of the breeding season before most of the females arrive. In a moderate-sized harem of perhaps fifty females, one individual, known as the alpha bull, is able to dominate and chase away all the other males. The "second-in-command" dominates all males except the alpha bull, and so on down to the lowest-ranking individuals. The top-ranking males get to mate with most of the females and exclude other males from doing so.

Dominance hierarchies in the northern elephant seal have been studied in detail over many breeding seasons by Burney Le Boeuf and his colleagues at the University of California at Santa Cruz. Names bleached onto the side of each breeding bull (as onto the weaned pup in Fig. 6.20) — using Lady Clairol Ultra Blue — make it easy to follow individual males to determine their position in the hierarchy. (Because elephant seals lose their names when they molt each year, permanent identification is determined from hindflipper tags.) Elephant seal dominance hierarchies are not always strictly linear, especially in crowded breeding areas like the main breeding beach (the Point Harem) on Año Nuevo Island, California, where large numbers of males compete to mate. A convoluted male dominance hierarchy at the Point Harem during the 1979 breeding season proceeded as follows: alpha bull Casey dominates Groucho, who bosses Twig, who is usually able to dominate Toranaga. But in certain situations, Toranaga can dominate Groucho and Shred can boss Toranaga. With so many males competing to mate, the hierarchy can become confusing at times. In addition, it can change throughout the breeding season. Bulls that ranked high during the first few weeks of the breeding season may find themselves underlings later in the season (Fig. 6.21).

A high-ranking elephant seal male can often move a subordinate male, even from a distance, simply by raising his head in the direction of the

Fig. 6.20. Northern elephant seal weaned pup with the name Gull bleached onto fur. *Photo by author.*

intruder or by rearing back and making a clicking threat vocalization. In fact, one can distinguish "foreign" bulls from another island by the pulse-repetition rate in the threat vocalization, which varies from one elephant seal colony to another (see chapter 9 for a more detailed discussion) (Le Boeuf and Peterson 1969; Le Boeuf and Petrinovich 1974). If vocal threats do not work, the male moves to attack the other bull. If both bulls refuse to back down, a long and bloody fight takes place. Each bull throws his weight against the other and slashes at his opponent's face and chest with long canines. Sometimes the two giants teeter back and forth, chest to chest, like mammoth rocking chairs, and for minutes neither budges an inch. At other times the fighting continues in the surf and out in the water for up to an hour, turning the water a bloody red. By the end of the fight, each bull has expended a tremendous amount of energy and both the winner and loser are obviously exhausted (Fig. 6.22). Deutsch (1987) was able to quantify the greater expenditure of energy by bulls with higher ranks.

When the harem increases to more than fifty females, one bull has a hard time keeping the other bulls out of the harem. This is especially true of the Point Harem on Año Nuevo Island. Because over a thousand females now breed in this area, one can imagine what a tremendous job it is for one male to prevent the two hundred or so other males in breeding

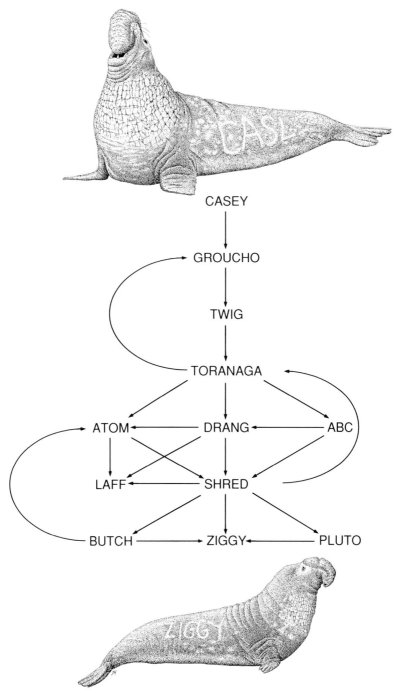

Fig. 6.21. Example of northern elephant seal male dominance hierarchy during one breeding season at Año Nuevo Island. *Drawing by Pieter Folkens.*

Fig. 6.22. Northern elephant seal bulls fighting during the breeding season on Año Nuevo Island. *Photo by W. B. Tyler.*

condition from mating with these females. And what a task it would be for the alpha bull to mate with all the females himself, much as he might want to! Furthermore, the large Point Harem is situated on a beach that is unprotected on two sides, so that subordinate males can readily sneak into the harem. It is easier for the alpha bull to defend females in a smaller, more sheltered area where any males approaching from the water can be detected quickly.

The number of bulls that can monopolize mating in the Point Harem has increased each year since 1968. In this year most of the males were excluded from mating. The five top-ranking bulls mated with 83 percent of the 193 females, and only 14 bulls mated at all. During the 1971–74 breeding seasons, the same highly successful alpha bull reigned, mating with at least 225 females. In 1973 close to one-third of the males succeeded in mating with 470 females in the Point Harem (Le Boeuf 1974). Within such a huge group of females, some of the high-ranking bulls tend to defend females in a particular area, blurring the distinction between dominance and territorial systems. In 1979 over half of the island's males mated with the 1,200 breeding females (Le Boeuf 1981). So in the crowded Point Harem the male monopoly has slowly crumbled.

Lifetime reproductive success in males has been studied by Le Boeuf and Reiter (1988), who found that male breeding success (measured by the number of different females inseminated each year) increased from age 6 to age 11 and declined to zero at 13 years. The prime breeding age for males was from 9 to 12 years. Males in this age range were responsible for mating with 93 percent of the females at the Año Nuevo colony. The variance in male reproductive success was extreme. Over the lifetime of a sample of male elephant seals followed from birth, only 8.8 percent mated at all. The three most successful males of known age mated with as many as 121, 97, and 63 females throughout their lives. Practically the only romantic aspect of the decidedly unromantic mating behavior of the elephant seal is that the peak in copulations on Año Nuevo Island, by coincidence, occurs on February 14 — Valentine's Day.

ALTERNATIVE MATING STRATEGIES IN MALE PINNIPEDS

Males of each pinniped species have evolved specialized and sometimes ingenious strategies for mating with as many females as possible. A number of variables help the top-ranking bulls and territorial males to compete successfully with other males: large size, experience, ability to bluff and fight, and timing. That is, a successful male must arrive early enough in the breeding season to establish a territory in a good spot or to secure a place for himself in the dominance hierarchy but not so early that he exhausts his energy reserves before the last females come into estrus. The correlation between large size and breeding success in males has been demonstrated in a number of pinnipeds, such as grey seals (Anderson and Fedak 1985), northern elephant seals (Le Boeuf 1974), and southern elephant seals (McCann 1981). The same link between large size and reproductive success in males generally holds true for most male vertebrates (Reiss 1982).

But how can the younger or subdominant males copulate at all in the face of such intense competition? Young subordinate males cannot even begin to compete effectively with the well-established older bulls. So rather than give up, they adopt an alternative mating strategy, often termed a satellite strategy. To achieve any mating success at all the young males employ a method of mating with a female that involves either a great deal of sneakiness or, less commonly, a group effort, as males "gang up" on the territorial or dominant bull.

Northern elephant seal "riff-raff" (subadult males), for example, are quiet and stealthy. These subadults, relatively small and less conspicuous than older males, "pretend" to be one of the females, pulling in their noses and keeping a "low profile." They do not, however, fool the females, who hiss and threaten them, catching the attention of a dominant bull. Apparently the chance to mate is worth the risk of discovery and attack by a dominant male. On occasion this sneaky strategy might work when the deceptive male finds a quiet, receptive female. But very likely she has already been inseminated repeatedly, and the subdominant male's sperm is too late to fertilize her egg, although "sperm competition" within the female's reproductive tract is always a possibility. New molecular genetic techniques have actually made it possible to determine paternity in pinnipeds such as grey seals (as well as in humans, birds, whales, and other animals) from blood samples, using DNA fingerprinting (see, e.g., Amos et al. 1987).

Young elephant seal males may pursue another strategy in their attempt to mate, trying to intercept females as they depart for sea. Sometimes a female on her way to the water is followed and harassed by as many as twenty subadult males at once. Unfortunately for the young males, these females almost certainly have been inseminated by a high-ranking male. The same subadults that try to mate with departing females also harass weaned pups, who occasionally sustain fatal injuries inflicted by huge subadults trying to mount them. Subadult grey seals also employ the strategy of attempting to mate with females as they depart for sea. And similar alternative male reproductive strategies are found in other highly polygynous animals, such as some species of frogs and fish.

In contrast to the elephant seals' alternative mating strategies, those of some subordinate male otariids (usually attempts to establish territories) are anything but sneaky. Many otariid males attempt to establish a territory by suddenly rushing onto the rookery. Often they time their move to coincide with the disturbance resulting from another male's attempt, acting simultaneously yet independently. An individual raider also stands less chance of injury from resident males when he is part of a group rather than on his own.

The subadult male strategy in two sea lion species seems predicated on the idea of strength as well as safety in numbers. Among Australian sea lions, for example, nonterritorial subadult males form groups that cruise around the breeding rookery at Dangerous Reef and sexually harass breeding females. Older group members try to copulate with estrous females until challenged by the territorial male (Marlow 1975). In some southern sea lion rookeries, the nonterritorial males band together in noisy raiding parties that invade female areas. This joint action was first documented on Lobos Island off Uruguay by Vaz-Ferreira (1975a, 1985), who found that subadult males raided the main breeding areas in groups of up to three hundred males. During these "surprise attacks," many group members attempted to mate with females or abduct mothers and pups.

A similar alternative reproductive tactic occurs among the southern sea lions along the Valdés Peninsula in Argentina (Campagna 1987; Campagna et al. 1988). In this area nonterritorial males are able to secure females by taking advantage of the confusion resulting from group raids. Most of the raiders (66 percent) are subadult males. The raids occur suddenly, causing the separation of many mothers and pups and creating chaos within the breeding groups. The resident territorial males cannot effectively repel all the raiders and keep the females within their territories at the same time. A mean of 144 raids was observed per breeding season, with most (57 percent) taking place during the peak time of copulation (January 14–26). Raids take place as often as every two hours at this time, and the raiding groups contain from two to forty males, with a mean group size of ten.

Sometimes a raiding southern sea lion is successful in abducting a female or achieving copulation, but often he is not. The younger subadult males are the least successful in securing females. Sometimes they grab pups (18 percent of the raiders) instead, biting and shaking them and

FIG. 6.23. Southern sea lion male, female, and pup. *Photo by John Francis.*

tossing them into the air. At other times they "kidnap" pups from the breeding area (57 percent of the time), holding them by force as if they were adult females. Some males attempt to mate with these pups. Not surprisingly, many pups sustain injuries during this rough treatment; according to Campagna et al. (1988) at least five pups were accidentally killed.

When a female is abducted, three outcomes are possible: a male loses the female within one hour and is driven out of the main breeding area (53 percent of the seizures); the attacks of the resident males force the raiding male to herd females to the periphery of the breeding area (8 percent); or the male remains as a resident in the breeding area (Fig. 6.23) with one or more females (39 percent) (Campagna 1987).

Both Vaz-Ferreira (1985) and Campagna et al. (1988) found that sub-dominant sea lion males employ other mating strategies as well. Some subadults hold wandering females away from the main breeding area on isolated beaches or mate with occasional females during the winter nonbreeding season when the breeding area is deserted. Other males attempt to intercept females that leave the main breeding group on the way to and from the water, as elephant seal and grey seal males do. In most other otariids, however, strategies of subadults are different and less conspicuous and aggressive than in the above-mentioned species. Among

Steller sea lions, for instance, the subadults never interrupt the copulations of territorial males, and nonterritorial males are never observed
mating in another male's territory (Schusterman 1981b).

COPULATORY BEHAVIOR

Mating takes place ashore among most, but not all, of the land-breeding
pinnipeds. The monk seals and harbor seals are the exceptions among
land breeders, since these species almost always copulate in the water. All
the otariids copulate on land, although mating may also take place in
shallow water or tidepool areas in some species, particularly those that
breed in warmer climates. In contrast, most of the seals that breed on ice
copulate in the water, including many of the Arctic phocids, walruses, and
the Antarctic Weddell seal (see Table 10). Apparently the crabeater seal,
however, usually mates on the ice. Stirling (1983) speculates that intensive
predation on crabeater seals by leopard seals and killer whales may have
promoted terrestrial copulation, since seals copulating in water would be
especially vulnerable. But Hawaiian monk seals mate aquatically even
though they are preyed on heavily by sharks. It is not known whether
copulation among the Ross seal and leopard seal usually takes place on
ice or in the water; these species have been less well studied. In at least one
case, however, a leopard seal has been observed mating on ice (Kooyman
1981a).

Among some phocid seals, copulation apparently can occur either in
the water or on land or ice. For instance, mating among harp seals usually
appears to take place on ice except in Newfoundland, where these seals
copulate aquatically (Ronald and Healy 1981). The Newfoundland seals
are the most distinctive and isolated of the three populations, which may
partly explain the practice of aquatic copulation among them. Hooded
seals also mate on ice at times. Although most harbor seals mate in the
water, Allen (1985) has seen at least four copulations on land among the
Phoca vitulina richardsi population. Among grey seals, copulation may also
occur either on land or in the water (Bonner 1981a).

Mate Choice and Sexual Behavior in the Female

As the preceding discussion has shown, the reproductive strategies of
males and females differ as males seek to mate with as many females as
possible, while females discriminate in selecting a mate. As Barash (1978)
puts it, "Males tend to be selected for salesmanship; females, for sales

resistance." Although a male can inseminate many females with little expenditure of energy, a female is fertilized by only one male during each breeding season. Females have only one chance to mate with a male who will provide high-quality genes to help produce a healthy, successful offspring. Therefore, the female's choice of a mate is especially important. But how can a female pinniped exercise choice when it seems as though all the competition for mates is left to the territorial or dominant males?

Perhaps because mate choice in females is less conspicuous than that in males, female choice has been less well studied in pinnipeds. Yet even in species such as elephant seals where the enormous dominant bulls seem to control all the action, females are able to have "some say" about who inseminates them. According to Cox and Le Boeuf (1977), when a female is mounted by a male, she behaves in a way that increases her chances of mating with an older, high-ranking bull. A female in early estrus that is mounted by a male always croaks noisily in protest, slapping her tail against the male's side and trying to squirm away. Her protest alerts other bulls that a copulation may be taking place. The highest-ranking male in the area interrupts the copulation attempt and often tries to mate with the female himself. The female's habit of vigorously protesting when mounted therefore makes it unlikely that she will mate with a young subadult male, since the top bull will undoubtedly win the competition she has incited among the males. According to Boness et al. (1982), a similar situation exists among grey seals. Female grey seals protest and attack bulls that attempt to mount them, thereby inciting competition among the males and increasing their chances of mating with a dominant bull. In all probability, the genes of such high-ranking males are good, since these genes have "allowed" their carrier to survive to adulthood and outcompete other males. When a female mates with one of these superbulls, she will likely receive some of these high-quality genes for her offspring.

Reiter et al. (1981) have found that a female elephant seal's location in the harem also influences her choice of mate. The older, more dominant females generally remain in the center of the harem, where they are more likely to mate with either the alpha bull or another high-ranking male. Conversely, females on the periphery of the harem are more apt to mate with one of the young subordinate males that are relegated to the harem edges.

Female walruses also appear to have a significant role in choosing a mate, which they do on the basis of intricate visual and vocal displays conducted by adult bulls in the water. Subdominant males do not display at all but are excluded to the peripheries of the lek. No other pinniped

FIG. 6.24. Walrus mother with her calf. *Photo by Lloyd Lowry.*

undergoes courtship displays as elaborate as those of the walrus. Males display to females by producing sounds both on the surface and under-water. Surface vocalizations consist of various barks, growls, and whistles; underwater vocalizations sound remarkably bell-like. The massive tusks of the male walrus (which can weigh up to 10 pounds) appear to play an important role during the breeding season as a symbol of rank (to repel other males) and as an epigamic signal to females, who may choose males partly by the size of their tusks. Displaying bulls often use their tusks to threaten one another. A bull with very large tusks can move another male simply by threatening at a distance, whereas small-tusked males cannot dominate other males so easily (Miller 1975c). Tusks in males are not strictly a secondary sexual trait but serve other functions as well (these are discussed in chapter 3). Females possess tusks too and may use them to dominate other females in minor squabbles and to defend their young against polar bears or hunters (Fig. 6.24) (Fay 1982).

Among otariids, female California sea lions and possibly other species appear to exercise some choice in mating (Heath and Francis 1987). The movement of female California sea lions to cool areas near water brings them closer to a number of different males and seems to allow them some degree of choice in their mates. Females respond differently to the mating attempts of various males. Females do not necessarily mate in the male's territory where they spend most of their time resting and caring for their pups.

At times, Steller sea lion females may actually solicit and initiate copulation with territorial males, but males tend to be responsible for most of the courtship interactions, and the extent of female choice is unclear (Fig. 6.25). One study of Steller sea lion female courtship displays showed that they involved a stereotyped sequence of movements such as lateral neck swings, dragging of hindquarters, and sinuous movements of the female's body against the male (Sandegren 1975, 1976). Gisiner (1985) found that females initiated 7 percent and 13 percent of the pre-copulatory interactions among Steller sea lions at Año Nuevo Island and Marmot Island (in Alaska), respectively. He was careful to distinguish female courtship behavior from female appeasement behavior, used to deflect aggressive bulls. He found that females directed most of their courtship behavior toward the older "proven" territorial males. One of the oldest and most successful males received over half of the courtship displays during his fifth, sixth, and seventh years of territorial tenure. At Marmot Island, many of the soliciting females were probably young nulliparous females who were less likely to come into contact with a male unless they actively solicited his attention. In at least three cases, female

Fig. 6.25. Steller sea lions copulating; note extreme sexual dimorphism. *Photo by Bob Gisiner.*

Guadalupe fur seals were also observed to initiate copulation by approaching the territorial male and repeatedly nipping at his neck and sides (Pierson 1987).

When mating activity is aggressive and rough, females sometimes sustain injuries. The cost of mating to females may be especially high among highly polygynous pinnipeds with pronounced sexual dimorphism, such as elephant seals. In fact, Bartholomew (1970) has pointed out that elephant seal males may have reached the upper limit of dimorphism; if they were any larger, they would probably kill the female just by attempting to copulate with her. According to Le Boeuf (in press), a number of breeding northern elephant seal females at the Año Nuevo rookery sustain some type of injury during attempts at copulation. Occasionally a female is even killed inadvertently by a bull who attempts to mate with her, but such an event is rare. The most traumatic incidents take place as females depart for sea, when many subadult males surround and attempt to mate with the female who is still in estrus and weakened after her 34-day fast. Females appear to minimize the risk of being accosted by the peripheral males by departing at very high tides, during another female's departure, or when males are engaged in fights.

Although the polygynous Hawaiian monk seals show little sexual dimorphism, the cost of mating may be high for some females because many males at once fight over them for mating privileges. During one incident, a female monk seal was severely injured by several males that bit her as they all tried at once to hold her down and copulate with her in the water (Kenyon 1981). Estrous females have even been killed by groups of up to thirty males attempting to mate with them; they die either from severe injuries or from subsequent shark attack. Such extreme competition among the males is more common around islands where the sex ratio is heavily skewed toward males (Gilmartin et al. 1987). Kooyman (1981b) also found a female leopard seal, another species with little dimorphism, that appeared to have been killed during mating.

Although the male generally pursues a good strategy in mating as often and with as many females as he can, the strategy does not always work. Male pinnipeds may indiscriminately attempt to copulate with females of different species. Although interspecific copulation has occurred mainly in captivity, in the wild a male New Zealand sea lion has been observed attempting to copulate with a dead female New Zealand fur seal (possibly killed by an earlier copulation attempt) (G. J. Wilson 1979), and a subadult male grey seal has been seen trying to mate with a young harbor seal (*Phoca vitulina concolor*) female on Sable Island (S. C. Wilson 1975). P. B. Best et al. (1981) even observed a southern elephant seal male trying to mate with a South African fur seal female. Male Steller sea lions and various male fur seals (*Arctocephalus* spp.) have also been observed copulating with females of another species. A male Guadalupe fur seal not only defended a territory among breeding California sea lions on San Nicolas Island but also mounted three females, with intromission occurring at least once (B. S. Stewart et al. 1987). Mixed breeding groups of Antarctic and subantarctic fur seals are seen on the Prince Edward Islands and Marion Island, and the existence of hybrid adult males (which can be distinguished both in the field and on the basis of skull characteristics) indicates that interbreeding sometimes takes place (Condy 1978; Kerley and Robinson 1987).

That male pinnipeds are often not too particular about what they mate with, as long as it remotely resembles a female, is exemplified in species such as northern elephant seals, who do not always mount an appropriate animal. Each year, for instance, a few weaned pups are crushed and sometimes killed by inexperienced subadults who indiscriminately attempt to mate with the relatively tiny weaners instead of adult females (Le Boeuf 1981). Researchers at the University of California at Santa Cruz have even used the male's indiscriminate taste in mates to their advantage in weighing the gigantic bulls. Chip Deutsch, a graduate student working with Burney Le Boeuf, constructed a life-sized model of a female northern elephant seal, using a surfboard for her "base." She was dubbed Raisin, being brownish and a little wrinkled (Fig. 6.26). A female "croak protest," typically emitted by females that are mounted, was recorded and played back near Raisin. As expected, the recording successfully attracted the interest of breeding males, who climbed onto the scale as they approached her. Often a male proceeded to the stage of actually mounting Raisin, and one persistent bull would not let her go!

Another humorous example of indiscriminate mating by male pinnipeds involves a Hawaiian monk seal male that approached with great interest the researcher Brian Johnson, who was resting on the sand. Brian apparently resembled a female closely enough that the male mouthed his leg and was ready to mount him in a prelude to copulation (C. Deutsch, pers. comm.). No doubt this has happened to other seal researchers who were keeping a "low profile" during the breeding season.

Social Organization

Social organization and behavior vary considerably among the pinnipeds, depending on the species as well as the time of year. Among the pinnipeds that breed in large colonies, social organization changes seasonally. During the breeding season, the rookery contains groups of

Fɪɢ. 6.26. Northern elephant seal male that attempted to mount Raisin, a model of a female northern elephant seal designed to lure adult males onto the scale so they can be weighed. *Photo by Chip Deutsch.*

females and pups in areas controlled by territorial or high-ranking bulls. Juvenile animals and subdominant males, excluded from breeding activities, often congregate in bachelor groups.

Even during the nonbreeding season, the limited number of islands and isolated mainland sites suitable for seals to haul out on promotes dense aggregations when animals come ashore to molt or rest. Moreover, there may be advantages for a seal that associates with a group onshore during the nonbreeding season. For example, seals in a group may benefit from the thermoregulatory effects of huddling together during cool weather. Huddling behavior is particularly pronounced during the walruses' pupping season, for instance, when the young calves are susceptible to heat loss. Groups also offer seals protection from predators. Many pairs of eyes and ears can detect danger more readily; the group enhances collective defense capabilities, and the seals find "safety in numbers" — the larger the group, the less the likelihood of one individual's being taken by a predator, as Hamilton (1971) theorizes in his model of the "selfish herd." Walruses grouped together on land or ice floes can detect predators such as polar bears more successfully. And if a polar bear should charge the walruses, the adults in the group defend the young calves (Fay 1982).

Harbor seals tend to haul out in groups of varying size and mixed age and sex classes, although females, dependent pups, and weaned pups sometimes form groups that are segregated from other animals (Newby 1973; Evans and Bastian 1969). The social organization within these groups appears to be loose or nonexistent. Unlike elephant seals and grey seals, harbor seals maintain a space of up to several feet between themselves and their neighbors. To maintain this distance, harbor seals use vocalizations such as growling and snorting as well as aggressive foreflipper waving and even head butting and biting (Bigg 1981).

Harbor seals probably group together primarily as an antipredator strategy, according to da Silva and Terhune (1985), who found that harbor seals formed groups in the Bay of Fundy in Canada even though there was no shortage of haul-out sites. Seals in the groups frequently scanned the area for danger (Fig. 6.27); as group size increased, however, individual seals were able to spend more time resting and less time being vigilant. In addition, the harbor seals appeared to derive thermoregulatory benefits from grouping: as wind speed increased, the distances between individual seals in a group tended to decrease.

During the nonbreeding season large pinniped aggregations often have a vague or ill-defined social structure, at least in comparison with the

FIG. 6.27. A group of harbor seals showing vigilance behavior. *Photo by W. E. Townsend, Jr.*

Fig. 6.28. Walruses resting together in a large group on land; a mother is nursing her calf. *Photo by Brendan Kelly.*

highly structured social organization characterizing many species during breeding activities. For instance, during the nonbreeding season Australian sea lions gather in small groups that include all ages and both sexes, although only one adult male is usually present. Hot weather causes a reduction in the grouping behavior, especially among adult males, who space themselves regularly along the water's edge during warm weather. Dominance generally increases with age among Australian sea lions, and high-ranking sea lions of both sexes often move subordinate animals during cold and rainy weather, assuming the vacated spots themselves (Walker and Ling 1981a).

Seals resting close together engage in positive thigmotaxic behavior. Walruses are one of the most positively thigmotactic of the pinnipeds, usually resting in large, dense groups (Fig. 6.28). Females and calves actually prefer to position themselves in the more crowded portion of the herd, and the calves often rest on top of a "bed" of adult animals. During the nonbreeding season in the summer, walruses travel north in sexually segregated herds of females and calves, followed by males. At this time, some males remain in the southern part of the range in the Bering Sea, where they haul out on beaches in huge herds. The males also rest together in compact groups, sometimes even lying on top of one another.

Even during the nonbreeding season, the largest, most aggressive walruses with the longest tusks are dominant. Since the most crowded positions at the center of the group are the preferred resting sites, the dominant males use their tremendous bulk to force their way through the walrus herd, moving animals in their way by prodding and jabbing with their long tusks (Miller 1975c).

Northern elephant seals are another pinniped whose social structure undergoes dramatic changes during the nonbreeding season. After the winter breeding season various groups, segregated by age or sex, haul out on islands to molt at different times of the year, beginning with adult females in the early spring, followed by juvenile and subadult animals later in the spring, adult males in the summer, and finally yearlings and two-year-olds in the fall. Agonistic behavior declines substantially among the animals in these groups, with adult and subadult males resting side by side in tight groups. Like grey seals, northern elephant seal females rest close to their neighbors throughout the year, with bodies touching in the denser colonies.

Gentry (1975) compared the spacing between otariid females and found that fur seal females tend to maintain a greater distance between themselves (one meter or more) than sea lion females. For instance, Steller sea lion females often rest in close bodily contact with one another, allowing many females to fit into one male's territory. Australian fur seal mother-pup pairs rest apart, as do northern fur seals, but the latter species does so only toward the end of the breeding season in the summer. When northern fur seals first arrive they are clustered together as densely as 2.1 females per square meter (R. Gentry, pers. comm.). Female spacing may also vary a great deal throughout the day, depending on the weather and the air temperature, in species such as California sea lions (Heath 1985), as I noted earlier in this chapter.

7

Reproduction and Life History

Reproducing in the Marine Environment

Several similar reproductive features in pinnipeds and other marine mammals allow them to bear and raise their young in the ocean. This austere and challenging environment amplifies a young mammal's need for rich milk and prolonged or intensive maternal care. (Maternal care and lactation strategies are discussed in detail in chapter 8.) The young of pinnipeds and other marine mammals are born after a relatively long gestation period, so that they begin life in the harsh marine world in a well-developed, precocial condition. All pinnipeds give birth on land or ice, unlike other marine mammals, which give birth aquatically (although sea otters sometimes give birth on land). Even precocial young, however, require a great deal of maternal care to survive in the ocean. In many species of marine mammals, including walruses, some otariids, and a number of cetaceans, the female cares for her offspring for a relatively long period, one to two years or even longer. Other species, including many phocid seals, adopt the more intensive strategy of pumping up their young with as much fat-rich milk as the mother can provide during a period of maternal care that lasts only a few weeks or even days.

Seals, like other marine mammals, give birth to a single offspring. Twin births are extremely rare (Spotte 1982). Because a mother would be unable to produce sufficient rich milk to care for two offspring at the same time, if she kept them, both twins would probably die or at least be extremely underweight when weaned. Although twins are uncommon among pinnipeds, histological, or tissue, examinations of female reproductive tracts have revealed the existence of twin fetuses or blastocysts in the following species: elephant seals, Weddell seals, crabeater seals, grey seals, harbor seals, ringed seals, California sea lions, Steller sea lions, southern sea lions, northern fur seals, South African fur seals, Australian fur seals, and walruses (King 1983). Unborn triplets have even been reported in a ringed seal (Kumlien 1879). Live births of twins have also been witnessed. Peterson and Reeder (1966), for instance, observed northern fur seal females giving birth to twins on three occasions. In each case, the mother soon rejected one of the twin pups. Bell (1979) observed an Antarctic fur seal female give birth to twins at South Georgia Island, but both pups died before reaching one month of age. A female California sea lion gave birth to twins in captivity, but one of the pups died soon after birth (Uchiyama 1965).

On rare occasions, however, twin pups have survived to weaning. A subantarctic fur seal mother was able to raise twins to weaning age (Bester and Kerley 1983). Doidge (1987), following two Antarctic fur seal mothers who gave birth to twins (female-female and female-male), found that only one twin (of the female-female twins) had a growth rate and weaning weight similar to that of a single pup. The other three twins had below-average growth and weaning weights. Curiously, both the amount of time the two mothers spent at sea and their pattern of attendance were similar to those of mothers with only one pup, but the mothers of twins could have been working harder when foraging at sea.

Female pinnipeds and other marine mammals that invest heavily in their offspring in terms of a long period of gestation and intensive maternal care give birth only annually or biennially. The interval between births may be two years or more in walruses and some sea lions and fur seals, as well as in sirenians. This low and relatively inflexible rate of reproduction makes pinnipeds and other marine mammals especially vulnerable to the effects of heavy commercial exploitation. Recovery has been slow for many marine mammal populations that were dramatically reduced during the nineteenth century by intensive commercial hunting. Paradoxically, the same K-selected* reproductive and behavioral features

* K-strategists (e.g., elephants) produce relatively few offspring, which are intensively nurtured to maximize resource utilization in stable, saturated habitats (where offspring that are heavily invested in are best able to compete); r-strategists (e.g., mice) produce large numbers of offspring, with little investment in any of them, in an environment that favors a rapid increase in population. The concept of K- versus r-selection is, of course, a relative one.

that have enabled marine mammals to adapt to ocean living have also made them especially sensitive to the effects of overexploitation. Like other K-strategists, marine mammals produce relatively few, intensively nurtured offspring. Other K-selected traits characteristic of marine mammals include a large body size, long life span, relatively low mortality of young, and delayed reproduction.

J. A. Estes (1979) suggests that intensive hunting of marine mammals has forced these K-strategists to deal with an artificial "r-selected" situation. The slow recovery rates of heavily exploited marine mammal populations reflect the difficulty of overcoming K-selected reproductive strategies by breeding earlier in life and more often, although a few heavily exploited species of seals and whales have shown some plasticity in reproductive patterns by breeding earlier in life. Still, the flexibility to alter reproductive rates is limited in pinnipeds—as in all marine mammals—in that they cannot produce more than one offspring at a time and must adhere to their rigid seasonal reproductive cycle.

Delayed Implantation and Gestation

All female pinnipeds have evolved an internal mechanism called delayed implantation, which enhances the precise scheduling of their reproductive cycle, allowing them to breed at regular yearly intervals, as well as to give birth and wean their pups during the more favorable times of the year. After a female mates and her egg is fertilized, the tiny embryo stops growing for a while at the blastocyst stage. It may remain completely inactive for up to several months—alive but not developing. This suspension of implantation ceases when the egg implants in the uterine wall and begins full fetal development. The delay postpones birth either until breeding conditions are favorable or until the female again hauls out on land the next year, so that mating and pupping take place within a restricted breeding season.

Delayed implantation is known to occur in one other marine mammal—the sea otter. A number of terrestrial mammals, including bears, marsupials, temperate-zone bats, armadillos, roe deer, various rodents, and mustelids such as badgers and mink also undergo a period of delayed implantation. The cetaceans, however, do not, although the growth of the embryo may slow during the earlier developmental stages and accelerate in a rapid spurt toward the end of gestation. At this time, a blue whale fetus gains weight at a rate of 75 pounds each day. The hormonal and other physiological mechanisms regulating delayed implantation remain one of the most fascinating and perplexing biological mysteries.

The entire gestation period, which consists of the interval between fertilization and birth, including the time when implantation is delayed, varies with each pinniped species (Table 11). During the period of active gestation the embryo is implanted and actually growing. The length of this period in part determines the duration of the inactive gestation period, when the embryo remains unattached.

In most pinnipeds that give birth annually, the total gestation period is roughly 10.5–11.75 months. For otariids that give birth annually, the total gestation period, like many other uniform reproductive features in the Otariidae, is consistent at 11.75 months for almost all species whose gestation periods are known. This consistency makes sense because nearly all female otariids except the California and Galápagos sea lions mate about one week after giving birth. This pattern means in effect that many female fur seals and sea lions are *not* pregnant for only about one week out of every year. The active gestation period among most otariids is from 7.75 to 8.5 months and the delay of implantation from 3.5 to 4 months on average. Unlike the total gestation period, the delay in implantation varies in otariids from 2 to 5 months.

Among phocids with a moderately long pup dependency period (a month or so), a female who mates toward the end of her lactation period gives birth around the same time the next year, after a gestation of roughly 11 months. Of course female phocids with shorter lactation periods consequently have longer total gestation periods and vice versa. The delay in implantation in most phocids ranges from 1 to 4 months, with the average delay lasting about 2.5 months, which is less than the average delay for otariids of 3.5–4 months.

Walruses have the longest gestation period of any pinniped: 15–16 months, with a delay in implantation of 4–5 months. A walrus nurses its calf for two or more years, mating several months after birth. The long gestation period is therefore related to the calf's lengthy dependency period. Unlike phocids or otariids, walruses are pregnant only during the last half of this dependency period. Unfortunately, researchers' estimates of the gestation period, delayed implantation period, and lactation period in many of the pinniped species vary widely in the literature, making it difficult to summarize the reproductive features listed in Table 11. The estimates given there are based on the most recent data available for each species or on the information I judge to be most accurate.

TABLE 11

Reproductive Biology of Female Pinnipeds

SPECIES	DURATION OF LACTATION*	TIMING OF OVULATION†	TOTAL GESTATION‡ (in Months)	DELAYED IMPLANTATION§ (in Months)	REFERENCES
Hawaiian monk seal	35–42 days (39)	End of lactation ~70 days >birth?			Kenyon & Rice 1959; Johnson & Johnson 1984; Johanos 1984; Johanos 1984; Henderson 1986; Gerrodette, pers. comm.
Mediterranean monk seal	42–49 days (46) 120 days (may have been isolated case)#				Sergeant et al. 1978; Boulva 1979; Kenyon 1981 Mursaloglu, in press#
Northern elephant seal‖	22–29 days (27)	Late in lactation 25 days>birth	11	4	Le Boeuf et al. 1972; Reiter et al. 1981
Southern elephant seal	20–25 days (23)	Late in lactation 18 days>birth	11.25	4	Laws 1953b, 1956b, 1960; Carrick et al. 1962a; McCann 1980a
Weddell seal	45–50 days	Late in lactation 48 days>birth	10.25	1.6	Lindsey 1937; Bertram 1940; Kaufman et al. 1975; Laws 1981, 1984; Siniff 1982
Ross seal	~28 days	Est. end of lactation	est. 11	est. 2.5–3	Tikhomirov 1975; Kovacs & Lavigne 1986b

Crabeater seal	28 days	End of lactation 30 days>birth	11	6	Siniff et al. 1979; Bengtson & Siniff 1981; Laws 1981, 1984
	14–21 days (17)#				Shaughnessy & Kerry 1989#
Leopard seal	30 days?	End of lactation 48 days>birth	11?	1.6	Sinha & Erickson 1972; Siniff & Stone 1985; Laws 1984
Hooded seal	3–5 days# 7–12 days** (10)	End of lactation 12 days>birth	11.5	up to 3.7	Bowen et al. 1985# Shepeleva 1973;** Øritsland & Benjaminsen 1975; Reeves & Ling 1981
Bearded seal	12–18 days (15)	Late in or at end of lactation	11.5	2	Burns & Frost 1979; Burns 1981a
Harp seal	9–15 days (10–12)	End of lactation 14 days>birth	11.5	2.6	King 1964a; Ronald & Healey 1981; R.E.A. Stewart & Lavigne 1980; Kovacs & Lavigne 1986c
Ribbon seal	21–28 days (25)	Late in or at end of lactation	11		Burns 1981b
Caspian seal	28–35 days (32)	Late in or at end of lactation ~30 days >birth	11		Ognev 1935; Fedoseev 1976; Frost & Lowry 1981; King 1983
Baikal seal	2–2.5 mos. (68 days)	Late in or at end of lactation	9.5		Popov 1979b

Continued on next page

TABLE II, *continued*

SPECIES	DURATION OF LACTATION*	TIMING OF OVULATION†	TOTAL GESTATION‡ (in Months)	DELAYED IMPLANTATION§ (in Months)	REFERENCES
Largha seal§§	28 days	End of lactation		1.5–3?	Bigg 1981
Harbor seal	21–42 days‡‡	End of lactation		1.5–3	Bigg 1969a, 1973, 1981; Bigg & Fisher 1975; Boulva & McLaren 1979; Stein & Jeffries 1985
Ringed seal††	35–49 days‡‡ (42)	Late lactation ~ 30 days >birth	11	3.5	McLaren 1958; Fedoseev 1975; Curry-Lindahl 1975; Frost & Lowry 1981
Grey seal	16–21 days‡‡ (19)	Late lactation	11.25	3.4	Hewer and Backhouse 1960; Cameron 1967; Bonner 1972, 1981a; Burton et al. 1975; Boyd 1983, 1984, 1985
California sea lion	6–12+ mos.‖	Early lactation 28 days > birth	11	3?	Odell 1972; Heath 1985
Galápagos sea lion	10–12+ mos.##	Early lactation*** 21 days > birth	11.25?		Trillmich 1986b
Steller sea lion	12–36+ mos.‡‡	Early lactation 12 days>birth	11.50	3.5	Harrison 1969; Gentry 1970; Sandegren 1970
Southern sea lion	6–24+ mos.	Early lactation 6 days>birth	11.75		Vaz-Ferreira 1981; Campagna 1987

Australian sea lion††	12–24 + mos.	Early lactation 6–7 days > birth	11.75		Stirling 1972; Marlow 1975; Walker & Ling 1981a; Higgins 1987
New Zealand sea lion	6–12 + mos.	Early lactation 6–7 days > birth	11.75		Gaskin 1972; Marlow 1975; Walker & Ling 1981b
Northern fur seal	4 mos.	Early lactation 6 days > birth	11.75	3.5–4	Gentry & Holt 1986
Guadalupe fur seal	8–11 mos.	Early lactation 7–8 days > birth	11.75		Fleischer 1978b; Gentry et al. 1986; Pierson 1987
Galápagos fur seal	18–36 mos.	Early lactation 8 days > birth	11.75		Trillmich 1986a
South American fur seal‡‡	8–12 mos. (Uruguay) 12–24 + mos. (Peru)	Early lactation 6–8 days > birth	11.75?	4	Trillmich & Majluf 1981; Trillmich et al. 1986
New Zealand fur seal	10–12 mos.	Early lactation 8–9 days > birth	11.75	4	Stirling 1971b; E. H. Miller 1975a; Crawley & Wilson 1976; Mattlin 1987
Antarctic fur seal	3.7–4 mos.	Early lactation 6–7 days > birth	11.75	4.2	McCann 1980b; Kerley 1983; Laws 1984; Costa et al. 1985; Doidge et al. 1986

Continued on next page

TABLE 11, *continued*

SPECIES	DURATION OF LACTATION*	TIMING OF OVULATION†	TOTAL GESTATION‡ (in Months)	DELAYED IMPLANTATION§ (in Months)	REFERENCES
Subantarctic fur seal	9–11 mos.	Early lactation 8–12 days > birth	11.75	4.3	Roux & Hes 1984; Gentry et al. 1986; Bester 1987; Kerley 1987
South African fur seal	8–18 mos. (most 9–11 mos.)‡‡‡	Early lactation 5–6 days > birth	11.75	~4	Rand 1959; David & Rand 1986
Australian fur seal	11–12 mos.§§§	Early lactation 5–6 days > birth	11.75	~3	Stirling & Warneke 1971; Shaughnessy & Warneke 1987
Walrus	24 + mos.‖‖‖	Mid-lactation	15	4–5	Krylov 1969; Fay 1981b, 1982

NOTES: Question mark indicates probable but inconclusive information; tilde (~) indicates approximation; > birth means "after birth."

* Lactation periods are given in days when less than 2 months and in months when more than 2 months (calculated using 30-day months). The average length of the lactation period is given in parentheses.

† The period between birth and estrus.

‡ Includes the period of delayed implantation, calculated in 30-day months.

§ The time between conception and implantation.

‖ Length of lactation is known to vary (increase) with the female's age.

Symbol indicates correspondence of data and reference.

** Symbol indicates correspondence of data and reference.

†† *Ph. ochotensis* breeds on more unstable ice and has a lactation period of only 21 days.

‡‡ Length of lactation varies geographically or with population or subspecies.

§§ Varies in the largha seal with the stability of the pack ice.

‖‖ Some California sea lion yearlings nursed.

Some Galápagos sea lion juveniles nursed 2 or more years if female had no new pup.

*** Twenty-one days is an estimate.

††† Australian sea lions appear to have a 510- to 540-day (17- to 18-month) reproductive cycle.

‡‡‡ In South African fur seals nursing may continue into second or third year if newborn pup dies.

§§§ In Australian fur seals nursing may continue up to 2–3 years.

‖‖‖ Some walrus calves are nursed for 3 years.

FIG. 7.1. Subantarctic fur seal female. *Photo by John Francis.*

Ovulation and Estrus

The timing of ovulation varies among the families of pinnipeds. As I have mentioned, almost all otariids ovulate and mate about 6–7 days after parturition. The timing of ovulation varies from 5 to 14 days after birth in most species, with the longer period of time elapsing in the sea lions. Subantarctic fur seals (Fig. 7.1) are the exception among fur seals in that estrus occurs 8–12 days after parturition, rather than the typical 6–7 days. Among the sea lions, *Zalophus* does not fit the pattern of mating about one week after birth. Instead, California sea lions mate about one month after giving birth (Heath 1985). Trillmich (1986b) speculates that the closely related Galápagos sea lion does not enter estrus until as long as three weeks after parturition, although he based this estimation on the time that males become attracted to females, since no copulations were actually observed. Most otariids remain onshore with their pup after giving birth, not going to sea to feed until after copulation. South African fur seal females (David and Rand 1986) and Galápagos fur seal females (Trillmich 1986a), however, leave their pup once before copulation to forage.

The timing of ovulation in phocid seals is more variable than in otariids. All phocids enter estrus and mate either during late lactation or soon after weaning their pup following lactation. In many species it is unclear whether the female mates toward the end of lactation or after

FIG. 7.2. A very rare photograph of a Mediterranean monk seal swimming underwater. *Photo by Jorge de Castro.*

weaning her pup. More information is needed to fully understand the intervals between these reproductive events. For instance, little is known of reproductive timing in Mediterranean monk seals (Fig. 7.2), Ross seals, ribbon seals, and other seals that are difficult to study. Because ovulation takes place during late lactation or after nursing, the interval elapsing between birth and estrus in phocids depends on the length of the female's lactation period. Very generally, ovulation in phocids tends to occur 2.5–6 weeks after parturition. Little is known about the length of estrus in most species. In the northern elephant seal, it lasts 3–4 days, beginning in late lactation and ending when the female departs for sea. Harbor seal females remain in estrus anywhere from 1 to 9 weeks.

Walruses are unusual among pinnipeds in that females enter a postpartum estrus in late summer after giving birth about four months earlier, in April or May. Usually a female is unable to mate at this time, however, because there are no fertile males; most males are fertile only from December to March (Fay et al. 1981b). She therefore waits another six months or so until another mating season, usually in February, before mating. As a result, a female lactates around ten months before her second estrus and conception. Because walruses nurse their pup for two years or longer, females ovulate and mate around the middle of their lactation period.

Sexual Maturity

Pinnipeds tend to be characterized by sexual bimaturism (one sex matures sooner than the other): females reach sexual maturity more quickly than males. In fact, many male mammals and birds mature later than females, among them grouse (Wiley 1974), red-winged blackbirds (Peek 1971), mountain sheep (Geist 1971), and even humans. Deferred maturity in males tends to be especially pronounced in polygynous species where competition among males is intense and males postpone breeding until they are old and large enough to compete effectively for mates and resources. Among phocids, females are generally capable of breeding (that is, mating for the first time) at an average age of 4 years, whereas males reach puberty at around 5 years. The age of sexual maturity varies between species and within each species from 2 to 6 years in female phocids (although some hooded seal females do not breed until 9 years), and from 3 to 7 years in males.

The degree of sexual bimaturism is more pronounced in otariids than in phocids. Female otariids reach sexual maturity at an average age of about 3.5 years, whereas males attain reproductive maturity about 2–3 years later, at 5–7 years of age. As with phocids, the age of sexual maturity varies, from 1 to 7 years in females (although most mate for the first time between 2 and 5 years), and from 3 to 12 years in males (most, however, reach puberty within 4 to 8 years). Curiously, New Zealand fur seal males (Fig. 7.3) do not mature sexually until the age of 10–12 years, whereas females of this species are sexually mature at 2–5 years.

Among the pinniped families these differences between the sexes in the timing of sexual maturity make sense in view of the more pronounced sexual dimorphism and highly polygynous breeding system characterizing all the otariids and odobenids. Only a few species of phocids, in contrast, are markedly sexually dimorphic and highly polygynous. Accordingly, in polygynous phocids, such as elephant seals and grey seals, the differences in sexual maturity between males and females are most pronounced.

Table 12 lists the age of sexual maturity in males and females of each pinniped species, that is, the age at which a female or male is physiologically capable of breeding. A female who gives birth at age 4 would be considered sexually mature at age 3, when she first mated. Unfortunately, some authors do not specify their definition of sexual maturity (saying only that it occurs when the seal first "breeds"), so that it is unclear in some cases whether breeding means mating or giving birth. Other authors define sexual maturity as the age when a female first gives birth,

Fig. 7.3. New Zealand fur seal adult male. *Photo by Gerry Kooyman.*

which is technically incorrect; hence, the literature on the topic of sexual maturity is somewhat confusing.

Most pinniped males do not become "socially mature" until several years — usually at least 3–4 years — after reaching sexual maturity. Although a young male may be capable of breeding at a certain age, he is rarely able to copulate successfully with a female or to compete effectively with the dominant or territorial bull until he is older. Although grey seal males, for instance, typically reach sexual maturity at 6 years, most do not become active breeders until 12–18 years of age. Walruses reach sexual maturity at 9–10 years, but most males probably do not actually begin to mate until 13–16 years (Fay 1982). Pubescent five-year-old northern elephant seal males must wait until they are at least 4–5 years older to join in the mating activities (Le Boeuf 1974).

Variation in Breeding Age

As one can see from Table 12, the age of sexual maturity can vary widely in a particular species. Although most females tend to give birth by a certain age (5–6 years, for example), some females may mate for the first time at 2 years, while others postpone initial breeding until several years later.

TABLE 12
Age of Sexual Maturity in Male and Female Pinnipeds

SPECIES	FEMALES (Age in Years)	MALES (Age in Years)	REFERENCES
Hawaiian monk seal	≥ 5–6		Johnson and Johnson 1978
Northern elephant seal*	2–6 (4)	5	Le Boeuf 1974; Reiter et al. 1981; Le Boeuf & Reiter 1988
Southern elephant seal*	2–7	4–6	Laws 1953b, 1956b; Carrick et al. 1962a
Weddell seal	2–6	3–6	Mansfield 1958b; Siniff 1982; DeMaster 1978
Ross seal	3–4 3–5 (4)	2–7 3–5 (4)	Øritsland 1970a,c Tikhomirov 1975
Crabeater seal	2–6 (4)	2–6 (4)	Bengtson & Siniff 1981; Laws 1984
Leopard seal	2–7 (4)	4–6 (4.5)	J. E. Hamilton 1939; Laws 1957; Øritsland 1970a,c; Tikhomirov 1975; Laws 1984
Hooded seal	2–9 (3)	4–6	Øritsland 1975; Øritsland & Benjaminsen 1975
Bearded seal	3–6 (5–6)	6–7	Burns 1981a
Harp seal*	3–7 (5.5)	3.7–5.1 (4.4)	Sergeant 1966, 1973a; Ronald & Healey 1981; Ni et al. 1987
Ribbon seal	2–4 (3–4)	3–5 (4–5)	Burns 1969, 1971
Caspian seal	4–6	6	Ognev 1935; Fedoseev 1976
Baikal seal	3–6	4	Pastukhov 1969
Ringed seal	3–7 (6–7)	5–7	McLaren 1958; Mansfield 1967; Tikhomirov 1968; Frost & Lowry 1981
Largha seal	2–5	3–6	Goltsev & Fedoseev 1970; Burns & Fay 1973
Harbor seal*	2–7 (4–5)	3–7 (5)	Bigg 1969b; Pitcher 1977; Pitcher & Calkins 1979; Burns & Goltsev 1981

NOTE: The age of sexual maturity given here is the age at which a male or female is physiologically capable of breeding. Thus for females the age of first ovulation is given, not the age of first parturition, or birth. Males generally do not participate actively in breeding until 3 or more years after they reach sexual maturity. The age at which most animals in a species become sexually mature is given in parentheses.

* Age may vary with the colony or the status of the population.

TABLE 12, *continued*

SPECIES	FEMALES (Age in Years)	MALES (Age in Years)	REFERENCES
Grey seal	3–5 (4–5)	6	Hewer 1964; Mansfield & Beck 1977; Harwood & Prime 1978; Prime 1978; Bonner 1981a
California sea lion	4–5	4–5	Mate 1978
Steller sea lion	2–7 (5)	3–8	Perlov 1971; Pitcher & Calkins 1981
Southern sea lion	3–4	5–6	Vaz-Ferreira 1981
Australian sea lion	3?		Stirling 1972
Northern fur seal*	3–7 (4–7)	5	York 1979, 1983; Gentry 1981a; Lander 1981
Galápagos fur seal	3–5?		Trillmich 1984, 1987b
South American fur seal	2–3 (3)	7	Vaz-Ferreira 1979; Vaz-Ferreira & Ponce de Leon 1987
New Zealand fur seal	2–5 4–6	10–12	Mattlin 1978 Crawley & Warneke 1979
Antarctic fur seal	3–4	3–4	Payne 1977, 1979b
Subantarctic fur seal	4–6	3–4	Bester 1987
South African fur seal	≥ 3–4	3–4	Rand 1955; Shaughnessy 1979, 1982; David 1987b
Australian fur seal	3–6	4–5	Warneke 1979, 1982
Walrus	4–12 (5–6)	7–10 (9–10)	Mansfield 1958a; Krylov 1966; Fay 1982

Walruses are a good example of a species with a wide variation in the age of sexual maturity among females. Some females mate for the first time at 4 years; others delay breeding until they are as old as 12 years. Most females, however, ovulate for the first time at the age of 5–6 years. The majority of males do not become sexually mature until 9–10 years, although a few are fertile at 7–8 years (Mansfield 1958a; Krylov 1966; Fay 1982).

Why does this variation exist? The age of sexual maturity often varies from one breeding colony to another in a particular species, depending on when the colony was established and how crowded it has become. The age of sexual maturity may also change after a population has been reduced by intensive commercial hunting. Sexual maturity tends to occur earlier in the recently colonized, uncrowded breeding rookeries that are recovering from exploitation. As the colony becomes well established and

FIG. 7.4. Southern elephant seal mother and pup. *Photo by John Ling.*

more crowded, both males and females tend to begin breeding later in life. In addition, the age composition of a colony changes, with a higher proportion of younger animals present in the colonies that have been exploited. Northern fur seals (York 1979), harp seals (Sergeant 1973a), and southern elephant seals (see, e.g., Carrick et al. 1962a) began to breed at an earlier age and their population growth rates increased after their numbers were reduced by commercial harvesting. The pregnancy rates of commercially exploited cetaceans, such as blue whales and fin whales, have increased as well (Laws 1962; Gambell 1973).

These increases have been especially well documented among southern elephant seals (Laws 1953a, 1956a; Carrick et al. 1962a). When exploited colonies of southern elephant seals (Fig. 7.4) at South Georgia were compared with unexploited colonies at Signy Island and Macquarie Island, breeding age was found to vary greatly with the density of the colony. The average age of breeding bulls was 10 years (for some it was as high as 20 years) at Signy Island and 6 years at Macquarie Island, whereas males reached sexual maturity at 4 years at South Georgia. At the unexploited high-density colony at Macquarie Island, where elephant seals had not been hunted for 20 years, the average age of primiparous females was 6 years. In contrast, most females at the exploited South Georgia colony reached sexual maturity at age 2 and gave birth at age 3. In fact, all three-year-old females on South Georgia were found to be pregnant. At Macquarie Island, however, only one-third of four-year-old females were pregnant, and one-quarter postponed pregnancy until they were as old as 7 years. Seals on Macquarie Island also grew more slowly than those on South Georgia. The earlier breeding age of animals at the exploited colonies may be associated with reduced competition for food, which allows more rapid growth rates. However, an increase in the available food base may also have been promoted by the reduction in krill-eating baleen whales and squid-eating sperm whales in the area, which in turn may have increased the availability of food to elephant seals (Bonner 1982).

Variation in breeding age within a particular population actually occurs among a wide variety of vertebrate animals, depending on the environment. In a saturated (K) environment, food resources are limited and there is little suitable breeding space or few reproductive opportunities available to young animals. An environment that has abundant resources and breeding opportunities has not yet reached carrying capacity; its population is still growing (r). When populations are rapidly growing and expanding into areas with abundant food and space, the age of sexual maturity tends to decline (Cole 1954; Lewontin 1965). These uncrowded and productive areas for breeding often occur in areas where range expansion is taking place, where younger males and females tend to emigrate.

Among northern elephant seals (Fig. 7.5), new breeding areas are first colonized by young males and females who have been less successful reproductively in high-density breeding areas. Younger bulls cannot compete with the many high-ranking bulls that control the harems, and young females are more apt than older females to lose a pup in crowded areas (Reiter et al. 1981). New elephant seal colonies are established by immigrants from crowded islands (Bartholomew and Hubbs 1960; Le Boeuf and Petrinovich 1974). The most recently established elephant seal colonies on the Farallon Islands and the spacious Año Nuevo mainland are made up of mostly younger seals from the now crowded Año Nuevo Island (Le Boeuf et al. 1974; Le Boeuf and Panken 1977). During the early 1970s when Año Nuevo Island was still being colonized (colonization there began in the early 1960s), most of the females giving birth were also young. Today, a high proportion of older females breeds on the island (Reiter et al. 1981).

At the less crowded breeding beaches, pup mortality is low and most of the young female elephant seals are able to wean their pups successfully. In contrast to northern elephant seals, young Weddell seal females forced to breed in less crowded peripheral areas experience a decline in re-

FIG. 7.5. Northern elephant seal adult male, female, and pup. *Drawing by Irene Campagna.*

productive success (Stirling 1971c). But even in the "crowded" colonies, female Weddell seals are spaced farther apart and pup mortality is lower than in dense northern elephant seal rookeries. Meanwhile, the northern elephant seal population continues to expand. When all available breeding beaches become congested so that emigration to suitable new breeding areas is impossible, elephant seals may postpone sexual maturity and breeding until they are older and better able to compete in a difficult, high-density environment (Fig. 7.6).

FIG. 7.6. High-density northern elephant seal breeding colony on Año Nuevo Island. The bull in foreground is attempting to mate with a female, who is protesting. *Photo by author.*

Yet why do seals reach sexual maturity at different ages within the same colony? Reiter et al. (1981) believe that female northern elephant seals that breed earliest in life are probably the healthiest and largest of their cohort (a cohort consists of all the seals in a particular colony that were born in the same year). This would suggest that seals need to reach a minimum level of growth before initial ovulation and pregnancy. Indirect evidence tends to support this hypothesis: females that are themselves born early in the breeding season reach sexual maturity earlier than females born later in the season; older, larger females pup earlier in the season than younger, smaller-sized females; and the size of pups increases along with the mother's age and size.

The benefits of producing offspring at an early age are obvious: a female elephant seal may be able to increase the total number of offspring she produces over a lifetime (provided she lives a normal life span), and her offspring will themselves reproduce sooner. But are there disadvantages for females who breed early in life? Reiter (1984) was actually able to document the reduced longevity of early-breeding female northern elephant seals. Simply put, the earlier a female elephant seal reproduces, the sooner she will probably die. Females that pupped for the first time at age

four had a higher survival rate each year than females who were primiparous at three years of age. But breeding colony density also influences pupping success in females of all ages. All females are more likely to
wean a healthy pup in low density areas, although the breeding success of
young females improved the most in uncrowded rookeries as compared
to high-density areas. The optimal strategy for a female elephant seal is to
give birth for the first time in a low-density area when she is four years old;
next best is to pup in an uncrowded area for the first time at age three; the
worst is to give birth for the first time at age three in a high-density
rookery (a female that waits until she is four years old to give birth in a
high-density area would be a little better off). So a female that delays
pupping until four years and raises her pup in a spacious rookery survives
longer and has better success breeding.

R. E. A. Stewart (1983), measuring the possible cost of early breeding in
harp seals, found that young harp seal females invest more of their body
reserves in raising their pup than older females. Young mothers deplete
their blubber reserves to the minimum critical level needed for survival.
But it is not known whether this drain on their energy results in a reduced
lifespan, as it does in elephant seals.

Longevity and Mortality

The best way to determine a seal's age is by tagging it as a pup. But age can
be estimated by removing a canine tooth and counting the growth layers
in the roots. If the tooth is sectioned longitudinally, annually produced
ridges consisting of alternating dark and light rings of dentine can be
seen. The pattern and appearance of these rings actually represents
various events in a seal's life, such as the nursing and weaning period, the
breeding season, time of pregnancy, and so forth (see, e.g., Scheffer 1950).
In some seals, such as bearded seals, there also seems to be a significant
relationship between age and the annual ridges in the claws, at least until
the tip of the claw wears away in older animals.

The average lifespan for most pinnipeds is estimated to be from 15 to
25 years. The information on longevity presented by King (1983) suggests
that pinnipeds live slightly longer in the wild than in captivity and that
phocids live longer than otariids. The average life span for otariids,
including both wild and captive animals, is perhaps 17–18 years; for
phocids it is about 25 years. A maximum life span of 43 years has been
recorded for a wild female ringed seal (McLaren 1962) and a remarkable
46 years for a wild female grey seal (Bonner 1971).

Male pinnipeds of most species tend to have life spans shorter than those of females by several years, perhaps because males — especially territorial males or high-ranking bulls — generally live "hard and fast" in comparison with females. In fact, Le Boeuf and Reiter (1988) have found that about 77 percent of tagged northern elephant seal males die before reaching their breeding age of 5 years. As many as 86–93 percent of males are unable to survive to 8–9 years of age, at which time a male is socially mature and able to compete for females (Le Boeuf 1981). In harbor seals the mortality rate for males also exceeds that for females after the onset of sexual maturity (Bigg 1981).

Natality

The natality, or birth, rate (the number of pups produced each year), in a particular pinniped population varies greatly between species and even in the breeding colonies of the same species. The rate of pup production may vary dramatically with geographic location in ringed seals, for example; their natality rate ranges from 11 to 92 percent, depending on the area and year each population was studied (see, e.g., T. G. Smith 1973b; Stirling 1977). In general, the natality rate is relatively high in most pinnipeds, ranging from 80 to 90 percent or more; but it is notably low in at least three species: Hawaiian monk seals (Fig. 7.7), Mediterranean monk seals, and walruses.

The low natality rate in both species of monk seal seems to be associated with the small size and fragility of these endangered populations as well as the sensitivity of females to human disturbances. Kenyon (1981) estimates that the natality rate in Mediterranean monk seals is only 11.5 percent. Hawaiian monk seals similarly have a birth rate of only 16.3 percent (Rice 1960). Many pregnant Mediterranean monk seals abort their fetuses, probably in response to human disturbances. Pregnant Hawaiian monk seals disturbed by humans abandon preferred breeding beaches and give birth in less desirable areas where pup mortality is much higher (Kenyon 1972).

The natality rate in walruses ranges from only 7 percent to 15 percent, depending on the population (Chapskii 1936; Mansfield 1958a, 1966; Burns 1965; Krylov 1968). This low rate of pup production is related to the long gestation period (15–16 months) and the long calf dependency period — 2 or more years. The pregnancy rate in walruses declines significantly with age. Younger females tend to give birth every 2 years, whereas older females may give birth at intervals of 4 or more years. Therefore,

FIG. 7.7. Hawaiian monk seal adult male. *Photo by Chip Deutsch.*

walrus populations containing a high proportion of older females will be much less productive than a population of younger females (Fay 1981b).

The natality rate usually varies with different age classes in other pinnipeds as well. In some species, fecundity rises with increasing age in the female, so that the birth rates of the older age classes are higher than those of younger females. This is true of hooded seals, for example, whose natality rate is 98 percent for all females 6 years or older but only 78 percent for three-year-old females (Øritsland 1964). In Antarctic fur seals (Fig. 7.8), 55 percent of the females give birth by the time they are 3 years old, 90 percent by age 6 (Payne 1977, 1979b). The natality rate in older northern elephant seal females (8–11 years of age) is as high as that in younger females (3–5 years of age): 97.8 percent (Le Boeuf and Reiter 1988).

But what about very old seals? Are aged females able to produce offspring? The answer seems to be both yes and no, depending on the species. In many mammals natality increases to a point and then falls as the female ages. This appears to be the case for northern fur seals, among which extremely old and very young females are least successful re-productively. The natality rate is highest in fur seal females 8–16 years of age, declining to nearly zero among females aged 23 or older (see, e.g., Lander 1981). Yet aged southern elephant seal females do not appear to cease breeding. Two "elderly" female elephant seals at Macquarie Island,

FIG. 7.8. Antarctic fur seal females nursing pups. *Photo by Dan Costa.*

both 23 years of age, gave birth to pups that they raised successfully (Hindell and Little 1988).

Pupping Season

Most, but not all, pinnipeds give birth during the spring and summer months of the hemisphere they inhabit. Table 13 summarizes the timing of the breeding season for each species. For many of the pinnipeds breeding in Arctic and subarctic areas, the pupping season begins in spring when ice conditions are most favorable for raising a pup. Phocids such as largha seals, ribbon seals, and ringed seals pup earlier in the spring when the ice is most stable and there is the greatest accumulation of snow. Bearded seals and walruses give birth later in the spring when the ice is breaking up and leads and open water are more extensive, partly because pups of these species are able to enter the water soon after birth. The food supply also begins to increase for all species later in the spring, when pups are usually weaned (except for walrus calves, which are cared for from one to two years). The northern otariids in the eastern North Pacific also give birth during late spring or early summer (May–July). At this time of year, when the seasonal upwelling promotes a rich food

TABLE 13
Timing of the Breeding Season

SPECIES	PUPPING SEASON	TIME OF YEAR*	PEAK PUPPING	PEAK MATING	REFERENCES
Hawaiian monk seal	Year-round (least often Aug.–Nov.)	All year	Apr.–May	June–July?	Kenyon 1966, 1973; T. Gerrodette, pers. comm.
Mediterranean monk seal	May–Nov.	Spring, summer, fall	Sept.–Oct.	?	Sergeant et al. 1978; Kenyon 1981
Northern elephant seal	Dec.–mid-Mar.	Winter	Late Jan.	Mid-Feb.	Le Boeuf 1974; Le Boeuf et al. 1972
Southern elephant seal	Mid-Sept.–Oct.	Spring	Mid-Oct.	Early Nov.	Laws 1956b; Carrick et al. 1962a
Weddell seal†	Late Aug.–Nov.	Spring	Sept.–Nov.	Nov.–Dec.	Lindsey 1937; Mansfield 1958b; Stirling 1969a; Siniff et al. 1971a
Ross seal	Nov.–Dec.	Late spring–early summer	Nov.	Dec.	Øritsland 1970a; Tikhomirov 1975; Thomas et al. 1980
Crabeater seal	Late Sept.– early Nov.	Spring	Mid-Oct.	Late Oct.–Nov.	Siniff et al. 1979
Leopard seal†	Sept.–Dec.	Spring–early summer	Nov.	Late Dec.	J. E. Hamilton 1939; K. G. Brown 1957; Siniff et al. 1980; Siniff & Stone 1985
Hooded seal†	Mar.–Apr.	Early spring	Mid-late Mar.	Early Apr.	B. Rasmussen 1960; Øritsland 1964; Sergeant 1976
Bearded seal†	Mid-Mar.–mid-May	Spring	Apr.	May	Chapskii 1938; McLaren 1958; Tikhomirov 1966; M. L. Johnson et al. 1966; Potelov 1975b; Burns 1981a

Harp seal†	Late Jan.–early Apr.	Winter–spring	Late Feb.–mid-Mar.	Mar.	Mansfield 1967; Sergeant 1976; Lavigne 1979; Ronald & Healey 1981
Ribbon seal	Early Apr.–mid-May	Spring	Apr. 5–15	Late Apr.–early May	Tikhomirov 1964b; Burns 1981b
Caspian seal	Mid-Jan.–late Feb.	Winter	Late Jan.	Late Feb.–early Mar.	Ognev 1935; Fedoseev 1976
Baikal seal	Feb.–Mar.	Late winter–early spring	Mid-Mar.	May–June	King 1964a; Pastukhov 1969; Popov 1979b
Ringed seal†	Mar.–Apr.	Spring	Early Apr.	Late Apr.–early May	McLaren 1958
Largha seal†	Feb.–May	Winter–spring	Variable	Variable	Shaughnessy & Fay 1977; Bigg 1981
Harbor seal†	Feb.–Sept.	Winter, spring, early fall	Variable	Variable	Bigg 1969a; Scheffer 1974; Shaughnessy & Fay 1977; Hoover 1983
Grey seal†					
Baltic population	Late Feb.–mid-Mar.	Late winter–early spring	Early Mar.	Late Mar.	Davies 1957; Coulson & Hickling 1964; Bonner & Hickling 1971; Summers 1974; Curry-Lindahl 1975; Anderson 1977; Mansfield & Beck 1977
Eastern Atlantic (U.K.)	Sept.–mid-Dec.	Fall–winter	Late Sept.–mid-Nov.	Oct.–Dec.	
Western Atlantic (Canada)	Mid-Dec.–mid-Feb.	Winter	Mid-Jan.	Early Feb.	
California sea lion	Mid-May–late June	Late spring–early summer	Mid-June	Early July	Peterson & Bartholomew 1967; Odell 1972, 1975

Continued on next page

TABLE 13, *continued*

SPECIES	PUPPING SEASON	TIME OF YEAR*	PEAK PUPPING	PEAK MATING	REFERENCES
Galápagos sea lion[†]	Year-round (rarely in April/May)	All year	Aug.–Oct. (Jan. on South Plaza Island)	Variable	Eibl-Eibesfeldt 1984b
Steller sea lion	Early June to early July	Summer	Mid-June	Late June–early July	Mathisen et al. 1962; Gentry 1970; Sandegren 1970; Pitcher & Calkins 1981; Gisiner 1985
Southern sea lion[†]	Mid-Dec.–early Feb.	Summer	Mid-Jan.	Mid–late Jan.	Carrara 1952; Vaz-Ferreira 1975b; Campagna 1985, 1987
Australian sea lion[†]	Year-round (1.5 yrs. may elapse between breeding seasons)	All year	June, Oct.–Dec. (variable)	Variable	Stirling 1972; Marlow 1975; Ling & Walker 1978
New Zealand sea lion	Early Dec.–early Jan.	Summer	Late Dec.	Early Jan.	Marlow 1975
Northern fur seal[†]	June–July[‡]	Summer	Mid-June–mid-July	Late June–late July	Gentry 1981a; DeLong 1982; Gentry & Holt 1986
Guadalupe fur seal	June–late July	Summer	Mid-June	Early July	Brownell et al. 1974; Pierson 1978, 1987; Fleischer 1978a, 1987
Juan Fernández fur seal	Mid-Nov. to late Jan.	Late spring–summer	Late Nov.–early Dec.	?	Torres 1987
Galápagos fur seal	Mid-Aug.–mid-Nov.	Spring	Late Sept.–early Oct.	Mid-Oct.	Trillmich 1984, 1987b

South American fur seal†	Mid-Oct.–Dec.	Spring–early summer	Nov.–early Dec.	Dec.	Vaz-Ferreira 1979; Trillmich et al. 1986; Vaz-Ferreira & Ponce de Leon 1987; Majluf 1987
New Zealand fur seal	Mid-Nov.-mid-Jan.	Late spring–summer	Mid–late Dec.	Late Dec.–early Jan.	Stirling 1971a,b; Crawley & Wilson 1976; Mattlin 1981
Antarctic fur seal	Late Nov.–Dec.	Late spring–early summer	Early Dec.	Mid-Dec.	M. R. Payne 1977, 1979b; Laws 1984
Subantarctic fur seal	Late Nov.–Jan.	Late spring–summer	Mid-Dec.	Late Dec.–Jan.	Rand 1956; Condy 1978; Roux 1987b; Bester 1987; Kerley 1987
South African fur seal	Nov.–Dec.	Late spring–early summer	Late Nov.–early Dec.	Mid-Dec.	Rand 1967; Shaughnessy & Best 1976; Shaughnessy 1979; David 1987b
Australian fur seal	Nov.–Dec.§	Late spring–early summer	Late Nov.–early Dec.	Early–mid-Dec.	Warneke 1979; Warneke & Shaughnessy 1985
Walrus	Mid-Apr.–mid-June	Spring–early summer	May	Jan.–Mar. (Feb. peak)	Mansfield 1958a; Fay et al. 1981b; Fay 1981b, 1982

NOTE: Question mark indicates probable but inconclusive information.
* Time of year refers to the pupping season of the appropriate hemisphere.
† Timing varies geographically or latitudinally.
‡ Pupping season is about 2 weeks earlier on San Miguel Island, California.
§ Some pups born between late October and early January.

supply, mothers may find food easier to obtain when they leave their pups to set off on feeding trips.

The northern elephant seal and grey seal are the major exceptions among the northern phocids in that their pupping season extends throughout the winter and fall. Northern elephant seals breed in winter from December to early March. The weaned pups therefore go out to sea at an optimal time, during the annual spring upwelling. Unlike the eastern Pacific otariids, elephant seal mothers fast throughout the pup dependency period, and do not go to sea to feed until the spring upwelling.

The grey seal of the North Atlantic Ocean breeds during the fall and winter (from September to early April), depending on location. There are three geographically separate grey seal populations: western Atlantic (or Canadian), eastern Atlantic (British), and Baltic Sea. The timing of the pupping season varies both from one population to another and within each population. The eastern Atlantic seals tend to pup in the fall (from September to November), whereas the western Atlantic seals give birth during the winter (from December to February) (Mansfield 1966). Curiously, Baltic Sea grey seals give birth in the early spring (from late February to March) although they are geographically much closer to the eastern Atlantic population (Curry-Lindahl 1970, 1975). The timing of the relatively brief (three-week) pupping season of Baltic Sea grey seals may be related to their tendency to breed on pack ice. If so, the pupping season in the Baltic would coincide with the most favorable ice conditions for raising and weaning a pup. Although the pupping season varies little with latitude in the Baltic Sea, there may be a clinal (that is, changing with latitude) variation in breeding season among the British seals, with births occurring earlier in the south and later in the northern colonies (Bonner 1981a).

Davies (1957) developed three alternative hypotheses to explain the variation in breeding season and the reproductive differences among the three major populations: (1) that all three groups have deviated from a pupping season that in the past occurred in spring and summer, as it still does for most other pinnipeds; (2) that grey seals originally bred in the eastern Atlantic, and the seals that dispersed to the Baltic and western Atlantic adapted to breeding on winter sea ice by pupping at the end of winter rather than before and during winter; or (3) that the eastern Atlantic seals colonized a new area and changed their breeding season from that of the grey seals that originally inhabited pack ice. Bonner (1981a) suggests that the timing of breeding among grey seals may vary widely as a result of the predation of prehistoric humans on these vulnerable seals.

Like grey seals, harbor seals of the eastern North Pacific (*Phoca vitulina richardsi*) have a prolonged pupping season, from early spring to early fall (March to September). There is a gradual clinal variation from Mexico to the Gulf of Alaska, however, so that the timing of pupping varies widely with latitude (Bigg 1969a; Scheffer 1974; Shaughnessy and Fay 1977). The pupping season is about one to two months in any particular area, generally beginning later as one moves north. For instance, on Cedros Island off Baja California, the southern end of the range, harbor seal pups are born in early February. The pupping season is March and April on the California Channel Islands, late June to September in British Columbia, and June to mid-July on the Pribilof Islands (Map 20). Bigg (1973) and Bigg and Fisher (1975) have conducted experimental studies that indicate that an annual endogenous reproductive rhythm and a specific response to photoperiod enable each harbor seal population to breed on schedule.

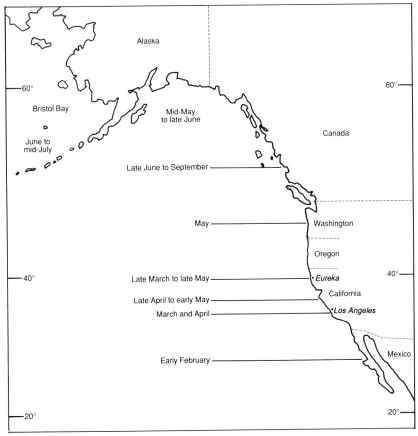

MAP 20. Timing of the pupping season of various harbor seal populations along the western coast of North America. *Monterey Bay Aquarium Graphics.*

The prolonged breeding season of the Galápagos sea lion (*Zalophus californianus wollebaeki*) also varies slightly from island to island as well as from year to year. The pupping season tends to begin earlier in the western part of the archipelago and later in the southeastern part. In 1976, for instance, female sea lions began to give birth around mid-May on Santiago Island, whereas in 1977 pupping began almost a month later, on June 15 (Trillmich 1986b).

The length of the pupping season in any particular colony of pinnipeds varies considerably with respect to a number of environmental variables. Table 13 gives the duration of pupping and breeding seasons for each pinniped. In general, pupping seasons are prolonged in areas characterized by mild aseasonal climates, of moderate length in temperate areas or areas of stable ice, and very brief in areas of unstable pack ice.

Pinnipeds with extended pupping seasons include the Hawaiian monk seal, Mediterranean monk seal, Australian sea lion, Galápagos sea lion, and Galápagos fur seal. Hawaiian monk seals breeding on the French Frigate Shoals have an eight-month pupping season, from late December to mid-August, although pupping has been recorded in every month of the year (Kenyon 1981). Mediterranean monk seals also have a prolonged pupping season, about seven months, from May to November (Sergeant et al. 1978). The Australian sea lion gives birth throughout the year, although the pupping season may peak from October to December (Walker and Ling 1981a). The Galápagos sea lion has an extended six- to seven-month period of pupping from June to November or December, whereas the sympatric Galápagos fur seal has only a three-month pupping season between mid-August and mid-November (Trillmich 1986b). Although the Galápagos fur seal's pupping season of three months is relatively long, it is short in comparison with the pupping seasons of six months or longer that characterize the Hawaiian and Mediterranean monk seals and the Australian and Galápagos sea lions—a puzzling difference.

Pinnipeds living in temperate, moderately seasonal climes (harbor seals, grey seals, elephant seals, and many otariids) and in areas of stable fast ice (such as Weddell seals [Fig. 7.9]) tend to have a pupping season that lasts roughly one to three months. The two species of otariids inhabiting environments of the highest latitude and seasonality (northern and Antarctic fur seals) have the briefest pupping seasons (McCann 1987a). In contrast, pinnipeds that give birth on unstable pack ice in sharply seasonal climes are characterized by a brief pupping season,

FIG. 7.9. Female Weddell seal nuzzling newborn pup. *Photo by Peter Anderson.*

sometimes lasting only a few days. Females give birth during the short period when ice conditions are most stable; breakup of the pack ice would be dangerous to young pups. Many of the Arctic phocids that breed on pack ice have pupping seasons lasting less than a month. Most of the hooded seal pups, for instance, are born within a two-week period at the end of March.

Molting

Breeding is not the only event that ties pinnipeds to land and ice. Seals return seasonally to their breeding grounds to molt. Although all pinnipeds molt, the process is more obvious in some species than in others. The fur seals and sea lions shed most of their fur each year, but the molting process is gradual and not apparent to an observer. A gradual molting process is especially important in fur seals, which need to maintain their dense, waterproof coat for temperature regulation. In fact, it takes as long as three years for northern fur seals to undergo a complete molt (R. Gentry, pers. comm.).

Some of the phocids undergo a drastic and very obvious molt, shedding skin and hair in great patches. The epidermis, or skin, actually peels off. This type of annual molt characterizes both species of elephant seals (Fig. 7.10) (Ling 1968, 1970, 1974) and Hawaiian monk seals (Kenyon and Rice 1959), all of which generally haul out on land during a certain time of year (depending on age and sex) and remain there much of the time until molting is completed. For instance, northern elephant seal yearlings

FIG. 7.10. Southern elephant seals molting skin in patches. *Photo by John Ling.*

haul out for about two months in the fall to molt; adult females, after feeding at sea following the breeding season, return to land for a month in the spring to molt; and adult males gather on land to molt in the summer. Elephant seals appear to fast while molting.

A less drastic yet similar type of annual molting occurs in grey seals (Bonner 1981a), captive harp seals (Ronald et al. 1970), and to some extent walruses (Fay 1982). Although molting can last several months in walruses, in males the peak period of hair shedding is from June to July and of hair replacement from July to August (Mansfield 1958a; Fay 1982). The timing of molting and of breeding activities may be intimately connected in some species. In grey seal females, for instance, there is evidence that the fertilized egg at the blastocyst stage is implanted at the end of the molting period; implantation may consequently be triggered by hormonal or other physiological cues associated with the cessation of molting (Bonner 1981a).

Pups molt their natal coat, or lanugo, either in utero or up to several months after birth. When harbor seals, for instance, molt before birth, the lanugo can sometimes be seen in the afterbirth. Since the lanugo would provide poor insulation against the cold ocean, the harbor seal pup is born with a warm and functional adult pelage. Elephant seals and Hawaiian monk seals, in contrast, shed their black natal fur during and after weaning; it is replaced by a silvery-grey coat of fur characteristic of recently weaned pups (Fig. 7.11). Molting of the more concealing dark natal fur (as well as the eruption of canine teeth) in northern elephant seals occurs later in males than in females. The delay may help males avoid detection when they steal milk and may ultimately benefit male weaners more than female weaners in terms of survival and future reproductive success (Reiter et al. 1978).

Migrations

Many pinnipeds as well as cetaceans make long-distance seasonal migrations to breeding rookeries or warm-water birthing grounds. Reproductive events and migratory movements are often timed to coincide with seasonal changes in the availability of food for the adults and newly weaned young. In pinnipeds, migrations are associated not only with the movement to good feeding grounds but also with the seasonal ice drift. The migratory patterns of many Arctic pinnipeds are closely linked to annual movements of the ice pack. Harp seals, for instance, travel both

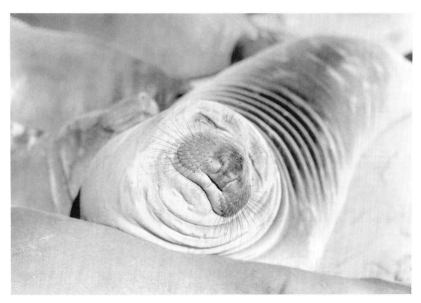

FIG. 7.11. Recently molted northern elephant seal weaned pup, showing new silvery fur. *Photo by author.*

actively and passively with the ice pack, as I noted in chapter 3 (Sergeant 1965).

Unlike pinnipeds, most of the baleen whales in both hemispheres have evolved a cyclic pattern of feeding in plankton-rich polar seas during the summer months and traveling toward the equator to give birth in temperate or tropical waters. Many migrating and breeding baleen whales either fast or consume little food during the six to eight months they spend away from their feeding grounds. Migratory movements of many odontocetes, such as pilot whales and Dall's porpoises, coincide with the seasonal distribution of schools of fish and squid.

Besides seasonal shifts in the abundance of prey, annual breeding cycles also help regulate the migratory patterns of pinnipeds. Reproductive events in a seal's life revolve around a rigid annual schedule. At the same time each year, land-breeding females return to the breeding rookery to give birth, raise young, and mate, while males return to establish territories or dominance hierarchies. Among species that breed on pack ice, the location of the pupping grounds may vary from year to year, although the beginning of the breeding season each year is generally consistent. Some pinnipeds, such as grey seals and elephant seals, also migrate to their rookeries at the same time each year to molt.

Biologists have long wondered how pinnipeds, as well as other vertebrates such as cetaceans, sea turtles, fish, salamanders, and birds, find their way back to their place of birth for breeding. Some of these animals migrate thousands of miles to reach their breeding grounds. Photoperiod, the proportion of the total day the animal is exposed to light, may be an environmental cue that triggers the migratory cycle. In addition, the synthesis of certain chemicals sensitive to photoperiod in the pineal gland may tick off the passage of set intervals and in so doing act as an internal biological clock. Of course, numerous birds undertake extensive seasonal migrations; many that travel at night navigate by celestial cues, using the position of the stars to find their way.

It is intriguing to consider that an "internal compass," using biomagnetism, may also direct animals back to their breeding grounds. In fact, a diverse group of animals, such as bacteria (Frankel et al. 1981), honeybees (Gould et al. 1978), homing pigeons (Walcott et al. 1979), and cetaceans (humpback whales, bottlenose dolphins, Dall's porpoises, pygmy sperm whales, Cuvier's beaked whales) (Bauer et al. 1983; Credle 1987), have been found to possess tiny lodestones, or magnetic material, in their brain tissues. These could well act as "compasses." Magnetite, a highly magnetic iron oxide synthesized within these animals, appears to act as a sensory receptor for transducing the geomagnetic field to the nervous system (Kirschvink 1982, 1987). Migrating marine mammals may therefore use marine magnetic lineations (created by sea-floor spreading and geomagnetic reversals) for navigation and orientation. In addition, Klinowska (1985, 1986) has demonstrated a fascinating correlation between areas in Great Britain where local magnetic field lines intersect the coast and the frequency with which live cetaceans are stranded. That live strandings of cetaceans are associated with such geomagnetic disturbances further suggests that biomagnetism may be used for navigation. Biomagnetism may also help pinnipeds to navigate, although magnetite has not yet been discovered in the brain tissue of seals.

Migrating pinnipeds and other marine mammals may also be guided by certain features of their underwater world, such as water and wind currents, the contours of the sea floor, the position of the sun and moon, and even the taste and temperature of the water. The harp seal, for instance, undergoes an amazingly accurate northward migration of up to 5,000 kilometers. How it accomplishes such a precise migration is not entirely understood, but some marine biologists believe that the harp seal may orient into the prevailing winds (Sergeant 1970) or follow the temperature profile of moving ice or that the seal's eye is "attuned" to the turquoise color of coastal waters (Lavigne and Ronald 1975). Juvenile seals can successfully find their way north during their first migration unaccompanied by adults that have made the trip before, a feat that suggests their navigational abilities may be innate (Table 14) (Ronald and Healey 1981).

TABLE 14
Migratory Patterns in Pinnipeds

SPECIES	MIGRATORY PATTERN	COMMENTS	REFERENCES
Hawaiian monk seal	Nonmigratory	Known to travel distances of up to 1,165 km in open ocean; movements as far as main Hawaiian Islands and Johnston Atoll	Kenyon & Rice 1959; Kenyon 1981
Mediterranean monk seal	Nonmigratory		Sergeant et al. 1978; Kenyon 1981
Northern elephant seal*	Long-distance seasonal migrations	Movements north from breeding rookeries up to 1,000 km. Extralimital sightings at Midway Island and Gulf of California	Condit & Le Boeuf 1984; Le Boeuf & Reiter 1988; Le Boeuf, pers. comm.
Southern elephant seal*	Seasonal movements	Little known. Random dispersal northward to sea (to South America) in winter and south to ice front in summer	Carrick & Ingham 1962; Ling & Bryden 1981; Laws 1984
Weddell seal	Seasonal movements	Some populations travel short–moderate distances in winter in association with ice front after breeding	Stirling 1969a; Kooyman 1968, 1975
Ross seal	Seasonal movements	Little known. Movements probably associated with pack ice and food distribution	G. C. Ray 1970, 1981; Gilbert & Erickson 1977

NOTE: Migration is defined as long-distance seasonal movements of a species; dispersion consists of less structured movements out to sea. Movement generally takes place between breeding rookeries and feeding grounds.

* Philopatric individuals are those known to return to their natal area to breed and/or molt.

† Indicates species that is often but not always philopatric.

‡ The San Miguel Island population of northern fur seals appears to be nonmigratory.

TABLE 14, *continued*

SPECIES	MIGRATORY PATTERN	COMMENTS	REFERENCES
Crabeater seal	Seasonal movements with pack ice	Movements into Ross Sea in summer; extra-limital sightings in Tasmania, New Zealand, South-ern Australia, South America	E. A. Wilson 1907; Lindsey 1937; Scheffer 1958; Stirling & Kooyman 1971; Kooyman 1981a
Leopard seal	Nonmigratory	Periodic dispersals, possibly associ-ated with distri-bution of food resources & pack ice; north in winter, south in summer with retreating pack ice. Extralimital sightings in New Zealand & Australia and at Cape Horn	K. G. Brown 1957; Scheffer 1958; Kooyman 1975, 1981b
Hooded seal	Long-distance seasonal migrations	Wide dispersal after spring breeding and summer molt, probably to feeding grounds. Extralimital sight-ings as far south as Florida and Portugal	Rasmussen 1960; Sergeant 1974; Kapel 1975; Reeves & Ling 1981
Bearded seal	Seasonal movements with pack ice	Movements north and south in close association with advance and retreat of pack ice; movements vary with population	Burns 1970, 1981a; Fay 1974a,b
Harp seal*	Long-distance seasonal migrations	Movements north in summer to feed-ing grounds; south in winter/spring to breed. Movements up to 5,000 km	Sergeant 1965, 1970, 1973b; Lavigne & Ronald 1975; Ronald & Healey 1981

Continued on next page

SPECIES	MIGRATORY PATTERN	COMMENTS	REFERENCES
Ribbon seal	Seasonal movements with pack ice	Dispersion after breeding & molting. Extreme dispersion south & north associated with unusual ice conditions	Tikhomirov 1961; Roest 1964; Burns 1969, 1970, 1981b
Caspian seal	Seasonal movements with pack ice	North in winter, south in summer to cooler, deeper waters	Frost & Lowry 1981
Baikal seal	Seasonal movements with pack ice	North in winter/ early spring (except for pregnant females)	Frost & Lowry 1981
Ringed seal	Long-distance seasonal migrations with pack ice	North in summer; south in fall with expanding ice pack	Burns & Harbo 1972; T. G. Smith 1973b; Frost & Lowry 1981
Largha seal	Seasonal movements with pack ice	Coastal in summer; offshore to edge of ice pack in fall/winter. Varies with population	Fay 1974b; Shaughnessy & Fay 1977
Harbor seal	Nonmigratory	Local movements associated with distribution of food resources and breeding activities; up to 300 km	Spalding 1964; Paulbitski & Maguire 1972; Bonner & Whitthames 1974; Pitcher & Calkins 1979; Bigg 1981; R. Brown & Harvey 1981; Beach & Jeffries 1981
Grey seal†	Nonmigratory	Wide dispersion after breeding season; pups undergo long-distance dispersal (up to 1,280 km)	E. A. Smith 1966; Bonner 1972, 1981a; Mansfield & Beck 1977; Boness 1979

SPECIES	MIGRATORY PATTERN	COMMENTS	REFERENCES
California sea lion	Long-distance seasonal migrations	Males travel north after breeding; females do not move beyond southern California Channel Islands	Orr & Poulter 1965; Mate 1973, 1975; Odell 1972, 1975; Ainley et al. 1977a,b, 1978; Gallo-R. & Ortega-O. 1986
Galápagos sea lion	Nonmigratory	Confined to Galápagos Islands	Trillmich 1979; Eibl-Eibesfeldt 1984b
Steller sea lion*	Long-distance seasonal movements	After breeding season, California population moves north, Alaska population south. Males leave first, followed by females	Mate 1973, 1975; Schusterman 1981b; Loughlin & Livingston 1986
Southern sea lion	Nonmigratory	Some dispersion after breeding; distances up to 1,600 km	Vaz-Ferreira 1975a, 1981
Australian sea lion	Nonmigratory	Some local movements 20–40 km in association with feeding; up to 300 km	Wood Jones 1925; Ling & Walker 1976, 1977
New Zealand sea lion	Nonmigratory	Some dispersion; nonbreeding males to Macquarie Island & South Island	Marlow 1975; Walker & Ling 1981b
Northern fur seal*	Long-distance seasonal migrations‡	Movements out to sea in summer after breeding (adult males) and in fall/winter (adult females & juveniles); south to Mexican border & Honshu coast. Adult males at sea up to 9 months. Summer: Bering Sea; winter: North Pacific	Kenyon & Wilke 1953; Lander & Kajimura 1976; Fiscus 1978; Gentry & Holt 1986

Continued on next page

SPECIES	MIGRATORY PATTERN	COMMENTS	REFERENCES
Guadalupe fur seal	Nonmigratory	Small numbers disperse to California Channel Islands	Bonner 1981b; Fleischer 1978a,b
Juan Fernández fur seal	Nonmigratory	Confined to islands of the Juan Fernández archipelago	Aguayo 1971, 1973; Bonner 1981b
Galápagos fur seal	Nonmigratory	Confined to Galápagos Islands	Trillmich 1979, 1984
South American fur seal	Nonmigratory	Some dispersion in winter; most females remain near the breeding rookery all year	Vaz-Ferreira 1982a; Trillmich et al. 1986
New Zealand fur seal	Nonmigratory	Some seasonal dispersion northward after breeding and south again before the breeding season	G. J. Wilson 1974; Crawley & Warneke 1979; Bonner 1981b
Antarctic fur seal	Seasonal migrations	Females leave breeding rookeries in winter and migrate to unknown area; males remain in rookery. Movements north out to sea in winter, south to ice front in summer	Erickson & Hofman 1974; Payne 1979a; Doidge et al. 1986
Subantarctic fur seal	Seasonal movements	Males move out to sea after breeding; females & pups remain in rookery	Payne 1979a; Bonner 1981b
South African fur seal	Nonmigratory	Seasonal movements occur; some long-distance dispersion from breeding rookeries in winter; up to 1,500 km	Rand 1956, 1959; David & Rand 1986

SPECIES	MIGRATORY PATTERN	COMMENTS	REFERENCES
Australian fur seal	Nonmigratory	Remains near breeding islands; local movements offshore to feed	Rand 1959; Warneke 1979
Walrus	Long-distance seasonal migrations in association with ice pack	Movements south in fall with advancing ice; north in spring as ice recedes; up to 3,000 km. (Several thousand males remain in Bering Sea throughout the summer)	Nikulin 1941; Brooks 1953; Fay 1957, 1981b, 1982

8

Maternal Care and Lactation Strategies

A seal's breast milk will raise an inch of fat.
IRISH PROVERB

Although human females might not like to produce milk and nurse babies nearly continuously throughout their lives, this seems to be the fate of most sea lion females, who may be in a virtually continuous state of lactation once they have begun to bear pups. Most phocid seals, in contrast, have a relatively short but highly efficient lactation period in which the young grow rapidly. Despite differences in maternal care tactics among the various species, maternal investment is substantial in all pinnipeds.

This chapter explores the fascinating diversity of lactation strategies and mothering systems in pinnipeds. To start we might ask the question, How does maternal care in pinnipeds differ from that in land mammals? Maternal care in seals revolves primarily around milk. The mother seal's key task is to transfer large quantities of extremely fat-rich milk to her offspring. But successful mothering among many species of seals involves more than lounging on the beach with nipples available and ready for suckling. Although it sometimes seems that seal mothers do little else, in many of the crowded breeding rookeries mothering can be a complex task, and pup loss is high. A good mother must be able to protect her pup, deter milk-stealing orphans, and defend her breeding space from other females. Furthermore, sea lion and fur seal mothers, each time they return from feeding at sea, must find their own pup among others at crowded breeding rookeries.

As far as we know, the mother-pup bond in pinnipeds (with the possible exception of walruses and some otariids) persists only as long as the pup is able to obtain milk. In other marine mammals, including some cetaceans, the association between mother and young may continue for quite some time after lactation has ceased. Weaned bottlenose dolphins, for instance, sometimes remain in the same group as their mothers, and the mother and her adult offspring turn to each other during stressful times (Tavolga and Essapian 1957). In many highly social land mammals, such as elephants, primates, and lions, the bond between mother and female offspring can persist for as long as a lifetime, with older daughters sometimes helping to care for their siblings or half-sibs (Riedman 1982). There is much we do not know, however, about the ability of pinniped mother and young to recognize each other after weaning has taken place. While northern fur seal females do not recognize their pups after weaning (Gentry 1981a), some fascinating research on captive California sea lions (discussed in chapter 9) shows that an adult and its mother may recognize and preferentially associate with one another.

All pinnipeds follow the pattern of returning to land or ice to give birth and raise their young. A marine mammal on land encounters some obvious problems, the most serious among them the inability to escape predators efficiently or protect young from them. Maternal behavior and lactation strategies in pinnipeds are therefore influenced by the respective breeding systems that arose in relation to the constraints of terrestrial breeding as well as the breeding environment: islands, pack ice, or landfast ice.

All marine mammals produce milk that is extremely rich in fat. Depending on the species and stage of lactation, milk-fat content generally ranges from 30–60 percent. On average, perhaps half of the milk is made up of fat. Marine mammal milk also contains relatively large amounts of protein (5–15 percent or more), important for rapid tissue growth. In most species of pinnipeds and cetaceans, the young grow quite rapidly on a diet of fat-rich milk. A blue whale calf gains weight at a rate of 200 pounds each day, or nine pounds per hour, which amounts to 50,000 pounds throughout the six- to seven-month nursing period (Slijper 1962). The milk of marine mammals generally contains far more fat and protein than that of terrestrial mammals. Cow's milk and the milk of human mothers contains only about 2–4 percent fat and 1–3 percent protein. Even the 16 percent fat content of the richest ice cream pales in comparison to that of the milk produced by seals and whales. The milk of seals and other marine mammals contains little or no lactose, or sugar, whereas that of terrestrial mammals typically contains 3–5 percent sugar. The sugar content of human milk is as high as 6–8 percent.

In the sections that follow I review the maternal care and lactation strategies among the pinniped groups: odobenids, sea lions, fur seals, ice-breeding phocids, and land-breeding phocids. Very generally, female phocids have short lactation periods, during which the pups ingest only milk and the mothers fast and remain with their pup. Phocid milk is extraordinarily rich in fat, and the young grow rapidly. Otariid mothers have a relatively long nursing period, leave their pup to feed during lactation, and produce less rich milk. Their pups grow more slowly and may begin to eat solid food later in the lactation period. Weaning is abrupt in phocids but may proceed more gradually in otariids and walruses. Walrus pups may accompany their mothers on feeding trips and continue to be nursed for one to two years or longer after reaching nutritional independence (Table 15). Excellent reviews of lactation strategies in pinnipeds are provided by Bonner (1984) and Oftedal et al. (1987a).

Although the growth rates of phocid pups are much higher than those of otariids, the total investment of energy on the part of otariid mothers may in some cases be greater than that for phocid mothers. Northern fur seals, for example, provide a total milk energy investment of about 1.3–1.8 times that of northern elephant seals (over a 27-day lactation period) and grey seals (over an 18-day period). Although northern fur seals have a 4-month lactation, they spend only about 21–28 days onshore nursing their pup, a period similar to that of some nursing phocids. Otariid females may be able to invest more energy because they feed rather than fast during lactation (Costa 1985). Another way of looking at the investment strategies of phocids and otariids is to note the differences in pup growth: rapid blubber deposition in phocids, slow lean body growth in otariids. While some otariid pups begin to forage and eat solid food before weaning, the extensive blubber layer of phocid pups provides them with a buffer to tide them over while they gain feeding experience after weaning.

Phocids

All phocids fast or feed very little during lactation. Maternal care in phocid seals revolves around a comparatively brief but intensive lactation period, during which the pup is nursed several times each day. Lactation may continue only a few days, as in the hooded seal, or may last as long as two and one-half months, as in the Baikal seal (see Table 11). The practice of fasting in itself naturally shortens the lactation period: the fasting mother lives off her own reserves in addition to nursing her pup, and there is a limit to the amount of weight a female can lose and still survive. Phocid pups, moreover, are larger at birth and grow more rapidly than otariid or odobenid pups.

Phocid milk contains a higher proportion of fat than the milk of otariids or walruses, especially toward the end of the lactation period (Table 16). Even the milk produced by whales generally contains less fat than the milk of true seals. Phocid milk typically contains 40–50 percent fat and 5–14 percent protein; that of northern elephant seals contains up to 63 percent fat at its richest (Riedman and Ortiz 1979). Because creamy

TABLE 15
Summary of Lactation Strategies in Pinnipeds

PHOCIDS	OTARIIDS	WALRUSES
Short lactation (4 days–2.5 mos.)	Longer lactation (4 mos.–3 yrs.)	Longer lactation (2–3 yrs.)
Tend to fast or feed very little during lactation	Feed during lactation*	Feed during lactation
High milk fat content ~40–50% fat ~5–13% protein	Lower milk fat content ~20–35% fat ~10–14% protein	Lower milk fat content ~30% fat ~5–11% protein
Milk composition changes throughout lactation (fat content increases in 4 species that fast)	Milk composition relatively constant throughout lactation (some fluctuations may occur)	Milk composition probably relatively constant throughout lactation
Newborn pups larger sized and more precocial	Newborn pups comparatively smaller sized and less precocial	Newborn pups comparatively smaller sized and less precocial
Pups usually eat no solid food during lactation	Milk diet often supplemented with food later in the lactation period	Milk diet supplemented with solid food later in the lactation period
Rapid growth of pups	Slower growth of pups	Slower growth of calves
Weaning occurs abruptly	Weaning occurs gradually†	Weaning occurs gradually

NOTE: Tilde (~) indicates approximation.

* Except for interval between parturition and departure for first feeding trip to sea and during subsequent suckling bouts on land.

† Except in the northern fur seal and Antarctic fur seal, in which weaning occurs abruptly; the pups leave the breeding grounds before their mothers wean them.

TABLE 16

Milk Composition and Lactation Patterns of Pinnipeds

SPECIES	PERCENTAGE OF FAT	PERCENTAGE OF WATER	PERCENTAGE OF PROTEIN	CHANGE IN FAT CONTENT	REFERENCES
Northern elephant seal	12–55 (43.9)	35–75 (43.7)	5–12* (7.7) 9–14	Increase	Le Boeuf & Ortiz 1977; Riedman & Ortiz 1979; Ortiz et al. 1984 Ortiz, pers. comm.
Southern elephant seal	9.5–43.9 (~40) 13–49	55–35 (43.6)	9.1–13.5 4–13.8	Increase Increase†	Peaker & Goode 1978 Bryden 1968
Weddell seal	30–60 (42.2)			Increase	Kooyman & Drabek 1968; Kooyman 1981c,d
Hooded seal	45–65				Shepeleva 1973
Bearded seal§	40.4‡ 49.5‡	49.8‡ 46.4‡	6.7‡ 6.8‡		Ridgway et al. 1975a Fay 1982
Harp seal	23–40	50–60	6.6	Increase	Lavigne et al. 1982
Harbor seal (*Pv. vitulina*)	45‡	45.8‡	9‡		Harrison 1960
Grey seal	52.2‡	46.4‡	6.8‡		Amoroso & Matthews 1951; Amoroso et al. 1950
California sea lion	36.5‡ 34.9‡	47.3‡	13.8‡ 13.6‡		Pilson & Kelly 1962; Ridgway et al. 1975a
Galápagos sea lion	17		9		Trillmich & Lechner 1986
Steller sea lion	20‡	61.8‡	11‡		Poulter et al. 1965

Species	Fat	Water	Protein	Composition	Reference
Northern fur seal	37.5–46.3 (41.5)	41.7–48.4 (44.4)	11.7–15.5 (14.2)‖	Relatively constant (fluctuates throughout lactation)#***	Costa & Gentry 1986
Galápagos fur seal	25–32 (29)		12	Relatively constant**	Trillmich & Lechner 1986
South American fur seal	38.3–57.1††	29.0–59.3††	8.1–12.5	Variable††	Vaz-Ferreira and Ponce de Leon 1987
Antarctic fur seal	40.3	42.4	17.4	Constant	Costa et al. 1985 (N = 63 samples)
	26.4‡	51.1‡	22.4‡		Bonner 1968 (N = 1 sample)
South African fur seal	18.6*		10.0*		Rand 1956
Walrus	14–32‡‡ (~30)	54–68‡‡ (~60)	5–11‡‡	Unknown§§	Fay 1981b, 1982

NOTES: Minimum and maximum percentages are given for fat, water, and protein from the beginning to the end of lactation. The average percentage is given in parentheses.

Tilde (~) indicates approximation.

* Older females may produce milk higher in protein than the milk of younger females (Riedman & Ortiz 1979).

† Bryden (1968) found that although fat content generally increased during lactation, it declined suddenly just before weaning.

‡ Only one value given by the source (see reference column) or milk samples not collected throughout the nursing period. Since milk composition may change throughout lactation, this percentage may not represent the average composition.

§ Milk samples from a female with a two- to three-week-old pup.

‖ Protein content increased significantly during the 7-day period between birth and a female's first feeding trip but remained steady thereafter.

Fat content declined during the 7-day period between birth and a female's first feeding trip and in each 2-day suckling period on land (the percentage of fat was highest when the female arrived after her feeding trip and lowest when she departed for sea).

** In otariids, fat content remains relatively constant but decreases slightly after the mother begins to forage for the first time following parturition.

†† Fat and water content in the South American fur seal in Uruguay varied during the year as follows: fat content, 28.3–32.3% (December), 35.3–48.8% (April), 53.7–57.1% (September), and 51.7% (October); water content, 45.6–59.3% (December), 41.3–50.4% (April), 29.0–32.3% (September), and 38.7% (October). The pupping season in Uruguay occurs during November and December.

‡‡ Samples not necessarily collected from beginning to end of lactation.

§§ Fat may decrease slightly after birth, then remain constant.

white elephant seal milk reminded me of rich vanilla ice cream, I decided to taste it as I collected samples for analysis. As it turned out, the milk did not taste like ice cream or anything else I had eaten before. It had no sweetness but rather a nutty blandness and a waxy texture, probably due to its abundant fat.

Phocid pups grow rapidly on their diet of fat-rich milk, while their mothers lose a large amount of weight. A phocid pup may gain an average of one-quarter (23 percent) of its birth weight each day (Bonner 1984). During a brief (10- to 12-day) lactation period, harp seal pups more than triple their birth weight, from 10.8 to 34.4 kilograms (a rate of 2.5 kilograms per day) (R.E.A. Stewart and Lavigne 1980). Northern elephant seal pups quadruple their birth weight, from an average of 34 kilograms at birth to 136 kilograms at weaning, during a one-month lactation period (Le Boeuf and Ortiz 1977). A few especially sneaky elephant seal pups manage to grow to 7 times their birth weight (to 238 kilograms) by suckling other lactating females during the 1- to 3-month period between weaning and departure for sea (Reiter et al. 1978).

CHANGES IN MILK COMPOSITION DURING LACTATION

As Table 16 illustrates, the water and fat content of the milk produced by many phocid seals changes dramatically during lactation. Milk composition may change throughout nursing in all phocids that fast, but we know for certain only that such changes take place in four species from which milk samples have been taken throughout the course of lactation: southern elephant seals (Bryden 1968; Peaker and Goode 1978), northern elephant seals (Riedman and Ortiz 1979), Weddell seals (Kooyman and Drabek 1968) and the harp seal (Lavigne et al. 1982; Yurkowski 1985). During lactation in all these species, the fat content of the milk gradually increases while the water content declines. The amount of protein in the milk remains relatively constant. In northern elephant seals, for instance, fat content increases from a low of 12 percent at birth to a plateau of 55 percent during the last week of lactation (Fig 8.1), while the water content correspondingly decreases (Riedman and Ortiz 1979).

It is difficult to determine changes in seal milk composition over time simply because it is difficult to obtain the successive milk samples in the first place. Milking a domestic cow is relatively easy, but how do you milk a wild seal? Weddell seal females are docile enough to allow researchers to kneel beside them while sucking their milk into a "giant straw" consisting of a short length of plastic tubing. A much longer "milking stick" and a more cautious approach are needed to collect milk samples from the

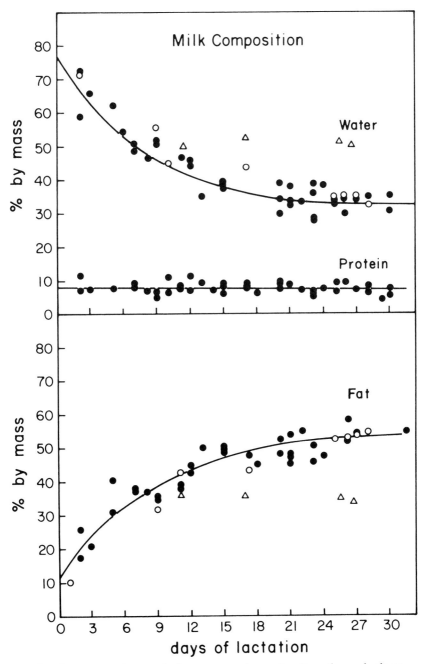

FIG. 8.1. Percentage changes in fat, water, and protein of northern elephant seal milk throughout lactation *(from Riedman and Ortiz 1979)*.

more aggressive northern elephant seal females (Fig. 8.2). Leo Ortiz and I devised a 9-foot wooden pole, attaching to one end a funnel connected to vacuum tubing and a small milk trap. Wearing a vacuum pump (actually an old fire extinguisher) on my back like a scuba tank, I could suck up a little milk by pressing certain valves. The whole contraption looked silly, but it worked.

Speed was the key in taking a milk sample successfully. After sneaking up to a female and applying my stick to her teat, I had only a few seconds to collect milk before the mother realized that what was suckling her was no pup. Enraged, she would invariably hiss threateningly or charge me, prompting me to retreat quickly to a safer location. I repeatedly sampled one female, whom we named Fran, who was situated away from the main group of females and easier to milk. But she and her pup soon caught on to my methods. The pup learned to recognize the threatening milking stick and would stop suckling as I approached, thus signaling Fran that trouble was close at hand. Suddenly very alert, Fran of course saw me stalking her with my stick. After a while it became difficult to get close to this observant mother-pup team.

It is remarkable that female phocids can provide such high-calorie milk for their pups while fasting. Among terrestrial mammals, most females must double or even triple their food intake to compensate for the expense of lactation. The changes in milk composition in phocid seals reflect the mother's severe physiological stress as well as the changing metabolic needs of her pup as it grows. This varying composition of the milk is a highly efficient nursing strategy. The increasingly high fat content of the milk allows the female to conserve water (water needed during fasting is derived metabolically); the need to conserve water increases as the female remains fasting on land.

The progressive substitution of fat for water in seal milk also benefits the pup. Newborn pups need to ingest much more water than older pups. An older pup, like a fasting adult, can derive water metabolically from its fat stores, but a young pup with little fat can obtain water only by ingesting its mother's watery milk. The high fat content in late lactation helps the pup to develop a thick blubber layer to sustain it after weaning.

In some species of phocids the weaners, or weaned pups, fast for up to three months after weaning. During this period they may lose a considerable amount of weight while surviving off their fat stores. After weaning, northern elephant seal pups spend two to three months fasting on land before going out to sea. Bonner (1984) suggests that the physiology of these fasting pups "requires a certain period from birth to nutritional independence." The interval between weaning and the pup's departure to

a. Milking the female (inset shows milk sample).

b. Female detects human milker while pup looks on.

FIG. 8.2. Northern elephant seal milking stick. *Photos by Bob Gisiner.*

feed at sea may also represent a period of learning for the weaned pup, as it practices swimming, foraging, and social skills in shallow water and tidepool areas.

ICE-BREEDING PHOCIDS

As I noted in chapter 6, the breeding environment, mating system, and lactation strategy of females are intimately related in pinnipeds. The length of lactation in each species depends in large part on the habitat used for breeding: moving pack ice, fast ice, or land. Among seals that breed on pack ice there is strong selection pressure for a short, efficient nursing period. The faster a pup born on ice gains weight and is weaned, the better, for two reasons. First, pack ice is an unstable substrate. A pup may be crushed or drowned as strong winds and currents cause the ice to raft or break up. There is less risk of harm to a pup weaned quickly. Second, pack ice is a cold, unsheltered environment for newborn pups that have very little blubber. A seal pup needs to develop a layer of insulating blubber as quickly as possible.

[handwritten margin note: reasons for fast weaning]

It follows that the shortest and most intensive lactation periods are generally found among the five species of Arctic seals that breed on pack ice: bearded seals, harp seals, hooded seals (Fig. 8.3), ribbon seals, and largha seals. Harp seals and bearded seals have nursing periods of only ten to twelve days and twelve to eighteen days, respectively. When the ice is most unstable, largha seals have only a two-week lactation period; largha seals pupping on more stable ice may nurse their pups for up to six weeks.

Hooded seals are characterized by the shortest lactation period known for any mammal: three to five days (Bowen et al. 1985, 1987). In the past, estimates of hooded seal lactation ranged from seven to twelve days (Shepeleva 1973). It is possible that the hooded seal, like the largha seal, may be able to vary the length of lactation in relation to changing environmental conditions. Bowen et al. (1985) found that hooded seal pups nearly double their birth weight of 22 kilograms, reaching 42.6 kilograms in an average of 4 days. This is a weight gain 2.5–6 times that of other phocids, relative to body weight. According to Shepeleva (1973), the pup's intestine remains closed throughout the nursing period, presumably to absorb all possible nutrients and fuel. In addition, a dense plug of fur and feces is thought to form near the anus before the pup is born, to promote maximum retention and use of the mother's milk (Khuzin and Yablokov 1963). Hooded seal mothers lose at least 35 kilograms of their initial body weight in the process of nursing their pup with milk that contains 45–65 percent fat. In relation to other phocids, weaned hooded

FIG. 8.3. Hooded seal mother and pup. *Photo by Kit Kovacs.*

seal pups are able to attain a similar weight in only one-third to one-tenth the lactation time because they are heavier at birth and gain weight at an extremely rapid rate. Hooded seal pups are in fact the only pinniped possessing such a well-developed blubber layer at birth.

Arctic seals that pup on stable land-fast ice have longer lactation periods. Lactation continues in the Caspian seal for four to five weeks and in the ringed seal for five to seven weeks. The Baikal seal has one of the longest lactation periods known for any phocid: four to six weeks. A subspecies of ringed seal (*Phoca hispida ochotensis*) that breeds on pack ice has a shorter lactation than other ringed seals, as might be expected since this subspecies breeds on less stable ice. *P.h. ochotensis* mothers nurse their pups for only three weeks, whereas other ringed seals nurse for six.

In the Antarctic, leopard seals and crabeater seals breed on pack ice, but the southern ice is less susceptible to breakup, and the problem of unstable ice is less severe in this area than in the Northern Hemisphere. Both leopard and crabeater seals have a lactation period of roughly one month. The other Antarctic seals—the Ross seal and Weddell seal—inhabit relatively stable ice. The Ross seal breeds on the heavy consolidated interior pack ice. Its lactation period is thought to be about a month. Weddell seals, which give birth on fast ice, nurse their pups for a relatively long period, six to seven weeks (Fig. 8.4) (Laws 1981, 1984).

Fɪɢ. 8.4. Weddell seal female. *Photo by Dan Costa.*

LAND-BREEDING PHOCIDS

All six species of land-breeding phocids have long lactation periods in comparison with those of seals that breed on unstable pack ice. Lactation periods range from about three to six weeks in the land breeders, which include two species of elephant seals, two species of monk seals, harbor seals, and grey seals. The various harbor seal subspecies either fast or feed little during lactation, and grey seals may sometimes enter the water but apparently do not feed much, if at all. Other species remain onshore and do not eat or drink throughout the nursing period (Table 17). Southern elephant seals nurse their pups for slightly more than three weeks, northern elephant seals for four weeks. Southern elephant seal pups gain about 6.8 kilograms per day, or a total of 70–120 kilograms throughout lactation (Bonner 1984). The enormous transfer of energy from mother to pup is obvious to anyone who simply watches an elephant seal mother-pup pair from birth to weaning: the mother gradually "deflates" while her pup balloons into a little barrel of blubber (Fig. 8.5). This transformation and the energy invested in a pup by its mother have been quantified in the northern elephant seal. Ortiz et al. (1984) found that mothers transferred 138 kilograms of milk (604,000 kilocalories) during the four-week nursing period, with a mean rate of energy transfer equivalent to 21,600 kilocalories per day (five times the metabolic requirements of the pup). Costa et al. (1986) showed that by the end of lactation, a mother has lost an incredible

42 percent of her initial body weight, while a pup has gained about 55 percent of the mass lost by its mother—a very efficient transfer of energy. Most of the female's weight loss is fat rather than muscle tissue. Milk production accounts for 60 percent of a female's energy expenditure during her 26.5-day lactation period.

Grey seal mothers also transfer a substantial amount of energy in the form of milk to their pup, but among land-breeding phocids their lactation period is the shortest—only sixteen to twenty-one days (Fig. 8.6). (Although some grey seal populations in the Baltic Sea and the western North Atlantic breed on fast and drift ice, most grey seals breed on land.) Grey seal pups gain about 1.6 kilograms per day, while mothers lose 3.6 kilograms per day, a net weight gain of about 30 kilograms for the pup and a net weight loss of 65 kilograms for the mother (84 percent of her stored energy reserves). A pup gains 46 percent of the mass lost by its mother, who transfers about 17,000 kilocalories per day (or about 3 liters of milk) to her pup, who assimilates 14,000 kilocalories per day. However, the mother uses an enormous amount of energy—30,000 kilocalories per day—to maintain herself as well as produce milk (Fedak and Anderson 1982). Put another way, the mother expends about fifteen times the calories humans need to eat each day (2,000) to maintain body weight—equivalent to over 12 gallons of ice cream per day for the seal!

The tropical monk seals have one of the longest nursing periods among the phocids. As the oldest living phocid seal in terms of evolutionary origins, the monk seal may have a lactation pattern that represents the ancestral form from which all modern forms evolved. The little-known and very rare Mediterranean monk seal probably lactates for six to seven weeks (Boulva 1979), although Mursaloglu (in press) observed one case in which a mother suckled her pup in a cave for at least four months. Whether this single observation represents atypical behavior is unclear. The better-known Hawaiian monk seals have a lactation period of five to

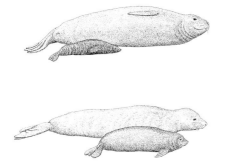

FIG. 8.5. During lactation the northern elephant seal mother loses weight while her pup gains correspondingly (top drawing shows beginning and bottom shows end of lactation period). *Drawing by Pieter Folkens.*

TABLE 17

Fasting and Feeding among Pinnipeds during the Breeding Season

FEMALES			MALES		
Fast	*Substantially Reduced Feeding*	*Feed*	*Fast*	*Substantially Reduced Feeding*	*Feed*
Hawaiian monk seal (5–6 wks.)	Harp seal*†	Ribbon seal*†	Northern elephant seal* (2–3 mos.)	Harp seal*?	Hawaiian monk seal
Northern elephant seal* (5 wks.)	Ringed seal*†	Harbor seal	Southern elephant seal* (2–3 mos.)	Ringed seal*?	Ribbon seal*
Southern elephant seal* (4 wks.)	Bearded seal*†	Walrus‡	Crabeater seal (~4 wks.)	Bearded seal*?	Harbor seal
Weddell seal†§ (6–7 wks.)		All sea lions‖	Ross seal* (unknown)		Walrus
Ross seal* (unknown)		All fur seals‖	Leopard seal (unknown)		Weddell seal
Crabeater seal (~4 wks.)			Hooded seal* (~4 wks.)		
Leopard seal (unknown)			Grey seal* (3–8 wks.)		
Hooded seal* (1.5–2 wks.)			All sea lions** (2.5 mos.)		
Grey seal*†# (2.5–3 wks.)			All fur seals (2.5 mos.)		

NOTES: The approximate duration of the periods of fasting or reduced feeding is given in parentheses. The maximum period of fasting is given for the male.

A question mark indicates probable but uncertain information. A tilde (~) indicates approximation.

* Also known or suspected to fast or feed very little during the molting season.

† Phocid females that leave their pup unattended when they enter the water; they may or may not be feeding (this practice is not noted for otariids since all sea lion and fur seal females leave their pups).

‡ Although walruses feed during their long lactation, the females appear to fast or considerably reduce food intake during the first days or weeks after giving birth (Fay 1982).

§ No evidence of feeding, although female Weddell seals enter the water frequently, and the possibility exists that some feeding may occur.

‖ Although otariid females feed during most of lactation, they fast for several days (usually 1–2 weeks) between birth and departure for sea as well as when ashore nursing their pup.

Some grey seal females leave their pup unattended whereas others do not, depending on the mother's location in the breeding colony and the accessibility of the water. (Reduced feeding may take place occasionally.)

** California sea lion males enter the water periodically (every 2 weeks) to feed.

Fig. 8.6. Grey seal pup. *Photo by Sheila Anderson.*

six weeks, during which time females fast completely (Kenyon and Rice 1959). Monk seal pups quadruple their weight from birth to weaning, from 17 to 64 kilograms (Kenyon 1981), after which they appear to fast for a period of two to three months (T. Gerrodette, pers. comm.). Yearling monk seals weigh only about 45 kilograms—even less than recently weaned pups.

The harbor seal is the only phocid that nurses and cares for pups in the water as well as on land. The lactation period varies from four to six weeks, but it is not known if the mother forages during this time. If so, her food intake is probably reduced. Perhaps to facilitate this strategy, harbor seal pups are highly precocial; they can swim at birth and dive for up to two minutes when only two to three days old (Finch 1966). A harbor seal mother provides a great deal of care for her pup, carrying it in the water, interacting and playing with it, and protecting it by diving with it or clasping it to her at the slightest sign of danger. Bonner (1984) suggests that the widespread harbor seal may be evolving toward a fully aquatic mode of birth and suckling, an adaptation that would minimize its dependence on land and allow pups to feed while still attended by their mothers (Fig. 8.7).

Otarioids: Sea Lions, Fur Seals, and Walruses

Among otariids and walruses, mothering and lactation strategies have taken an evolutionary pathway less extreme than that of the phocid seals.

Maternal behavior in the eared seals and walruses more closely resembles that of large terrestrial mammals in that female otarioids feed while nursing and have a relatively protracted lactation period, during which the pups gain weight slowly. In addition, pups of many species begin to eat solid food later during the nursing period and are therefore weaned more gradually than phocid pups. Sometimes sea lion and fur seal pups even manage to continue suckling for one or more years past the normal weaning time. Some sea lion pups and walrus calves begin to feed on their own when they are a few months old. But among the subpolar fur seals with comparatively short lactations (the northern and Antarctic fur seals), pups do not begin to forage independently until after weaning. A sub-polar fur seal pup often weans itself, leaving the breeding grounds before its mother returns from foraging at sea. Because some otariid females nurse their pup until they give birth to a new pup, the phrase "Once a mother, always a mother" could certainly be applied to them. That is, a primiparous female (one having given birth for the first time) essentially lactates or produces milk continuously for the rest of her life.

MILK COMPOSITION

Fat and protein are present in otariid and walrus milk in much higher proportions than in the milk of land mammals, although the fat content is

FIG. 8.7. Harbor seal mother nursing pup. *Photo by Dan Costa.*

somewhat lower than that found in the milk of phocid seals. With the exception of the two high-latitude fur seals, the milk of most otariids typically consists of about 20–35 percent fat and 10–14 percent protein. The fat and protein content of northern fur seal and Antarctic fur seal milk, however, tends to be higher—about 40 percent fat and 14–18 percent protein (Costa et al. 1985; Costa and Gentry 1986) (see Table 16). Bonner (1968) analyzed one sample of Antarctic fur seal milk that contained a remarkable 22 percent protein (although this high percentage does not seem to characterize most samples).

Otariid mothers fast for no more than one to two days at a time, except for the period of several days immediately following birth. Otariid pups, however, must fast as long as seven days while their mothers are away on feeding trips. The duration of these trips and the consequent length of the pup's fast vary with each species. Gentry et al. (1986) suggest that the fat content in otariid milk may be related to the length of the mother's feeding trip and the amount of time she spends nursing her pup. That is, the longer her absence from her pup, the higher the milk fat content. For instance, Galápagos fur seals and South African fur seals provide milk of lower fat content (29 and 19 percent, respectively), but they nurse their pups at more frequent intervals than other otariids. The two high-latitude species—northern fur seals and Antarctic fur seals—produce milk with a higher fat content (about 40 percent), but females nurse their pups less frequently over a shorter, four-month, lactation, and may leave their pup for long periods while foraging at sea. Therefore the overall energy delivered to fur seal pups of various species could be similar, even though the milk-fat content and attendance patterns vary from species to species.

Trillmich and Lechner (1986) have shown that the correlation between the length of a female's feeding trip (or absence from her pup) and the fat content of her milk holds true for sea lions (*Zalophus*) as well. They suggest that the milk-fat content is modified to satisfy the pup's physiological requirements while fasting. They cite a dramatic example involving the two subspecies of California sea lion. In California, sea lions have a milk-fat content of 35 percent and their feeding trips last for about three days. Yet on the Galápagos where sea lion mothers nurse their young nearly every day, the milk fat content is only 17 percent.

The composition of otariid and walrus milk appears to vary less dramatically than that of phocid milk, as one would expect of a mammal that feeds throughout lactation. If anything, milk fat may decrease slightly as lactation progresses. Studying northern fur seals, Costa and Gentry (1986) found that although their milk composition does not change dramatically as in phocids, it does fluctuate in a predictable fashion throughout lactation. The percentage of milk fat is highest just after birth, when it is 46.3 percent; it declines to 39 percent about seven days later when the mother departs for sea. The fat content is also high (about 44 percent) when a female returns from her feeding trip; it then decreases throughout the two-day nursing period (to 37.5 percent). Protein content increases significantly during the period from birth to first departure for sea (from 11.7 percent to 15.5 percent) but then remains steady until the end of lactation. Costa and Gentry (1986) suggest that the temporary decline in fat content during the two-day nursing period indicates that the female's fat reserves are depleted. That is, fur seals may be limited by their available fat stores, although unlike phocids, they have sufficient water during lactation.

Trillmich and Lechner (1986) discovered a fascinating lactation pattern among Galápagos fur seals: during the mother's fast immediately after birth, the fat content of her milk is higher than it is later during lactation when she begins to forage at sea. Milk fat during the postbirth fast (eight days) was about 32 percent; on or after the twelfth day following birth, it fell to 25 percent. During the fasting period mothers conserve not only body water, by producing richer milk, but also bone, by producing milk with less calcium. The high-fat milk also enables the newborns to accumulate a blubber layer before their mothers leave for sea. Later in lactation the high water and calcium content of the milk helps the pup contend with high temperatures on land and allows for faster bone growth.

OTARIID MOTHERING SYSTEMS

All otariids breed in temperate or tropical regions of the world. Because sea lions and fur seals do not have to contend with the problems associated with raising young on unstable ice in a frigid climate, as pagophilic phocids must, there is little environmental pressure to encourage a short, intensive lactation. The otariid strategy of feeding rather than fasting throughout the period of pup dependency also allows females to nurse their pup for a relatively long period—several months or more (see Table 11).

Maternal care systems, like otariid mating systems, are relatively similar among most species of fur seals and sea lions, although there are slight differences in attendance patterns from species to species. Female otariids give birth soon after arriving at the breeding rookery (usually within the first two days and always within a week) and remain with their pup continuously for one to two weeks before leaving to feed at sea. The perinatal attendance period in most fur seals is about one week. Just before the mother departs for sea, she comes into estrus and mates.

Females fast only during the first few days of lactation and when they return to shore to nurse their pup. When a mother is ashore, she nurses her pup several times each day. In many species the duration of a female's foraging trips to sea gradually increases during the lactation period, and the time she spends onshore nursing her pup gradually declines (although a female's time onshore is consistently two days in the northern fur seal). Between birth and the first feeding trip to sea a female otariid spends her longest continuous period with her pup. Thereafter, she remains onshore nursing it only about one to two days between feeding trips. Female attendance behavior in otariids—the cyclic pattern of onshore nursing and feeding at sea—has been reviewed by Gentry and Kooyman (1986), Croxall and Gentry (1987) and McCann (1987a).

Otariid pups gain weight at a rate of between 45 grams per day, in the New Zealand fur seal (Mattlin 1981, 1987), to 76–90 kilograms per day, in the Antarctic fur seal (Doidge et al. 1984). Not only do otariid pups gain weight slowly relative to phocids, but otariids also tend to weigh less both at birth and at weaning. Among the otariids (as well as the polygynous phocids) male pups tend to weigh more at birth and at weaning than female pups. The energy intake of California sea lion pups has been measured by Oftedal et al. (1985, 1987b). They found that during the first month of life, a pup's intake of energy averaged 2,296 kilocalories per day, which is about 60 percent higher than that estimated for neonates of terrestrial mammals. After daily maintenance requirements have been met, approximately one-third of the energy ingested is deposited in body tissue.

The reproductive energetics and maternal investment strategy of the Antarctic fur seal (Fig. 8.8) have been particularly well documented among the otariids (Costa et al. 1985; Bonner 1968). The female Antarctic fur seal nurses her pup for about four months, one of the shortest otariid nursing periods. After giving birth, the female remains with her pup for seven to eight days, nursing it every six hours. After each of her seventeen or so feeding trips to sea, lasting four to nine days each, the mother returns for about two days to nurse her pup. Female Antarctic fur seal pups increase their weight from 5.7 kilograms at birth to 14.5 kilograms at weaning, a rate of weight gain equivalent to 76 grams per day. Male pups gain weight at a rate of 90 grams per day, growing from 6.7 kilograms at birth to 17.1 kilograms at weaning (Doidge et al. 1984). Since intervals of four to nine days elapse between feedings, pups gain weight sporadically, unlike phocid pups, which gain weight steadily and rapidly. Weight gain becomes steadier if otariid pups begin to supplement their milk diet with solid food, but Antarctic fur seal pups do not eat solid food until after weaning, which takes place when the pup abandons the beach

FIG. 8.8. Antarctic fur seal mother and large pup. *Photo by Dan Costa.*

while its mother is away at sea on a feeding trip.

Pup dependency periods in otariids are generally long, often lasting a year, sometimes longer. Sea lions usually nurse their pups for six to twelve months, although sometimes females nurse yearlings or two-year-olds. Pup dependency periods in fur seals seem to increase with decreasing latitude. The two subpolar fur seals have the shortest lactation periods (four months), followed by the temperate species (eight to twelve months). Galápagos fur seals inhabiting equatorial areas have the longest nursing periods, twelve to thirty-six months. The length of lactation may vary considerably even within the same species, however. Southern sea lions and California sea lions may nurse their pups for as little as six months or up to one year or longer (see Table 11). This variation may be related to food resources, the female's age and health, the pup's sex, and the birth of a new pup.

Some otariid mothers may continue to nurse and care for their offspring well past the typical weaning time—until the youngster reaches two, three, or even four to five years of age. Observers have recorded the nursing of yearlings in ten of the fifteen otariid species, including all five sea lions (although the behavior is rare in southern sea lions) as well as four fur seals: Galápagos fur seals, South American fur seals, South African fur seals (*Arctocephalus pusillus pusillus*), and Australian fur seals (*A.p. doriferus*; Fig. 8.9) (Bonner 1984). In all four fur seals as well as in

TABLE 18

Otariid Females That Care for Juveniles or More Than One Pup

	PUP & YEARLING	TWO JUVENILES	YEARLING	TWO-YEAR-OLD	THREE-YEAR-OLD	FOUR-YEAR-OLD OR OLDER	REFERENCES
South American fur seal	X	X*	X†	X	X*	X*	Trillmich & Majluf 1981; Trillmich et al. 1986
Galápagos fur seal‡	X	X§	X	X	X	X‖	Trillmich 1984, 1986a
South African fur seal			X#				Rand 1959; David & Rand 1986
Australian fur seal			X	X	X		Rand 1955; Stirling & Warneke 1971
New Zealand sea lion	X		X	X			Marlow 1975
Australian sea lion	X	X**	X	X			Stirling 1972; Marlow 1975; Walker & Ling 1981a
Southern sea lion			X	X			Vaz-Ferreira 1981
Steller sea lion							
Calif.	X		2%				Gentry 1970; Sandegren 1970; Pitcher & Calkins 1981; Schusterman 1981b
Alaska	X	X††	80%	X	X		
California sea lion	X		X			X‡‡	Peterson & Bartholomew 1967; Odell 1972; Boness et al. 1985
Galápagos sea lion	X		X	X			Eibl-Eibesfeldt 1984b; Trillmich 1986b
Walrus			X	X	X		Fay 1982

NOTE: Nursing of juveniles 2 years or older is relatively uncommon except among Galápagos fur seals. But in many species a mother continues to nurse her offspring for an extended period if she does not give birth to a new pup or if her newborn dies.

* One female nursed a two-year-old male and a three- or four-year-old male simultaneously.   § Female nursed yearling and three-year-old.
† South American fur seals are often seen nursing pups a year old or older in Peru.   ‖ One four- to five-year-old juvenile still nursed.
‡ Galápagos fur seal pups commonly nursed for 2–3 years.   # One pup was known to be still suckling at 21 months.
** Female nursed one yearling and one 1.5-year-old juvenile.   †† Two juveniles ages 1–3 years nursed.
‡‡ One adult female nursed another adult female who was simultaneously nursing a juvenile.

FIG. 8.9. Australian fur seal. *Photo by Dan Costa.*

Steller sea lions and New Zealand sea lions (Table 18), juveniles older than yearlings are routinely or occasionally nursed.

Among Galápagos fur seals it is in fact normal for females to care for their young for two to three years. Although pups begin to forage occasionally as yearlings, their mothers continue to provide milk, spending 70 percent of their time onshore nursing the yearlings. The pups' demand for milk even increases substantially during their first year. The percentage of time that females spend nursing declines to 40–50 percent only when pups reach two years of age, although milk is still an important part of the diet for two-year-olds. The prolonged period of maternal care may contribute significantly to the pups' survival by reducing the time they spend foraging in the water, where shark predation is an important cause of mortality in young seals (Trillmich 1984). Only about 15 percent of the females accompanied by a yearling or two-year-old give birth (compared to a 70 percent natality rate among pupless females), although all females come into estrus and copulate after parturition. Only the largest and probably the oldest mothers give birth to a new pup while still caring for an older offspring. But if a female with a yearling does bear another pup, the newborn almost always dies of starvation, losing out in the competition for milk to its stronger and more aggressive sibling. If the newborn's sibling is a two-year-old, its chances of surviving are improved to 50 percent—still not very high. Apparently, there is not enough milk for two pups, and the older is better able to monopolize its mother's milk resources, even though the mother usually protects the newborn against its older sibling (Fig. 8.10). The birth of the younger pup may act as "insurance" in case the older offspring dies, so that the mother does not lose an entire year before reproducing again (Trillmich 1984).

Trillmich (1987c) suggests that the extended pup dependency period of Galápagos fur seals may be related to the six-month warm season that occurs from December to June. Every year at this time, the currents shift so that less cold water enters the Galápagos archipelago, and local upwelling and food resources for the fur seals decline—almost like an annual mini El Niño. A mother therefore must spend more time away from her pup searching for food at sea. Pups five to six months old grow little during this time and could not survive on their own if weaned. In addition, the stress of foraging under difficult conditions while nursing one or two pups may mean that the female is not able to sustain a developing fetus, so she often skips a season in giving birth. A mother still caring for her eighteen-month-old juvenile can skip a second season as well. If the female does produce a new pup, the yearling will probably outcompete it. An El Niño year makes the situation worse, whereas in an especially cool year females may be able to wean their pups as yearlings

FIG. 8.10. Galápagos fur seal mother with yearling. *Photo by Phil Thorson.*

before their next pup is born. The females' flexible reproductive cycle therefore helps them to cope with an unpredictable environment.

The Steller sea lion in Alaska may exhibit some of the most unusual nursing behavior: one large adult female was observed nursing a smaller adult female, who in turn was nursing a juvenile herself—all at the same time (Pitcher and Calkins 1981). Although this type of nursing triad is not typical of Steller sea lions, females often care for their pups as yearlings, those of the northern population in Alaska caring for their pups longer than those of the southern population at Año Nuevo Island in California. At Año Nuevo Island, only 2 percent of the females nurse their pups as yearlings, whereas in Alaska, 80 percent of the mothers care for yearlings or two-year-olds (Fig. 8.11) (Pitcher and Calkins 1981).

WALRUSES

Maternal care in walruses is unique among the pinnipeds with respect to the duration and intensity of the mother-pup bond. Although walruses breed on pack ice, as do the Arctic phocids with short, intensive lactations, the odobenid lactation strategy is more like that of the otariids than the phocids. The relatively unstable pack-ice breeding environment may not affect the walrus lactation strategy as it does that of Arctic phocids, since walrus calves are precocial and even very young calves are able to

FIG. 8.11. Steller sea lion mother with pup. *Photo by Bob Gisiner.*

swim in the water with their mothers if the ice breaks up. Moreover, other adult males and females aid young calves in the water.

Female walruses care for their calf for two years or more — a long nursing period not only among pinnipeds but in any mammal. If they do not give birth to a new pup, some walrus mothers continue to nurse until their young are three years of age. The interval between successive births increases as the female ages: younger females may produce a calf every two years; older females may give birth only every three or more years (Fay 1981b). Calves begin to forage on invertebrates with their mothers at about six months of age and are usually able to feed proficiently on their own by two years. Many researchers and other observers have commented on the striking solicitude shown to the calf by its mother, who constantly interacts with her offspring and fiercely protects it from danger. Fay (1982) observed one cow approach to within 2 meters of hunters to retrieve her calf, an act that unfortunately resulted in the mother's death. When danger threatens, mothers usually push calves into the water before jumping in themselves (Burns 1965), and other adults often behave protectively toward the young. Distressed and barking calves seem to attract older animals who "herd" the alarmed calves into the water. And orphaned calves are sometimes rescued and cared for by adult or subadult males as well as by females (Fig. 8.12) (Fay 1982).

The Mother-Pup Bond

Many pinniped mothers that bear and raise their pups in crowded breeding rookeries face two substantial problems: how to tell their pup apart from so many others and how to keep their pup with them in a densely packed and often chaotic breeding colony. A sea lion or fur seal female must relocate her pup among similar-looking young every time she returns from a feeding trip. The task of finding her pup is considerably easier because a female always returns to the same site, and her pup remains close to the area where it was last suckled. When she returns, the mother at least knows where to begin searching for her hungry pup. In a South American fur seal colony of over two thousand seals, mother-pup pairs show a distinct preference for a particular site, which encompasses a relatively small area (Pejoves 1985). Northern fur seals exhibit the same preference (R. Gentry, pers. comm.).

A mother finds her pup by a combination of vocal, chemoreceptive (olfactory and gustatory), and, to a lesser extent, visual cues. The bond that allows a mother to recognize her pup is established immediately after birth. During this critical period, much nuzzling and vocalizing take place as mother and pup "imprint" on one another. Female otariids just coming

Fig. 8.12. Walrus calf riding on its mother's back in the water. *Photo by Kathy Frost.*

ashore after a foraging trip generally give a characteristic warbling call (often termed a pup-attraction call). Generally, any pup in the vicinity will rush toward the vocalizing female, but she responds only to her own pup's call. After mother and pup are reunited, there is a great deal of mutual nuzzling and touching as the mother confirms the pup's identity by smell. Visual recognition seems less important than smell and vocal cues in many otariids. For instance, if a northern fur seal mother encounters her pup and it does not vocalize, she actually bites or tosses her own offspring as she would any strange pup that ventured too close (Gentry 1980). Vocalization and chemoreception also play an important role in mother-pup recognition among Galápagos sea lions (Figs. 8.13, 8.14) and fur seals (Trillmich 1984; Eibl-Eibesfeldt 1984b) and New Zealand sea lions. A vocalizing New Zealand sea lion mother and her pup can "home in" on each other when they are separated by distances of over 100 meters (Walker and Ling 1981b).

Most phocid seals do not leave their pup for extended periods while feeding at sea, but even keeping a pup safely beside her can be a difficult task for a mother in a crowded rookery. Land-breeding phocids, such as elephant seals and grey seals, often raise their pups in especially dense aggregations. Separations are common in the more crowded elephant

Fig. 8.13. Galápagos sea lion mother nuzzling her pup. *Photo by W. E. Townsend, Jr.*

seal and grey seal colonies, where a pup's life is constantly at risk. Mother-pup separations in northern elephant seals are frequently caused by storms and high surf that pounds the breeding beach, sometimes washing away scores of mothers and pups during high tide (Fig. 8.15). Young elephant seal pups that are separated from their mothers are often crushed by huge battling bulls that charge violently through the rookery, oblivious to other animals in their path. Many adult females aggressively attack and bite pups that happen to wander away from their mothers. Hungry orphaned pups and weaners constantly prowl the harem stealing milk from unsuspecting females, who must distinguish the thief from their own pup. As many as 24–57 percent of the pups born at the crowded Año Nuevo rookery were separated from their mothers during the breeding seasons from 1977 to 1980 (Riedman and Le Boeuf 1982). Well-developed mother-pup recognition abilities can therefore save a pup's life.

During the 1983 El Niño storms that happened to hit crowded Ano Nuevo Island at the peak of the northern elephant seal pupping season, however, even mothers and pups were separated that in normal conditions would probably have stayed together. Three consecutive days of high tides and devastatingly large surf at the end of January not only demolished the island's catwalks, piers, and retaining walls, all constructed in the 1930s, but contributed to an astonishing 70 percent mortality among the 975 pups born in 1983 (Le Boeuf and Condit 1983).

In phocid seals as in otariids, acoustic, olfactory, and visual cues all appear important in maintaining and restoring the mother-pup bond. Many reunions of northern elephant seal mothers and pups have been observed that involve active visual seaching on the part of the mother, with both mother and pup vocalizing to each other. The mother emits a characteristic pup-attraction call during these searches. After finding her pup, she usually spends some time nuzzling and smelling it. Sometimes a female northern elephant seal even seems to entice her newly found or wandering pup to remain with her by rolling over into a nursing position and initiating a nursing bout. Although most mother-pup reunions take place soon after the separation, before the pup has had a chance to travel far, one elephant seal mother found her pup on the other side of the breeding rookery after a separation of four days—an amazing accomplishment considering the hundreds of seal bodies and aggressive mothers that separated the pair (Riedman and Le Boeuf 1982). A similar sequence of vocalization by both mother and pup, followed by final identification of the pup by smell, has also been observed in many other phocids, including southern elephant seals, grey seals, harp seals, and harbor seals. Unlike many other phocids, harp seal mothers remain with their pup continuously only during the first two days following birth, after which they spend much of their time in the water. Still, females keep close track of their pup, hauling out periodically to nurse and even checking pups from ice holes or leads (Kovacs 1987).

What Makes a Good Seal Mother?

Whether a particular seal pup grows fatter than others or survives at all depends on the quality of maternal care it receives. A pup's chances of survival both before and after weaning are also related to its sex and, in some species, the environment in which it is raised. Uncrowded, sheltered rookeries are easier to grow up in than high-density colonies in areas exposed to storms. Each of the variables that influence growth and survival of young is discussed in the sections that follow, with an emphasis on the better-studied pinniped species in which variation in maternal care and pup survival has been examined in detail.

OLDER MOTHERS MAY BE BETTER MOTHERS

Why are some females better mothers than others? In at least some species of pinnipeds, a female's ability to mother depends on her age, size, previous mothering experience, and social status, all of which are intimately related. Older females are generally larger as well as socially dominant over younger females. The relationship between increasing age and larger body size holds true for several pinnipeds, including largha seals, ringed seals, ribbon seals, bearded seals, northern elephant seals, and southern elephant seals (Tikhomirov 1968; Reiter et al. 1981). The older multiparous females (those having given birth more than once) have more mothering experience than younger primiparous mothers (those having given birth for the first time). The relationship between mothering ability and these factors may hold for all or most pinnipeds, but it has been extensively documented only in the better studied phocids, such as northern elephant seals and grey seals. In fact, it is a general rule in mammals that large multiparous females are usually more successful mothers than smaller primiparous females. That is, bigger mothers are better mothers. This is true of many land mammals, including numerous primate species, ungulates, and rodents (Ralls 1976; Riedman 1982).

In northern elephant seals, older females (six or more years of age) are larger and better mothers than younger females. More older females than younger ones successfully wean their pups. Reiter et al. (1981) found that

FIG. 8.14. Galápagos sea lion pup. *Photo by W. E. Townsend, Jr.*

young three- to five-year-old females weaned only 38 percent of their pups, whereas females six years or older weaned 73 percent of their pups. Pups of older mothers were also healthier and fatter than those raised by young mothers. The larger female elephant seals have greater body reserves, allowing them to remain longer at the breeding rookery and to provide more fat-rich milk to their pup. The number of days a female nursed her pup increased with her age. Young mothers three years of age nursed their pups for an average of only 24 days; nine-year-old mothers nursed their pup for an average of 29 days. A pup's weight at weaning also increased with the age of its mother. In addition, it is possible that older females may produce milk higher in protein content than that of younger females, although the number of females sampled whose age was known was small (Riedman and Ortiz 1979).

Older and larger female elephant seals also seem to be socially dominant over younger mothers. High social status is definitely a plus for mothers breeding in a crowded harem, since a big and aggressive mother can secure the best sites to give birth and raise her pup. She can also defend her pup better against other aggressive mothers in the rookery. Furthermore, older mothers arrive and give birth a little earlier in the season, so they can find the safest, most sheltered spot in the center of the

FIG. 8.15. Dense Año Nuevo breeding colony during a storm when the tide is high, showing mothers and pups awash in the surf. *Photo by author.*

harem (Reiter et al. 1981). A similar pattern appears to characterize Antarctic fur seals. Doidge et al. (1986) found that older females tended to give birth earlier in the season and therefore had a longer lactation period and presumably weaned heavier pups than younger females.

Finally, older females have simply had more mothering experience than younger females. Mothering skills are important in successfully rearing young not only among northern elephant seals but also in many species of primates and birds and possibly other pinnipeds (Riedman 1982; Riedman and Le Boeuf 1982).

How can a young elephant seal mother compete with the older females and successfully wean a pup in a high-density breeding area? She can wait until she is older and larger to give birth for the first time. Or she can move to a less crowded area to raise her pup. New breeding areas, such as the Año Nuevo mainland and Farallon Islands, are in fact colonized mainly by young females who emigrate from densely packed areas.

WHO RECEIVES MORE CARE — FEMALE OR MALE?

According to recently developed sex-ratio theories, a parent should vary the amount of maternal investment in relation to its offspring's sex, investing more in the sex with the highest variance in reproductive success. The assumption behind these theories is that parental investment influences the offspring's body size, health, and breeding success later in life (Trivers 1972, 1985; Trivers and Willard 1973; Maynard Smith 1980; Charnov 1982). In highly polygynous species, males show the highest variance: most produce few or no offspring, while a few may sire hundreds. Among polygynous and sexually dimorphic mammals, a female theoretically should produce a male offspring when she is in optimal breeding condition—that is, healthy and big enough to give her male offspring an edge in size and survival as well as in reproductive success after weaning. If this superior male offspring then goes on to become an alpha or high-ranking male, his mother's ultimate breeding success increases dramatically. In contrast, mothers in poor physical condition or of subordinate social status may be better off producing female offspring, since these female young will probably grow up to produce grandchildren, slowly but steadily, regardless of size or dominance. In other words, in polygynous societies, female mammals rarely have trouble finding a mate, whereas most males have a great deal of trouble securing access to receptive females.

These predictions of differential partitioning of investment according to the sex of offspring are borne out by studies of some but not all vertebrates. Do they hold true for pinnipeds? The answer seems to be a conditional yes for most of the polygynous species that have been studied in detail. Maternal investment is often unequal in the species studied, with male pups receiving a slightly greater share of mothering as measured in milk intake (see, e.g., Kovacs and Lavigne 1986b). Exceptions occur among southern elephant seals, Steller sea lions, and California sea lions.

In most polygynous and sexually dimorphic pinnipeds, male pups weigh more at birth and at weaning and take in more milk than female pups. They also grow faster than females during the lactation period (Kovacs and Lavigne 1986b). This difference between male and female young is especially pronounced among the highly polygynous species, where it pays to wean the largest male possible. In northern elephant seals and grey seals, mothers invest more energy in raising males than females, as measured in milk transfer. According to Anderson and Fedak (1985), grey seal mothers invest 10 percent more energy in raising male pups than female pups, with 95 percent of the investment expense occurring postpartum. Although their sample of mother-pup pairs was small, they found that more males are born to large mothers (13 males to 9 females) than to small mothers (9 males to 14 females). Mothers of male pups weigh

8 kilograms more than mothers of female pups, and the male pups themselves gain weight (1.9 kilograms per day) more rapidly than female pups (1.6 kilograms per day). In addition, males are born five days earlier than female pups. Anderson and Fedak (1985) conclude that adult females above average in size are better equipped to produce a male offspring without depleting their body stores to a level that might affect their future ability to reproduce. Of course, to test whether a female produces fewer of the more "expensive" sex — males — she would have to be followed throughout her lifetime, after which her investment in either sex could be weighed against her entire reproductive output.

Northern elephant seal male pups are nursed a full day longer than female pups. Some of the recently weaned elephant seal pups try to prolong their nursing period after their mother has gone by stealing milk from other lactating mothers. A few lucky weaners even manage to be adopted by females that have lost their own pups. More of the milk stealers are males, who are more persistent and successful at this risky behavior than females (Reiter et al. 1978). Although it can be dangerous for a weaner to venture into a crowded harem full of aggressive females, for some males the benefits of gaining weight by stealing milk apparently outweigh the risk of injury. Because the eruption of canine teeth and the molting of the black natal fur occur later in male than in female pups, male weaners may look and feel more puplike when they sneak a suckle. Most superweaners, or weaned pups adopted by a foster mother, are males.

Perhaps surprisingly, Fedak and McCann (1987) found no difference between the maternal investment in male and female offspring among southern elephant seals. Although male pups weighed more at birth, both male and female pups grew at the same rate (3.6 kilograms per day), were nursed for an equal period (twenty-two days), and were weaned at similar weights relative to birthweight (119 kilograms for males; 112 kilograms for females). Mothers lost 7.8–7.9 kilograms per day throughout lactation irrespective of their pup's sex.

Among phocids that are not highly polygynous, the difference in maternal investment for males and females is less pronounced than in polygynous species, as one would predict. At least this appears to be the case among the monomorphic harp seals. For instance, male and female harp seals are the same size at birth, grow at similar rates, and are close to the same weight at weaning (Kovacs 1987). Among serially monogamous phocids that show reverse sexual dimorphism, however, female pups tend to grow more rapidly and are larger at weaning than male pups (Kovacs and Lavigne 1986b).

Among the polygynous otariids, maternal investment has been studied in the northern fur seal (Costa and Gentry 1986), Galápagos fur seal (Trillmich 1986c), New Zealand fur seal (Mattlin 1981), Antarctic fur seal (Doidge et al. 1984), Steller sea lion (Higgins et al. 1988), and California sea lion (Oftedal et al. 1987; Ono et al. 1985; Francis and Heath 1985; Boness et al. 1985). In all these species, male pups weighed more at birth, grew at a faster rate, or ingested more milk than female pups. Northern fur seal mothers, for instance, invested more energy in male pups, whose milk intake was 61 percent higher than that of female pups. Galápagos fur seal mothers invested more in male offspring in suckling time and quantity of milk provided. Male pups were heavier at birth and grew more rapidly than females. Yearling and two-year-old males (which supplemented their diet with independent foraging) were also nursed more frequently than female offspring of the same age, although there were no differences in attendance patterns. The quantity of milk a pup received was directly related to the amount of time it spent suckling. Weaning for both sexes was estimated to take place at two years of age, although more male than female three-year-olds were observed suckling.

Yet sea lion mothers do not always seem to invest more in male offspring. Among Steller sea lions studied at Año Nuevo Island, the sex of the pup had no significant effect on maternal attendance patterns or suckling behavior. Although male pups spent more time suckling, their mothers were also absent for longer periods than the mothers of female pups. Among California sea lions breeding on the Channel Islands, Francis and Heath (1985) found that maternal care continues longer for female than for male offspring. Most of the yearlings that suckled past one year of age were female, whereas most of the juveniles sighted away from their natal island were male. Francis and Heath (1985) explain what appears to be an exception to the sex-ratio theory in a highly polygynous species by pointing out that female juveniles may derive greater benefits by remaining with their mothers near their natal site, including the opportunity to learn the skills of mothering and to become familiar with the local distribution of food resources upon which adult females depend more heavily. Perhaps territorial bulls also begin to exclude older juvenile males from the breeding area.

Early in the California sea lion's lactation period, however, Boness et al. (1985) found no differences in the attendance behavior of females with respect to the sex of their offspring. That is, females did not nurse pups of one sex longer than pups of the other sex during their time onshore. Moreover, they did not find a yearling sex-ratio bias toward females during the 1985 breeding season on San Nicolas Island. Over a three-week period, the estimated milk intake was higher for male pups (an average of 723 ± 31.0 grams per day) than in female pups (an average of 609 ± 24.0 grams per day), although for male and female pups there was no difference in milk intake as a percentage of body weight (Oftedal et al. 1987b).

Environmental and Physical Factors

POPULATION DENSITY AND TOPOGRAPHICAL FEATURES OF THE BREEDING AREA

As I have mentioned, among phocid pinnipeds that breed in colonies, a pup's chances of surviving are related to the density of females as well as the topography of the breeding area. The more crowded a breeding colony is, the greater the probability that a pup will be separated from its mother and die. A mother is also more likely to lose her pup if she raises it in an area that is especially exposed to surf and high tides. This is true of northern elephant seals (Riedman and Le Boeuf 1982; Le Boeuf and Condit 1983) and grey seals (Bonner and Hickling 1971). Human activity that disturbs grey seals on crowded beaches also causes mother-pup separations and pup mortality (Fogden 1968, 1971; Smith 1968).

OCEANOGRAPHIC AND CLIMATIC CONDITIONS AND THE AVAILABILITY OF FOOD RESOURCES

In many pinnipeds, breeding females as a group may change their maternal investment strategies from year to year in response to a changing environment. The availability of food resources in particular may determine how much a female can afford to invest in her pup. In general, a decline in the availability of food, whatever the environmental cause, causes a reduction in pinniped reproductive rates. Such decreases in relation to reduced food resources have been documented or are thought to occur in Weddell seals (Testa 1985) as well as in the species affected by El Niño conditions: northern fur seals, South American fur seals, southern sea lions, Galápagos sea lions, Galápagos fur seals, and possibly harbor seals and northern elephant seals in California.

The 1982–83 El Niño Southern Oscillation (ENSO) event, which occurred throughout the eastern Pacific Ocean, affected reproduction among many pinniped species. During this unusual oceanographic event—the strongest recorded in the twentieth century—anomalously warm water temperatures prevailed in areas inhabited by many pinnipeds, causing reductions in the abundance of local food resources and changes in the availability of various species of prey. In turn, these changes in food availability adversely and sometimes profoundly affected a female's reproductive success, the relationship between mother and pup, and population dynamics among several species of pinnipeds that breed along the western coasts of North and South America. The impact on sea lions and fur seals was especially great since these species continue to forage during the long period of pup dependency. In addition, pinnipeds breeding close to the equator were affected more than pinnipeds inhabiting higher latitudes. As I noted in chapter 3, the effect of El Niño is graded poleward from the tropics so that it is strongest near the equator. The diving and attendance behavior of the northern fur seal in the Bering Sea, for example, was unaffected by the 1982–83 El Niño (Gentry, pers. comm.).

The El Niño event caused a decline of about 60 percent from the 1982 level in pup production among northern fur seals breeding on the Channel Islands in California. Pup production remained suppressed throughout 1984. In addition, fur seal mothers spent significantly longer periods of time feeding at sea, and pup growth substantially decreased (DeLong and Antonelis 1985). California sea lions in the Channel Islands experienced similar effects during 1983 and 1984. Ono et al. (1985) found that the decline in food resources for lactating females appeared to result in a slower growth rate for pups and a higher pup mortality during the first two months of life. During the El Niño year, mothers not only left their new pups earlier to feed at sea but also remained at sea feeding longer than during the year preceding El Niño. Feeding trips in the year following El Niño were intermediate in duration in comparison to those of 1982 and 1983. Females spent less time nursing their pups, and the milk intake of pups was also substantially lower in 1983 and 1984 than in 1982 (Ono et al. 1985; Boness et al. 1985; Oftedal et al. 1987b).

Southern otariids in the eastern Pacific were also profoundly affected by the ENSO event. Although El Niño events occur irregularly every few years in the eastern tropical Pacific, the El Niño that took place between 1982 and 1983 in this area was exceptionally strong. Among southern sea lions and South American fur seals in Peru, the mortality of pups as well as adults was especially high during 1982 and 1983 (Limberger et al. 1983). Observations of South American fur seals in Peru indicate that females were unable to forage efficiently during 1983, when their foraging dives were longer and deeper than usual and they spent more time feeding at sea. Mothers suffered from severe malnutrition, and many pups and yearlings starved to death (Majluf et al. 1985; Trillmich et al. 1986).

The dramatic effects of the 1982–83 El Niño on Galápagos otariids were documented by Limberger et al. (1983) and Trillmich and Limberger (1985). Some populations of Galápagos fur seals and Galápagos sea lions suffered nearly complete reproductive failure. Among the Galápagos fur seals, almost all the territorial males apparently died during the El Niño year, probably of starvation. During the 1983 breeding season following El Niño, there were no large adult males holding territories. The smaller males (five of which were able to establish territories), however, seemed no more subject to mortality than females during El Niño. Perhaps even more damaging than the loss of territorial males in the Galápagos fur seal population was the loss of most members of the four youngest age classes (those born in 1980–83), and the death of about 30 percent of the adult females and nonterritorial males. From 1979 to 1981, about 80 percent of marked nonlactating females gave birth; in 1983, immediately following El Niño, pup production was only 11 percent of average. Pups produced during and after El Niño also weighed less at birth than pups produced before. In addition, during El Niño, because mothers stayed at sea longer on feeding trips, pups were undernourished.

Among the Galápagos sea lions, almost all the pups born during the El Niño year died. Many were abandoned by their mothers and starved to death. Following El Niño, yearling sea lions were rarely seen on several of the Galápagos Islands. The yearling age class was probably reduced to 5–20 percent of its typical size. During the breeding season following El Niño, pup production fell to less than 30 percent of normal, the decline varying from colony to colony.

The foraging efficiency and reproductive rates of phocid seals were also apparently affected by El Niño to some degree. The diet of harbor seals in the Channel Islands changed during the 1983–85 period in comparison to that of 1980–82. The overall number of species preyed on per seal (measured in the seals' scat) decreased, and the proportion of various prey eaten changed. For instance, harbor seals ate more octopus and less rockfish during the El Niño period (Stewart and Yochem 1985). Among northern elephant seals breeding on the Farallon Islands in northern California, a decline in the number of pups born during the 1984–85 breeding season may be attributable to the effects of El Niño, according to Huber (1985).

Adoption and Fostering Behavior

Fostering behavior is surprisingly common among pinnipeds, especially among the phocid seals. Incidents of such behavior and adoption have been reported in thirteen pinnipeds, 40 percent of the thirty-three species (Table 19). Some form of fostering behavior has often been observed in the walrus, Hawaiian monk seal, northern elephant seal, southern elephant seal, grey seal, Weddell seal, Australian sea lion, and Antarctic fur seal (although fostering incidents in the Antarctic fur seal occur mainly at high-density beaches). It has been reported but is probably uncommon in the harbor seal, largha seal, northern fur seal, South African fur seal, Guadalupe fur seal, California sea lion, and Steller sea lion.

Although the majority of female seals, with the possible exception of the Hawaiian monk seal, care only for their own young, enough females nurse and protect a pup other than their own to make us wonder about the reasons for this behavior. Many kinds of fostering are, in fact, widespread among numerous taxonomically and ecologically diverse species of mammals and birds (Riedman 1982). Parental care of nonfilial young is especially interesting in that it represents a form of altruism, in which the foster parent assists others at its own expense (Hamilton 1964). A female that provides parental care for young that are not her own is called an allomother (E. O. Wilson 1975). Allomaternal care and adoption would seem to be reproductively expensive forms of behavior. But are they? What benefits might accrue to the foster mother? Individuals that care for another's young may in fact derive selective advantages associated with caring for genetically related young (called increased inclusive fitness),

TABLE 19

Allomaternal Care, Adoption, and Fostering Behavior among Noncaptive Pinnipeds

PHOCIDS	OTARIIDS	ODOBENIDS
Hawaiian monk seal	Northern fur seal*	Walrus
Northern elephant seal	South African fur seal*	
Southern elephant seal	Guadalupe fur seal*	
Grey seal	California sea lion*	
Weddell seal	Steller sea lion*	
Harbor seal*	Australian sea lion	
Largha seal*†	Antarctic fur seal	

NOTE: References are given in the text.

* Fostering behavior appears to be uncommon in these species.

† Although the adoption occurred under noncaptive conditions, it was induced experimentally.

parental experience, reciprocal altruism (usually involving babysitting), and the exploitation (mainly among primates) of the fostered young. Maternal errors involving cases of mistaken identity may be responsible for other cases of adoption. Environmental constraints, such as a scarcity of breeding resources or food resources that require cooperative foraging strategies, also seem to have promoted the evolution of fostering behavior and parenting systems in which young are cared for communally (Riedman 1982).

Limited suitable breeding space on islands available to seals may have indirectly promoted allomaternal care and adoption in a number of pinniped species. A shortage of breeding sites often results in the formation of high-density breeding colonies where mother-pup separations frequently occur. The preponderance of orphaned pups and pupless females therefore set the stage for fostering behavior.

This behavior has been comprehensively studied in the northern elephant seal at the Año Nuevo Island and mainland rookeries (Reiter et al. 1981; Riedman and Le Boeuf 1982; Riedman 1983). Elephant seals engage in an interesting array of fostering behaviors, creating a virtual "soap opera" of female-pup exchanges (Table 20). The most common form of adoption occurs when a pupless female adopts a single orphaned pup. Less frequently, an allomother fosters a pup while it is still with its real mother. These lucky pups get twice as many nipples to suckle as other pups. Some females, especially those that have recently lost their own pup, even try to steal a pup away from its mother. And each year, a few females act like public "drinking fountains" for any orphans that solicit suckling. These "promiscuous mothers" nurse any pup and are often surrounded by several starving orphans at once. As I mentioned earlier in this chapter, some pupless females adopt a pup after it is weaned by its own mother, turning it into a superweaner (Fig. 8.16). These jumbo pups (called double-mother sucklers) are sometimes so stout that they have difficulty maneuvering and have been seen tobogganing and somersaulting helplessly down hills like oversized beach balls. The largest known superweaner (whom researchers named K-fat, after a Santa Cruz area radio station) weighed in at 525 pounds (240 kilograms), two to three times the size of the average weaned pup. It is conceivable that these enormous pups have a survival and, perhaps, reproductive advantage as adults due to their abundant body reserves, but this advantage has yet to be demonstrated.

For weighing, both super and regular weaners are secured inside a funnel-shaped weaner bag. The straitjacketed weaner is then attached to a scale and hoisted onto a sturdy metal tripod—a task that requires the

TABLE 20

Females Exhibiting Fostering Behavior on Año Nuevo Island
(The number of adoption cases is in parentheses)

FOSTERING BEHAVIOR	FREQUENCY			
	1977	*1978*	*1979*	*Total*
Point Harem				
Female loses her own pup and				
Adopts single pup	10	25	13	48
Adopts two pups	0	0	1 (2)	1 (2)
Adopts weaned pup	1	4	1	6
Adopts orphan with another female	1 (2)	1 (2)	0	2 (4)
Shares care of pup with its mother	5	5	1	11
Female nurses her own pup and				
Adopts single pup	3	4	6	13
Adopts two pups	1 (2)	0	0	1 (2)
SUBTOTALS	21 (23)	39 (40)	22 (23)	82 (86)
Cove Harem				
Female loses pup and				
Adopts single pup	3	0	0	3
Adopts two pups	0	1 (2)	0	1 (2)
Shares care of pup with its mother	3	0	0	3
Female keeps pup and				
Adopts single pup	1	1	0	2
SUBTOTALS	7 (7)	2 (3)	0*	9 (11)
Saddle Harem				
Female loses pup and				
Adopts weaned pup	0	1	0	1
TOTALS	28 (30)	42 (44)	22 (23)	92 (98)

* Area monitored infrequently.

efforts of at least four people. Even supposedly helpless weaners immobilized inside weaner bags, however, can be dangerous when swinging back and forth on a tripod, as weaner weigher Jim Estes discovered when one such weaner opened its mouth on a forward swing and latched onto Jim's leg, which happened to be in the way, giving him a nasty bruise.

A pupless female elephant seal that adopts or helps care for an alien pup may gain valuable maternal experience. This learning-to-mother hypothesis is supported by several lines of evidence. First, most of the females that adopt or help care for an alien pup have lost their own pup. On Año Nuevo Island 75 percent of the foster mothers had lost their own

Fɪɢ. 8.16. Superweaner northern elephant seal. *Photo by author.*

pup. On the less crowded Año Nuevo mainland rookery, 62 percent of the females that lost their pup adopted another. Second, most of the foster mothers are young and inexperienced. And third, mothering experience in itself enables a female to become a better mother.

The reproductive biology and behavior of all seals, including elephant seals, further points to the necessity of developing adequate mothering skills. Females give birth once a year to a single pup that requires a prolonged or energetically intensive period of dependency. The task of mothering can be complex and difficult, especially in some of the larger, more crowded breeding colonies. A female can produce only ten to twelve offspring at most in her lifetime. If a pup dies because of faulty maternal care, a female's lifetime reproductive success diminishes significantly.

Another potential, although untested, benefit to foster mothers that have lost their own pup is that regular nursing somehow induces or helps to induce ovulation. Females that adopt or nurse pups can therefore mate and give birth again the following year. Le Boeuf et al. (1972) found that most females in the small Año Nuevo Island Cove Harem that nursed a pup to weaning age copulated before departing for sea. In contrast, eight females that never or rarely nursed after losing their pups were not observed to copulate.

The proximate cause of some cases of adoption appears to be reproductive mistakes by the foster mother, or misplaced reproductive function, as G. C. Williams (1966) has called this misdirected maternal behavior. Adoption by a foster mother that keeps her own pup certainly involves a serious mistake in mothering since a female seal cannot raise more than one pup successfully. In attempting to do so, a female gives her own pup insufficient milk and inadequate care. Like other mammals, female elephant seals that have lost their young appear to be hormonally and behaviorally primed to provide parental care.

Often there may be a sensitive and critical period following birth in which a mother learns to recognize her young by chemoreception and vocalization, a process often referred to as imprinting (see, e.g., Lorenz 1935), which I discuss in chapter 9. Confusion about pup identity and mistaken bonding by females may be particularly likely to occur among mammals such as elephant seals and grey seals, which give birth seasonally with a relatively high degree of synchrony and rear their young under conditions of high female density. Most foster elephant seal mothers adopt soon after giving birth, and their foster charges are usually quite young (Fig. 8.17).

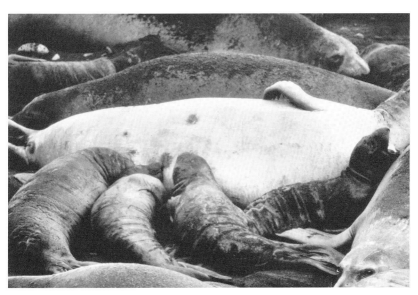

FIG. 8.17. Female northern elephant seal with orphans clustering around her trying to suckle. *Photo by Burney Le Boeuf.*

Fostering behavior is observed frequently not only in northern elephant seals but also in such other phocids as grey seals, southern elephant seals, Weddell seals, and especially Hawaiian monk seals. All these species breed in moderate to large pupping colonies rather than in isolated family units. Because mother-pup separations may take place frequently, there are many opportunities for adoptive behavior. Since the lifetime reproductive potential of all phocid seals is low, increased mothering experience may be a general benefit associated with fostering behavior in pinnipeds.

A wide range of fostering behavior similar to that observed in northern elephant seals occurs among female grey seals (see, e.g., Hewer and Backhouse 1960; Hickling 1962; Coulson and Hickling 1964; Fogden 1968, 1971; E. A. Smith 1968; Bonner 1975; Burton et al. 1975; S. S. Anderson 1979). Females with and without their own pups nurse alien pups and steal pups belonging to other females. Sometimes two females even share a pup. Boness (unpubl. data) has found that as many as 25–75 percent of female grey seals engage in some form of fostering in certain rookeries. Fostering behavior is more common in high-density areas where beach topography and human disturbances contribute to mother-pup separations. because females leave their pups and retreat into the water en masse when humans are present.

According to Carrick et al. (1962a), fostering behavior is relatively common in southern elephant seals, whose breeding biology and behavior are similar to those of the northern species. Many females are tolerant of alien pups, particularly toward the end of lactation, and it is not uncommon to observe females nursing alien pups or orphans. Cases of adoption in southern elephant seals have also been reported by Laws (1956a), Bryden (1968) and T. McCann (pers. comm.).

Fostering behavior has also been observed in Weddell seals, which breed in large well-spaced colonies on the Antarctic fast ice. According to Stirling (1975a), 7.8 percent of 179 females were observed nursing pups different from those they were originally recorded with, although the original filial relationship was not definitely known. Kaufman et al. (1975) found that maternal behavior toward alien pups was highly variable. Although most Weddell seal females appeared to reject strange pups, some females nursed orphaned young. Females that lost their own pup were observed adopting alien pups. Most of these adoptions, however, were induced by experimental manipulation.

Adoptive behavior is remarkably common among Hawaiian monk seal females, which nurse their pups for five to six weeks, fasting through-. out lactation (Kenyon and Rice 1959; Johnson and Johnson 1984). Most

FIG. 8.18. Hawaiian monk seal mother and pup. *Photo by Chip Deutsch.*

females give birth and rear their pups in close proximity to one another, although a moderate degree of spacing exists between mother-pup pairs. Surprisingly, the majority of the mothers engage in some form of fostering behavior, frequently switching pups with nearby females and weaning an alien pup (Johnson and Johnson 1978, 1984; Alcorn 1984; Boness unpubl. data). Such indiscriminate caretaking among monk seals is intriguing, and possible explanations for it are being investigated by D. Boness and others (Fig. 8.18).

Because harbor seal breeding systems vary from spatially isolated monogamous family groupings to large colonies comprising thousands of breeding females, a comparison of the occurrence and frequency of fostering behavior in several subspecies or populations of harbor seals would be especially interesting. At present, the frequency of fostering behavior in harbor seals and the closely related largha seal is not clear. Finch (1966) and Bishop (1967) observed adoptions in the moderately gregarious harbor seals (*Phoca vitulina richardsi*) in northern California and the Gulf of Alaska. However, Knudtson (1973) found that *P.v. richardsi* breeding in Humboldt Bay nursed only their own pups.

Although fostering behavior is probably uncommon among harbor seals and largha seals that breed alone or in small groups on glacial ice, A. Hoover (pers. comm.) observed two harbor seal (*P.v. richardsi*) mother-pup pairs on one ice floe and believes that the mothers may have nursed each other's pup on occasion. In addition, Burns et al. (1972) experimentally induced a noncaptive largha seal female of a monogamous family group on an ice floe to adopt a young pup.

Both allomaternal and allopaternal care of young have been documented in the walrus. In fact, only among walruses have males been observed to actively assist in the care and protection of the young. Orphaned walrus calves are protected and carried away from hunters by both male and female subadult and adult walruses (Bruce and Clarke 1899; Bel'kovich and Yablokov 1961; Fay 1982). Fay (1982) and Burns (unpubl. data) observed seven definite occurrences of temporary or permanent adoption by adult females. Other cases have been reported as well (see, e.g., Eley 1978b). Most of the adopted calves were relatively young, ranging from a few days to several weeks old. None of the foster mothers had a calf of her own at the time of adoption, although two females accompanied by calves that were several weeks old were known to have just given birth (Fay 1982). Although adopted calves have not been seen suckling the foster mothers, these calves appeared healthy and fat, suggesting that they were in fact nursing.

Fostering behavior in otariids, in contrast to phocids, is relatively rare, and females seem to have little difficulty locating and recognizing their own pup (Stirling 1975b). Still, isolated cases of adoption and fostering behavior, evidently caused by human interference or disturbance, have been observed in the following species: South African fur seals (Rand 1955), northern fur seals (Bartholomew 1959), Guadalupe fur seals (Riedman, unpubl. data), Steller sea lions (B. Gisiner, pers. comm.), California sea lions (Francis 1987; Boness, unpubl. data; Schusterman, unpubl. data), and Australian sea lions (Marlow 1972; Walker and Ling 1981b). Perhaps surprisingly, Antarctic fur seals are frequently observed nursing nonfilial pups at crowded breeding rookeries. Although the significance of such fostering behavior has not been investigated, pups seem to profit from the "illegal" suckling even if no long-term adoption takes place (McCann 1987b).

In otariids, fostering behavior often appears to be based on maternal mistakes. However, Walker and Ling (1981b) report the occurrence of a fascinating (although probably uncommon) form of "babysitting" among Australian sea lions at Seal Bay, which they termed nanny behavior. In this case, a particular mother appeared to be responsible for a group of pups, including her own, and she defended the entire group from researchers attempting to tag them. Eventually, she was relieved by another mother, who also protected the nursery group. Other forms of cooperative babysitting by females occur among a wide variety of mammals, including cetaceans, primates, ungulates, canids, lions, coatis, and bats. Such babysitting behavior appears to have evolved to provide increased foraging freedom for females with young (Riedman 1982). It should be pointed out, though, that during pup-tagging operations in other otariids, pups may gather around a female that "stands her ground" rather than fleeing from the taggers. Pups can see that researchers will not approach the female closely. But when pups are lassoed away from the female with a noose pull, she does not defend them. They are on their own (Fig. 8.19).

Differences between otariids and phocids in the frequency of adoption and allomothering behavior may relate to the divergent strategies of maternal care in each pinniped family: most phocids nurse their pups for a short, intensive period, whereas otariid females alternate between periods of feeding at sea and bouts of nursing on land over an extended period, four months to more than a year. Many phocids fast throughout lactation, while remaining continuously with their pup on land. In otariids, well-developed mother-pup recognition may relate to the female's constant need to relocate her pup after returning from feeding trips at sea.

FIG. 8.19. Researcher lassoing Antarctic fur seal pup. *Photo by Dan Costa.*

The occurrence or absence of fostering behavior in each group may also be associated with the timing of ovulation and mating. While most phocid seals copulate during late lactation or after weaning their pup, all otariids mate soon after parturition. An otariid female that has lost her pup does not have to remain at the breeding grounds after giving birth, as many phocids apparently must do, in order to mate. Consequently, any advantages associated with adopting a pup, such as increased mothering experience, must be measured against the virtual certainty that the pupless otariid female has already been inseminated, and therefore does not need to remain near the breeding area, where opportunities for fostering behavior exist, to insure next year's pregnancy.

9

Communication, Cognition, and Learning

Cognitive and Learning Abilities

The study of awareness, intelligence, and learning in seals and other animals is both exciting and challenging. To understand how another species "thinks" and perceives its world is perhaps the ultimate scientific and empathetic experience: it comes as close as we presently can to "seeing through the eyes" of another animal. What does a pinniped really feel? Humans have always wondered what it would be like to experience life as another species, whether a bird, cat, or seal. And since ancient times, we have been intrigued by the idea of communicating with animals. But the study of consciousness and cognition in other species is inherently difficult because we are limited by our view of our own world, by our thoughts and feelings, and by our attempts to measure the intelligence of another species by our own linguistic and cognitive skills. In his thought-provoking book *Animal Thinking*, Donald Griffin (1984) writes: "The prospect of using communication as a window on the feelings and thoughts of animals seems the most promising if only because it is so useful with our own companions. In one sense animals may already be using the window, as they succeed in conveying to one another their feelings and simple thoughts. If other animals can get these messages, cognitive ethologists with the advantage of the human brain should be able to do as well."

The pinnipeds tend to be intelligent and playful animals, although such attributes vary according to species. Studies on captive sea lions by

FIG. 9.1. Sea lion underwater performing a task for the navy (attaching line to a dummy missile). *Photo by Flip Nicklin.*

Ron Schusterman, Bob Gisiner, and co-workers at Hayward State University and UCSC's Long Marine Lab have shown that, like the dolphins and primates, sea lions are capable of mastering complex tasks. Sea lions even possess the conceptual abilities required to comprehend an artificial or simplified human language. As most people know, sea lions have long been exploited as show animals in zoos and aquariums because of their excellent trainability. The evolution of well-developed cognitive and learning abilities among pinnipeds would presumably enhance their ability to inhabit a complex or difficult environment. A seal must find and obtain prey that is hard to capture, engage in often complex social interactions during the breeding season, learn how to raise and wean a pup, disperse to new breeding areas when necessary, and so forth.

The U.S. Navy is also interested in how well sea lions can perform tasks for the military and has supported much of the research on cognition and learning in sea lions. For instance, California sea lions have been taught to help retrieve objects such as missiles by locating the object on the ocean bottom and clamping a line to it (Fig 9.1). Sea lions can recover such objects more easily and safely than human divers or submersibles (Conboy 1975).

At the Northwind Undersea Institute in New York City, harbor seals are even being trained so that eventually they can assist in a variety of police work. Instead of police dogs, there will be police seals. The harbor seals, which were raised as orphaned pups, are being taught to retrieve objects such as .38-caliber revolvers that are tossed into their tanks. Someday they might be able to pick up handguns thrown into New York waters. The institute's director, Michael Sandlofer, hopes that eventually the "Seal Search and Recovery Team" will be able to take underwater photographs (for such purposes as bridge inspection) and to locate submerged drugs and human bodies. For work like that, the "seal swat team" will no doubt demand that the trainers sweeten the deal with a bigger payoff of fresh herring.

Schusterman (1981a) has summarized several aspects of conditioning and learning in pinnipeds, from the relatively simple processes of habituation to the more complex cognitive skills involving abstraction and language. The term *habituation* refers to a pinniped's becoming accustomed to something—such as a frightening sound, a captive environment, or the presence of people. The rate of habituation to a particular stimulus varies widely among species, populations, and individuals. For instance, the presence or sound of humans near a pinniped rookery may or may not frighten the animals into the sea, depending on the species. Habituation to what is called the startle or flight response also varies within different populations with respect to sex, age, season, time of day, and location. For example, although many populations of *Zalophus* sea lions in California are afraid of humans, *Zalophus* on some of the Galápagos Islands is rarely frightened by people (Fig. 9.2) (Peterson and Bartholomew 1967). Sea lions on islands in the Galápagos frequently visited by people seem to be more tame than those inhabiting the less accessible, more remote islands. Some Australian sea lions at Seal Bay on Kangaroo Island are quite tame and do not react when people approach to within 1–2 meters, whereas Australian sea lions only 80 kilometers away on the same island immediately stampede into the water when a person appears (Stirling 1972). Harbor seals are among the most easily frightened and alert pinnipeds. They interrupt their sleep frequently to scan their surroundings, even in captivity (Schusterman 1968).

In contrast, species such as the Weddell seal, hooded seal, harp seal, Hawaiian monk seal, and elephant seal tend to be fairly tolerant of human presence and are among the most easily approached pinnipeds. Elephant seals of both sexes, for instance, rarely flee when humans come within a meter or less. During the breeding season, however, adult females and bulls threaten and sometimes even chase researchers, causing the humans to startle and flee, in a reversal of the usual situation. Biologists studying northern elephant seals on Año Nuevo Island encountered a bull, aptly named Jaw, who was unlike other bulls in that he preferred to charge researchers rather than other breeding males. While most bulls had an I-see-you-but-am-not-really-concerned-about-you look in their eyes, Jaw's ever-watchful gaze said: "I see you and am out to get you." During one breeding season, Jaw particularly liked to use the path to the island outhouse as a sleeping spot. Our nighttime trips to the outhouse were often aborted in terror that winter when Jaw the bull suddenly reared up out of the quiet darkness, only inches from our feet.

Because the harp seal does not fear humans, it is especially vulnerable to hunters, and this species has been heavily exploited by the commercial sealing industry. Monk seals evolved on isolated oceanic islands or unin-habited mainland coasts where they were not exposed to land predators; as a result, they appear to have developed a "genetic tameness" (Kenyon 1981). Despite their tendency not to flee when approached by man, they are sensitive to disturbance by humans and as a result have suffered an increase in pup mortality (see, e.g., A. M. Johnson et al. 1982).

LANGUAGE

Studies on captive California sea lions have shown that these animals are able to comprehend an artificial language in much the same way that chimpanzees and dolphins have learned to communicate with humans by means of symbols and sign language (Schusterman and Krieger 1984, 1986; Schusterman and Gisiner 1988). After summarizing the past thirty years of animal language research, Schusterman and Gisiner (1988) con-clude that the precursors of language are apt to exist in animals that are capable of combining the following learning and cognitive skills: (1) higher-order conditioning or paired-associate learning (i.e., the ability to learn A and associate it with B, or to understand that there is a rela-tionship between A and B); (2) the ability to categorize objects and events into class and relational concepts, each with its own subcategories (e.g., class = square vs. round objects; relational = big vs. bigger); and (3) the acquisition of conditional sequential discriminations, or discriminating the temporal order of the paired associates (i.e., the A-B association precedes the C-D association, and so forth). No one has yet demonstrated that a nonhuman animal possesses all the attributes of human language. However, two different species of marine mammal—the California sea lion and the bottlenose dolphin—have shown similar abilities to under-

FIG. 9.2. Galápagos sea lion inspecting tourist's daypack. *Photo by W. E. Townsend, Jr.*

stand a syntax made up of ordered sequences of signs relating to two objects. In other words, these animals can understand and act upon a series of commands that relate to two different objects in their environ-ment (Figs. 9.3, 9.4).

Schusterman and Krieger (1986) found that sea lions are able to think not only in broad perceptual categories but also in specific images at the same time. Much of their work was carried out on their "star pupil," an energetic female sea lion named Rocky. Rocky was taught by a trainer who sat at a pool's edge and made various gestural signals with her hands and legs. For instance, to make the signal for "go," the trainer dropped her foot. It took Rocky about 2.5 years to learn and respond to an abbreviated or pidgin language that included eleven signs for objects (such as a ball, bleach bottle, floating cone, or toy boat); five signs for object properties (that is, adjectives or modifiers), such as size and brightness; and six signs for action and behavior. The goal was not to build up Rocky's vocabulary but to develop ways of learning how she processed information.

Rocky is currently capable of comprehending over seven thousand "sentence" combinations, even novel ones, with a length of up to seven signs. Like bottlenose dolphins (Herman et al. 1984), Rocky can carry out novel instructions by integrating different behaviors learned through training according to rules based largely on the sequence of the various

FIG. 9.3. Bottlenose dolphin at Long Marine Lab. *Photo by Randy Wells.*

FIG. 9.4. Bottlenose dolphin performing tasks at Long Marine Lab. *Photo by Randy Wells.*

signs. For instance, the sign sequence (something like a sentence in human terms) "Ball, black pipe fetch" means "Take the black pipe to the ball," and the sentence "Black ball, pipe fetch" means "Take the pipe to the black ball."

Rocky appears to rely on learning and cognitive skills, not on linguistic skills in the human sense, in interpreting the sign sequences she is given. She can even shift adjectives for size from the absolute form (large or small) to the relational or comparative form (larger or smaller). Given a complex command with hand signals, for example, to "find the larger of two black balls and push it to the smaller one," she could correctly carry out the task. The discrimination between absolute and relative is actually a cognitive "rule shift" that Rocky employs when given the signal for "large." She uses one meaning or rule first, and if it doesn't fit, she shifts to a relative rule ("larger, not the large one").

During a testing session (Fig. 9.5), Rocky rests her head on the trainer's boot. As the command is being given, she raises her head and quickly scans the pool for the appropriate objects. For instance, at the word "black," she instantly scans the pool for all black objects. When "ball" is signed, Rocky looks around for the black ball, always returning her head to the same position on the trainer's boot. When the sign is given for Rocky to perform her task, she immediately swims to the appropriate objects and carries out the trainer's command. Rocky's search for the correct objects and her performance of the appropriate sequence of actions is usually executed with amazing speed! She often knows what the signs mean even before the trainer finishes them. But Rocky does not always get it right the first time. According to B. Gisiner (pers. comm.), when Rocky is given a sequence of commands she does not know how to execute, she begins to make fussy sounds and to breathe forcefully through her nose in apparent frustration. In summary, then, although the sea lion system shares several features of human language, Schusterman and Gisiner (1988) consider it "prelinguistic," since their experiments suggest that sea lions have many of the basic cognitive skills necessary for language but not language itself.

IMPRINTING

Among many animals there is a critical and sensitive period following birth when a mother and her young learn to recognize one another, a process referred to as imprinting (Lorenz 1935; Hess 1973; Hess and Petrovich 1977). In a number of mammals, such as chimpanzees, elephants, and of course humans, mother and young continue to recognize

a. Rocky rests her head on the trainer's boot as trainer gives hand signal.

b. Rocky scans the pool for the object designated by trainer's signal.

FIG. 9.5. Rocky (a female California sea lion) performing tasks at Long Marine Lab facilities, University of California at Santa Cruz. *Photos by Bob Gisiner.*

c. Trainer gives Rocky a command, telling her what to do with the object.

d. Rocky proceeds to carry out the command.

and interact with one another long after weaning. The female-pup bond created by the imprinting process at birth is especially important to pinnipeds that rear their pups synchronously in high-density breeding colonies. Well-developed mother-pup recognition abilities are also essential for female otariids that must relocate their own pup among a number of similar-looking pups in the rookery each time they return to shore from a feeding trip.

Imprinting in captive California sea lions has been extensively studied with respect to their social attachment to people, with some intriguing results (Schusterman 1985). Eighteen sea lions ranging in age from one day to nearly five years (fifty-eight months) were observed for evidence of bonding to their human caretakers. During the experiments, the sea lions were offered choices between their original human surrogate mothers, their current caretakers, strange humans, and other sea lions. All sea lions who were cared for and bottle-fed by a person within twelve to ninety-six hours of birth made stronger and more persistent efforts to interact with people than eleven control sea lions who were nursed by their biological mothers for at least three weeks following birth before being cared for by humans. In addition, three of the human-imprinted sea lions (a yearling female, yearling male, and three-year-old female) were given the choice between their surrogate mother and current keeper. All strongly preferred their surrogate mother, to whom the seals directed frequent calling, nuzzling, following, and "climbing-on" behavior. In fact, sea lions almost five years old recognized and responded to humans who took care of them as infants (Schusterman 1985; Schusterman and Hanggi, unpubl. data). A pup nursed and handled for as little as a week by two different caretakers formed a preferential and nearly exclusive bond to one of them (Fig. 9.6).

A fascinating study of sea lion social behavior has shown that familial relationships and mother-pup bonds in captive California sea lions persist for several years and possibly even a lifetime (Hanggi and Schusterman 1987). In a captive sea lion colony containing two adult males and seventeen females (ranging from two to twenty years in age), mothers interacted almost exclusively with their own offspring, who ranged in age from two years to eleven years. Siblings also interacted more often with each other than with unrelated sea lions. In contrast, females who had no kinship ties within the colony either kept to themselves or interacted aggressively with other animals. Family members also recognized one another's vocalizations even after a long period of separation.

One would think that a wild sea lion could also recognize its mother as well as its offspring after prolonged separation. But it is unclear whether a

FIG. 9.6. Juvenile California sea lion with human surrogate mother. *Photo by Bob Gisiner.*

seal mother and her offspring remember one another after weaning under natural conditions. There is no evidence yet that such mutual recognition occurs. But observing long-term associations between genetically related wild pinnipeds is very difficult, so perhaps researchers simply are not seeing these interactions when they take place. The studies that have been carried out on female aggression so far show that females tend to behave with equal nastiness toward all other females around them, rather than inhibiting aggression toward any particular individual.

In a number of sea lion and fur seal species, however, a female and her older unweaned offspring may recognize and associate with one another. Females may continue to care for their older offspring (sometimes while mothering a younger pup) for one to three years after birth in as many as ten of the fifteen otariid species as well as walruses. Galápagos fur seals routinely care for their pups for two to three years. Phocid females and their offspring may be less likely than otariids to recognize one another after weaning, given their brief lactation period and abrupt weaning.

Communication and Vocalization

Pinnipeds produce a strange and varied assortment of sounds both above and below the water. Some pinnipeds, such as harbor seals, seem almost silent in comparison with the sea lion, which barks incessantly in its rookery—an endless cacophony that has been known to drive novice seal biologists slightly crazy, especially when they try to sleep at night after arriving on an isolated island where they had expected peace and quiet. The more vocal pinnipeds tend to be highly social and often colonial, whereas the quieter species seem to be more solitary.

Recent research has indicated, however, that the supposedly quieter phocids (such as harbor seals and ringed seals) may not be so quiet after all, but harder for humans to hear (Ralls et al. 1985). With the exception of the highly social and obviously vocal elephant seals and grey seals, many phocids not only are less gregarious but also tend to mate in the water, where the vocalizing associated with breeding is largely inaudible to humans on land. Ralls et al. (1985) also suggest that some of the northern phocids may have evolved quieter vocalizations because of land predators. Captive phocids that are kept alone or in mixed-species groups seem to spend less time vocalizing than phocids in their native habitats, a change that is perhaps not surprising since a seal that vocalized with no conspecifics around would be "talking to itself." At the New England Aquarium, for instance, male harbor seals that vocalized frequently in the main group of harbor seals produced sounds rarely, if at all, when kept alone or with pups.

Pinnipeds vocalize not only in air but also underwater, perhaps far more frequently than most people realize. The production of sounds underwater and the possibility of the evolution of echolocation abilities are discussed in chapter 1. This chapter focuses on vocalizations produced in a social context. Detailed reviews of airborne and underwater vocalizations in pinnipeds are available in Poulter (1968), Schusterman (1978), and Winn and Schneider (1977).

UNDERWATER SOUNDS

Pinnipeds produce a variety of underwater sounds characterized by a broad range of pulse rates, frequencies, durations, and intensities (Watkins and Wartzok 1985). Schusterman (1978) graphically describes sea lion underwater vocalizations as "loud barks, whinnies, faint clicks, trills, moans or humming sounds, chirps, belches, growls, squeals, roars, roarlike growls, etc." Pinniped sound frequencies tend to range from 0.1

to 10 kilohertz (Schusterman 1978), although sounds produced by some species such as Weddell seals range as high as 70 kilohertz (Schevill and Watkins 1971). In contrast, clicks produced by dolphins are characterized by a broad frequency spectrum that ranges up to 256 kilohertz, with maximum energies between 20 and 60 kilohertz. Since the limit of human hearing is about 20 kilohertz, the sounds made by odontocetes and some pinnipeds extend into frequencies that are inaudible to humans. In addition, dolphin clicks, with repetition rates of over 500 per second, are produced much more rapidly than those of pinnipeds, whose clicks repeat at a rate of about 50–150 clicks per second (Schusterman 1978). Each pinniped species that has been heard underwater seems to have a distinctive vocalization.

Underwater vocalizations often appear to be related to breeding activities and social interactions, particularly among breeding males that are maintaining territories or otherwise competing with other males for mates. Pinnipeds that become more vocal underwater during the breeding season include the harp seal (Møhl et al. 1975; Watkins and Schevill 1979), ringed seal (Stirling 1973), California sea lion (Schusterman 1968; Schusterman and Dawson 1968; Schusterman and Balliet 1969), walrus (G. C. Ray and Watkins 1975), bearded seal (Chapskii 1938; C. Ray et al. 1969; Burns 1981a), and Weddell seal (Kooyman 1981b,c; Thomas and Kuechle 1982). Species such as bearded seals, Weddell seals, and walruses produce complex and noisy vocal repertoires or "songs" when displaying underwater during the breeding season. (Animal vocalizations are considered to be "songs" when they are repetitive and stereotyped, produced seasonally, variable according to sex, and associated with courtship or territorial maintenance.)

Intricate, musical underwater songs are produced by bearded seal males and possibly by females as well (Chapskii 1938; C. Ray et al. 1969). Eskimo hunters used to listen for bearded seal songs by placing a kayak paddle in the water and pressing an ear against the end of the handle. Portions of the underwater songs may even be heard in the air by humans at close range and presumably by other bearded seals as well. Singing bearded seals usually repeat a ritualized sequence of dives while vocalizing, sinking slowly in a loose spiral, releasing a burst of air bubbles, and then surfacing where they first submerged, over and over again. Singing seals take little notice of human presence and seem unconcerned about approaching researchers.

Bearded seal songs may last more than a minute. They consist of oscillating warbles that change frequencies, punctuated by a brief unmodulated low-frequency moan. Bearded seal songs are produced from

FIG. 9.7. Bearded seal pup in water. *Photo by Kathy Frost.*

March to July, although the singing shows a distinctive peak during the April–May breeding season. C. Ray et al. (1969) suggest that adult males produce songs during the mating season to maintain underwater territories and exclude other males from mating activities or to attract females. Burns (1981b), however, believes that female bearded seals may also sing in some areas (Fig. 9.7).

Weddell seals are especially vocal underwater and produce a wide variety of songlike vocalizations that sound like whistles, chirps, trills, and low-pitched buzzes (see, e.g., Schevill and Watkins 1965, 1971; Thomas 1979). Thomas and Kuechle (1982) classified twelve underwater calls of the Weddell seal into thirty-four distinctive call types, using descriptive names such as chi-chi-chi, teeth clatter, mew, gutteral glug, jaw claps, too-loo and what-chunk. They found that the call levels vary with the individual seal and the direction of the sound beam. The seal's call is directional, so that the beam is directed downward when the seal is in its typical swimming position. Although such a directional sound beam would be of use in echolocating, so far no one has found evidence to show that Weddell seals actually do echolocate (Schevill and Watkins 1971). The underwater calls seem to be related to social interactions and breeding activities. Male Weddell seals have a larger vocal repertoire and produce more complex calls than females. During the breeding season, males maintain underwater territories beneath the ice and appear to use many vocalizations in territorial displays (C. Ray 1967; Kooyman 1968). In fact, males produce many of the same vocalizations both underwater and on land (Thomas 1979). Thomas and Kuechle (1982) found that the calls are transmitted great distances underwater and can even penetrate the fast ice, apparently allowing seals underwater to communicate with seals on the surface of the ice.

Harp seals, which also have an extensive underwater vocal repertoire, are much more vocal underwater than on ice, where mothers and pups produce most of the calls. A total of fifteen underwater harp seal calls have been recorded, including trills, "dove" cooing, warbles, squeaks, and the "distressed blackbird." A particular call is often repeated 10–20 times or more. The underwater vocalizations are produced only during the breeding season, when harp seals congregate in large groups (Møhl et al. 1975). Terhune and Ronald (1985) recorded sound levels of 140 decibels and found that harp seal calls carried as far as 2 kilometers underwater. They recorded vocalizations continuously over a 60-kilometer under-water transect, finding that seals distributed along this transect were probably in acoustic contact with one another. This contact would appear to help a harp seal searching for relatively small breeding herds on vast expanses of ice by offering a large acoustic target.

Perhaps some of the most distinctive and unusual underwater pinniped sounds are produced by male walruses during the breeding season. While female walruses rest together on ice floes or in the water, adult males position themselves alongside the female herd and engage in continuous acoustical and visual displays, each lasting two to three min-utes. During the underwater part of the display, males produce a series of loud stereotyped gonglike sounds, often described as sounding like ring-ing bells (Schevill et al. 1966). (The underwater vocalizations of northern elephant seals are also thought to sound bell-like or like "random notes on a xylophone" [Poulter 1968].) A complete walrus song consists of both in-air and underwater calls. Male walruses rise out of the water, emitting a brief whistle and a number of clacks made with their teeth. Underwater, the male vocalizes in a series of pulses, followed by a distinctive gong sound (G. C. Ray and Watkins 1975).

Male displays appear to attract females as well as to repel other males. When hauled out in a large group, walrus bulls apparently may produce musical sounds by striking their inflated throat, or pharyngeal, pouches with foreflippers to generate air pulses (Poulter 1968). The pouches, pres-ent only on males, also serve as resonance chambers that help produce the gong sound underwater. The throat pouches seem to serve another function analogous to that of a scuba diver's buoyancy-compensation vest: to regulate buoyancy and provide "lift" in the water. Inflated pouches help the walrus to float on the water's surface as it sleeps.

Sometimes pinnipeds produce sounds underwater while foraging. For instance, Poulter (1968) found that leopard seals vocalize underwater while foraging on Adélie penguins near the penguin rookeries, producing almost continuous calls that consist of sounds such as isolated clicks and low-frequency chirping. According to Kooyman (1981a), the sounds are like "haunting, sonorous drones" and are sometimes so intense that the lowest-frequency call can be felt as well as heard on the surface of the ice.

AIRBORNE SOUNDS

Many researchers have studied the wide variety of vocalizations produced by pinnipeds on land. In-air vocalizations can be separated into those produced during the breeding season and those produced at other times of the year. Vocalizations can be further distinguished on the basis of sex and age classes, especially in polygynous species. Breeding males, for example, often produce a distinctive call associated with dominance or territorial maintenance. The vocalizations emitted by breeding bulls are very loud, repetitive, and directional, presumably enhancing the male's ability to communicate in a crowded, noisy breeding rookery with a high level of background noise. For instance, high-ranking northern elephant seal bulls produce a distinctive "clap-threat" call to advertise their high rank or to threaten other males. They "rear and trumpet," or "RT": with head thrown back, they emit a series of loud and distinctively drumlike pulses of sound which for one researcher resembled the sound of a flatulent tuba (Bartholomew and Collias 1962; Le Boeuf 1974).

California sea lion males also have a strong, directional bark that serves to threaten other males and advertise dominance during the breeding as well as nonbreeding season. Schusterman (1978) suggests that barking among males may also serve to attract mates. Schusterman and Gentry (1971) have found that captive California sea lions bark more in some seasons than in others despite the absence of females. Although the dominant territorial male barked infrequently during the breeding season, his presence appeared to inhibit barking by the subordinate males. During the nonbreeding season, barking was apparently associated with a less intense size-related dominance hierarchy, and subordinate males barked more frequently. Field observations of California sea lions have shown a similar pattern in male barking.

Fur seals appear to have a richer vocal repertoire than sea lions. Bonner (1981a) summarizes the vocalizations of the southern fur seals, describing such calls as bark and whimper (used in situations related to social status, territorial maintenance, sexual interest, and individual recognition); "puff" (low-frequency pulses followed by an explosion of air, used typically during high-intensity threats); "pulsed growl" (used during low-intensity threats); and "submissive call." (The various otariid display postures associated with these vocalizations are described in chapter 5.) Guadalupe fur seal males, for example, produce a distinctive and repetitive whickering sound in a variety of situations. Territorial bulls typically whicker-bark during boundary displays or when they are trying to prevent a female from departing for sea (Pierson 1978). As they move about on their territory, northern fur seal bulls emit a "whicker" call that sounds like a long exhalation without using vocal cords, with males interrupting the air flow by closing off the throat (R. Gentry, pers. comm.). When they perform boundary displays, they also give a loud "puff" sound (Peterson 1968).

During the breeding season territorial fur seal males often emit a characteristic undirectional vocalization, termed the territorial call (TC) by Roux and Jouventin (1987). With head and neck thrown back "coyote style," males emit this call either spontaneously or when responding to the vocalizations of other males. The TC apparently serves to advertise the vocalizing male's territorial status to other males. As Roux and Jouventin (1987) point out, the TC vocalization has been described in various terms by different authors: as a "trumpeted roar" in the northern fur seal (Peterson 1968; Antonelis and York 1985); a "full threat call" (Pierson 1978, 1987) and "high-pitched roar" (Peterson et al. 1968) in the Guadalupe fur seal; a "full-threat call" in the South American fur seal (Fig. 9.8) (Trillmich and Majluf 1981) and New Zealand fur seal (Stirling 1971a); and a "male threat call" (Bester 1977) and "male roar" (Paulian 1964) in the subantarctic fur seal.

Recent work has shown that male fur seals of several species have individually distinctive territorial calls. Individual variation probably helps territorial males to recognize one another. The TC of Guadalupe fur seal bulls varies greatly from male to male, sounding quite different even to the human ear (Pierson 1978, 1987). Antonelis and York (1985) recorded the "trumpeted roar" TC of territorial northern fur seal bulls and were able to differentiate the vocalizations of each male by consistent and distinctive acoustical characteristics. Individually distinctive vocalizations have also been found in South American fur seals (Trillmich and Majluf 1981), New Zealand fur seals (Stirling 1971a), and subantarctic fur

FIG. 9.8. South American fur seal male and female. *Photo by Gerry Kooyman.*

seals (Roux and Jouventin 1987). Roux and Jouventin (1987) showed that subantarctic fur seal males are able to discriminate between the TC's of their neighbors and those of strange males. Experimental playback of the recorded TC's of individual males indicated that territorial males respond much less often to the TC's of a neighbor male than to those of a strange male. Neighboring territory holders seem to habituate to one another.

Hauled-out harbor seals produce a chorus of grunts, burps, snorts, and loud belches. Someone unfamiliar with these rather rude-sounding noises might find them a little startling. In fact, while observing seals and otters in Monterey, I have occasionally been subject to the embarrassed or questioning looks of onlookers who are suddenly aware of the loud belching sounds being produced nearby. Other sounds produced by harbor seals include growling, yelping, snarling, and bubble blowing (Venables and Venables 1957; Sullivan 1981). The leopard seals described in chapter 4 who were threatening scuba divers underwater and blowing bubbles may have been producing threat vocalizations. Harbor seals also often slap the surface of the water with their foreflippers or hindflippers, producing a resounding whack that on a calm day can be heard from quite a distance. Although such flipper slapping is not entirely understood, it could help to flush prey for easier capture, or it could be related to sexual-agonistic displays.

Harbor seal vocalizations have been studied in nature by C. L. Hamilton (1981) and in the captive seals at the New England Aquarium by Ralls et al. (1985). The aquarium seals were pups and adults of both sexes and various ages, most of which had come to the aquarium as orphaned pups. Females (excluding mothers with pups) more than a year old rarely vocalized. The two adult males studied, in contrast, not only were quite vocal but also produced sounds that mimicked human speech. One of them in particular — the famous Hoover — was especially adept at imitating English words and phrases. (The other male was able to say only hello.) Hoover (Fig. 9.9) was found as an orphaned pup in Cundy Harbor, Maine, in May 1971. Raised by local residents for the first three months of his life, Hoover was given to the New England Aquarium in August 1971.

When Hoover was seven years old, he was first heard to "talk" with humans. According to one observer, "He says 'Hoover' in plain English. I have witnesses." Hoover also mimicked other words or phrases, such as "hello," "hello there," "hey," "how are you," "get out of there (here)," "come out of here." These same words were often repeated by humans caring for the seals. Hoover even imitated human laughter, by very rapidly repeating his "hey" sound: "heyheyheyheyheyheyhey." While "talking," Hoover always assumed a unique posture in the water, one he used only while mimicking human speech. With his head thrown back, nose pointing skywards, and

FIG. 9.9. Hoover, the harbor seal at the New England Aquarium that learned to mimic human speech patterns. *Photo courtesy of the New England Aquarium.*

neck retracted, Hoover spontaneously produced speechlike sounds that were directed toward no one in particular. He often quickly repeated words strung together in a nonsensical way, for example, "Hey, get out, hey, hello, get out, hello there, hello, hey, hey," punctuated by his version of laughter.

Sonograms comparing Hoover's speechlike sounds with the same words spoken by an adult male human showed two major differences: the seal separated the words less distinctly, and the seal's high frequency bands lacked the well-controlled modulation characteristic of the human's sonogram. Hoover pronounced vowel sounds more clearly than consonants. Ralls et al. (1985) felt that Hoover's human speech sounded like that of a male human with a Boston accent. In addition, he often sounded a bit drunk because of his slurring together of human words. Why did Hoover imitate human sounds so effectively and so often? Ralls et al. (1985) speculate that Hoover's "talkativeness" may be related to his having been raised by humans away from other harbor seals. When very young, he may have imprinted on his human foster parent's manner of speech. Male harbor seals may learn to vocalize by imitating other seal males, whereas Hoover tended to imitate human males. Such mimicry of the sounds produced by humans and other species has been widely reported in birds (Kroodsma 1983) as well as in cetaceans such as bottlenose dolphins (Lilly 1962, 1965; Richards et al. 1984), humpback whales (Guinee et al. 1983; K. Payne et al. 1983), and beluga whales (Eaton 1979).

Dialects

Dialects of some sort are found in at least three seals—northern elephant seals, Weddell seals, and bearded seals. A dialect may be thought of as geographic variation in the vocalizations of a species, with consistent differences in adult calls or songs between populations of the same species. Song and vocal dialects are also found in a number of birds (Krebs and Kroodsma 1980), humpback whales (R. S. Payne 1978; Winn and Winn 1978), and of course humans. Dialects in birds such as white-crowned sparrows, for instance, have been particularly well studied (Marler and Tamura 1962, 1964; Baker 1974, 1975; Baptista 1975, 1977; Baptista and Petrinovich 1986). Nonverbal, or gestural, dialects are even found among different groups of chimpanzees in various geographic localities (Goodall 1986).

Dialects in northern elephant seals have been especially well documented (Le Boeuf and Peterson 1969; Le Boeuf and Petrinovich 1975). The threat vocalizations of male elephant seals differ from one breeding rookery to another, showing the existence of distinctive regional dialects.

The entire population of northern elephant seals descended from a small group of less than a hundred animals on Guadalupe Island, off Mexico, that managed to survive the intensive hunting by sealers in the 1800s (Bartholomew and Hubbs 1960). As the elephant seals moved north (see Map 10), they recolonized San Miguel Island first (ca. 1938), then the San Benito Islands (ca. 1948), San Nicolas Island (ca. 1950), Año Nuevo Island (1961), and Southeast Farallon Island (1972) (Bartholomew and Hubbs 1960; Radford et al. 1965; Le Boeuf et al. 1974). Even in recent years, when elephant seals have dispersed from their birthplaces, they have tended to move north (Le Boeuf and Petrinovich 1975).

Le Boeuf and Peterson (1969) have shown that although the mean pulse rates of male threat vocalizations were similar on Guadalupe Island (1.77 per second) and San Miguel Island (1.88 per second), pulse rates on San Nicolas Island were much higher (2.53 per second) and those on Año Nuevo Island were slowest of all (1.02 per second). One can discern that the threat vocalizations of males on Año Nuevo Island are noticeably slower than those produced on a southern rookery such as San Nicolas Island. As Año Nuevo Island was colonized by immigrants from the "faster-pulsed" rookeries, the overall mean pulse rate on the island gradually increased, although the pulse rate of individuals remained consistent from year to year.

Presumably, the immigration of males from southern rookeries over several years was responsible for the increase in pulse rates at Año Nuevo Island. Because of the extensive immigration of young elephant seals, Le Boeuf and Petrinovich (1975) suggest that elephant seal dialects do not function to reduce gene flow and maintain population integrity, as dialects appear to do in some birds. For example, male and female birds may return to breed in the area of their birth because they are attracted to their "native" dialect. If females prefer mates who produce song dialects from their natal area, this preference would help to maintain a discrete population as well as to perpetuate the dialect (see, e.g., Marler and Tamura 1962; Konishi 1965; Nottebohm 1969, 1972, 1975; Marler 1970). In one case, however, Baptista and Morton (1982) found that dialects in montane white-crowned sparrows probably did not influence mate choice. They suggest that particular song dialects instead may play a role in territorial defense and the repulsion of other males. Although the reasons that birds learn dialects seem to be complex and in many cases unclear, dialects themselves appear to serve a different purpose for birds than for northern elephant seals.

Le Boeuf and Petrinovich (1974) also compared the threat vocalizations of the two elephant seal species—northern elephant seals and southern elephant seals. These two species probably diverged one to three million years ago. The threat vocalizations of southern elephant seals along the Valdés Peninsula in Argentina sounded very different from those of their northern counterparts. The southern vocalizations were irregular in pitch and time, and lasted much longer than the northern ones, an average of nineteen seconds. In contrast, the northern elephant seals emitted a short, regular train of pulsed sounds of six seconds or less.

The underwater vocalizations also differ in two Weddell seal breeding populations located on opposite sides of the Antarctic peninsula— McMurdo Sound and the Palmer peninsula, indicating that there are at least two dialects, or regional differences, in the vocal repertoire of this species (Thomas and Stirling 1984). Not only do the same vocalizations differ from one site to the other, but each population produces calls not heard at the other location. These geographic differences in vocalizations appear to be promoted and maintained by the Weddell seal's pronounced fidelity to a particular breeding colony (which would tend to isolate the two populations), their polygynous mating system, and learning. Male Weddell seals constantly vocalize underwater to defend their territories and attract mates. Thomas and Stirling (1984) suggest that these calls are learned and that new calls could be developed and transmitted to other seals within the colony. If breeding females prefer certain male calls (perhaps having imprinted on a particular type in their natal area), that preference could serve to further perpetuate the vocalization in the population. No one yet knows, however, if females do have such a preference. Interestingly, the vocal repertoire of one population of Weddell seals has changed gradually over the past twenty years (Thomas, unpubl. data).

The underwater vocal repertoire of bearded seals (recorded from March to June at six sites) also varies geographically. Each of the recording sites was separated from the others by a distance of from 150 to 350 kilometers. The presence of such regional differences in underwater calls suggests that bearded seals may be fairly sedentary, with little mixing occurring between the breeding populations (Cleator et al. 1987). Perhaps future research will show that dialects also exist in other pinniped species.

Mother-Pup Vocalizations

Almost all female pinnipeds emit a distinctive pup-attraction call, just as pups respond to or search for their mothers with a special vocalization. These mother-pup vocalizations, which appear to vary individually, help a female to recognize, locate, and maintain contact with her pup in crowded breeding colonies. Female northern elephant seals (Petrinovich 1974) and monk seals (Miller 1985), both of which remain with their pup throughout the dependency period, utter a distinctive pulsed bawling call when vocalizing to their pup. The pup's call, in contrast, sounds more like an explosive squawk. Harbor seal pups have individually distinctive vocalizations in air and possibly underwater. Such individual variation in the vocalizing of harbor seal pups almost certainly helps mothers to locate and maintain contact with their pup more easily and to distinguish their pup from others (Renouf 1984; Perry and Renouf 1985).

Female sea lions and fur seals initially emit a pup-attraction call when returning from feeding trips at sea. This call has been documented in many species, including Antarctic fur seals (Bonner 1968); subantarctic fur seals (Paulian 1964; Roux and Jouventin 1987); northern fur seals (Bartholomew 1959); South American fur seals (Trillmich and Majluf 1981); Galápagos fur seals (Trillmich 1981); New Zealand fur seals (Stirling 1971a), and South African and Australian fur seals (Stirling and Warneke 1971). Often several pups respond to a vocalizing female, but she is able to pick out her own pup from the others on the basis of vocal as well as olfactory and visual cues at close range. Pups also recognize and respond to their mother's vocalizations in the Galápagos fur seal (Trillmich 1981), subantarctic fur seal (Roux and Jouventin 1987), and probably other species as well. Sonograms of female subantarctic fur seal pup-attraction calls showed that they were individually distinctive enough to allow recognition between mother and pup (Roux and Jouventin 1987).

Female pinnipeds also frequently employ an aggressive threat vocalization when confronted by a human or when engaging in agonistic interactions with other females. Female northern elephant seal threat calls, for instance, sound almost like loud and raspy hissing. Monk seal females bellow when protecting their pup from human intruders. Regardless of the degree of sexual dimorphism in a particular species, females tend to have a higher-frequency vocalization than males. This sexual difference is apparent even in young pups (Poulter 1968).

Play

Play behavior is common among the pinnipeds, especially among juveniles, as it is in many young mammals (Meyer-Holzapfel 1956; Loizos 1966; Bekoff 1972; Fagen 1981). Although in some cases play may simply be play, having no purpose other than "fun," in other instances it seems to serve as practice for adult activities such as aggression or sex. In this sense, play helps a young animal to learn adult behavior patterns. For instance, juvenile male northern elephant seals often practice on each other the techniques used in copulating with females. They also frequently engage in what look like mock battles with one another, relatively uncoordinated attempts to fight like adult bulls. On land, male subadult and nonterritorial adult Steller sea lions and Australian sea lions also engage in social play, which usually resembles agonistic behavior in adult males (Farentinos 1971; Gentry 1974; Marlow 1975). The play of subadult Steller sea lion males differs from that of subadult females; the play bouts of males are longer, more complex, and rougher (Gentry 1974).

S. C. Wilson and Kleiman (1974) interpret some types of play in rodents and carnivores, including harbor seals, as antipredator behavior. As Eisenburg (1981) points out, although play in young animals may not be related to immediate survival, play behavior may be strongly associated with survival later in life. He suggests that mammals with relatively high EQ values, such as many pinnipeds, spend a substantial portion of their early development engaged in play. (EQ, or encephalization quotient, refers to a rather limited and rough measure of "biological intelligence" based on the ratio between the actual brain weight and the expected brain weight for a defined body weight [Jerison 1973].) Play seems to be especially prevalent even among adults in many of the "higher mammals" (Eibl-Eibesfeldt 1967).

Harbor seals engage in play behavior well into adulthood. Renouf and Lawson (1985) have found that wild harbor seals of all ages often play alone, sometimes with objects such as kelp or driftwood. Their observations of a harbor seal breeding colony showed thirteen distinctive forms of play; 80 percent of all play behavior was solitary, generally involving "locomotor rotational exercise," or playful repetitive rotation of the head and other body parts. While yearlings and juveniles engaged in much of the play behavior observed as well as in most of the social play, adult harbor seals also spent a considerable amount of time playing. Mothers and pups also spent some time playing together (S. Wilson 1974).

In the water, harbor seals lose much of their fear of people and can become curious about human swimmers or divers. Sometimes scuba divers have interesting encounters with these inquisitive seals (Fig. 9.10). At Hopkins Marine Station Refuge in Pacific Grove, the seals are relatively tame and unafraid of people. One seal in particular, nicknamed Slug, seemed endlessly fascinated by the underwater work my colleagues and I were doing, and we were surprised when he did not tag along as we counted and collected subtidal animals. Slug followed us everywhere, repeatedly zooming away and then reappearing out of the murky water to

FIG. 9.10. Harbor seal underwater. *Photo by Jim Watanabe.*

execute intricate acrobatic loops around the tall *Macrocystis* kelp plants. His curiosity was obvious as he nosed up and down my wetsuit, from black jet fins to the top of my hood. It was a spine-tingling experience to view a row of sharp seal teeth so close and feel strong jaws gently grasping my head and faceplate. Slug's inquisitive playfulness lost its appeal when he began to destroy our study site. He would often grab our clipboard and swim away with it, as our carefully marked water-proof data sheets scattered and floated away. I once felt a ripping tug at my yellow "goody" bag, and watched with dismay as my carefully packaged snail samples tumbled out of a gaping hole. Later, we were to find the painstakingly placed buoys that marked all our study sites either torn off or full of suspiciously seal-like tooth holes. In retrospect, however, all the damage was worth a chance to observe this exuberant seal in his own element.

Young bearded seals also spend time playing together as well as bodysurfing. Burns (1981b) gives the following account of seals at play:

> Two small seals kept up a constant chase which involved a repetition of riding swells into the beach, active rolling, mock fighting with the foreflippers and tail-chasing back out through the surf. This activity continued for 37 minutes. In another situation, two small seals moved along with a small outboard-powered boat and engaged in active tail-chasing, rolling, jumping partially clear of the water,

FIG. 9.11. California sea lion surfing. *Photo by Frans Lanting.*

slapping the water's surface with their foreflippers and swimming from one side to the other immediately beneath the boat.

Young sea lions of several species are perhaps the most playful of pinnipeds. Adults as well as juvenile sea lions may often be seen bodysurfing near shore and even in the wakes of boats. Surfing seems to be a favorite pastime of sea lions from the Galápagos Islands (Eibl-Eibesfeldt 1984b) to the Channel Islands. Sea lions are adept at surfing, as they glide seemingly without effort along the crest of the wave, cutting out just before it crashes to shore (Fig 9.11). Juvenile sea lions ride waves over and over again for no evident reason other than recreation. Later in life, however, adult sea lions use surfing to help them land high on the rocks during storms (R. Gentry, pers. comm.).

Juvenile sea lions engage in rough-and-tumble play while swimming underwater, romping, wrestling, twisting, and grabbing at one another's fur with their mouths (Fig. 9.12). They have been seen to chase and capture their own exhaled bubbles of air underwater. Like harbor seals, young sea lions also seem to be curious about and playful with human scuba divers. While I was diving with other researchers near a California sea lion rookery off one of California's Channel Islands, large groups of sea lions constantly "ganged up on us," engaging in high-speed chases and zooming at breakneck speeds just inches away from our faceplates, a game we thought of as "terrorize the diver." Other sea lions nosed our wetsuits and grabbed our swimfins with their mouths. The least popular sea lions among the divers were those that continued to chew up a long plastic tape that had been laid carefully across the bottom to mark subtidal sampling stations for the research we were doing (Fig. 9.13).

Eibl-Eibesfeldt (1984a,b), who has extensively studied the Galápagos sea lion, comments on its "notorious playfulness"; the young animals constantly engage in bouts of rough play together. He notes that when sea lions grasp one another's heads, they keep their mouths open, since an apparent inhibition exists toward biting that might break the skin (Fig. 9.14). Young Galápagos sea lions often chase each other in the water, with females and subadults sometimes participating as well. Adult females play with objects too, grabbing pieces of driftwood on the surface and tossing them into the air. They may also retrieve pebbles during a dive and play with them in a similar manner. Once Eibl-Eibesfeldt even had his swimfins stolen by sea lions, who then played tug-of-war with him. Occasionally a marine iguana on its way back to shore may fall victim to sea lions that playfully pull its tail. These iguanas are not necessarily safe from harassment on land, either. Their extremely long tails prove irresistible to young fur seal pups, who nibble and play with them as the iguanas pass by (Fig. 9.15).

FIG. 9.12. California sea lions playing underwater. *Photo by Doc White.*

FIG. 9.13. California sea lions interacting and playing on surface at Cedros Island. *Photo by W. E. Townsend, Jr.*

FIG. 9.14. Galápagos sea lions. *Photo by Steve Webster.*

FIG. 9.15. Galápagos sea lion resting with marine iguanas. *Photo by Phil Thorson.*

A number of pinnipeds even appear to play with their food or with live prey. For instance, young California sea lions and Guadalupe fur seals sometimes play with fish without eating much of it, repeatedly tossing their "toy" into the air and sometimes catching it again. (Some instances of fish tossing, however, may serve to prepare fish for consumption by "tenderizing" it or breaking it apart.) In Monterey Bay, California sea lions routinely play with small ocean sunfish (called molas), tossing them about like Frisbees, slapping them on the surface, and nibbling off their fins (A. Baldridge, pers. comm.). Little of the fish is actually eaten, however, for the flesh is essentially inedible. According to Bonner (1982), leopard seals also occasionally seem to be playing with live prey, such as a penguin or a crabeater seal, without really intending to eat it.

One of the predominant activities among pinniped pups and weaners is play—either social play, such as mock fighting; or individual play with natural objects, like kelp and rocks; or play with the most convenient and readily available "toy" of all: the pinniped's own body. Harp seal pups spend only a small portion of their time engaged in play, usually alone. Pups often play with chunks of ice or ice ledges as well as with their own bodies, exhibiting energetic jerky movements while continuously emitting growling or mumbling sounds (Kovacs 1987).

Elephant seal pups spend more time playing. Rasa (1971) and Reiter et al. (1978) have studied social interaction and play behavior in detail among northern elephant seal weaned pups. They found that the two primary social interactions on land are "space arguments" and "mock fights." Space arguments occur when one weaner moves and disturbs another; head sparring and nipping are generally involved. Female weaners engage in these interactions more often than males. In contrast, male weaners participate in mock fights far more frequently than females. These involve behaviors remarkably similar to that of real fights between adult males: neck presses, neck grabs, and rear-and-slams (so named for the adult male's act of rearing up high for leverage and show, then slamming down hard with the chest onto his opponent).

A variety of seemingly dull and uninteresting objects, such as rocks, kelp, and driftwood, are often the target of a weaner's "fun" (Fig. 9.16). Weaners are particularly attracted to moving objects: seagulls, anything that dangles, people walking past them, and their own bodies. When no other interesting objects are around, a weaner sometimes inspects and flexes its own foreflippers for several minutes or bends its head upside down over its body and tries to bite its tail (Reiter et al. 1978). One particular weaner we observed seemed to enjoy endlessly rocking back

Fig. 9.16. Northern elephant seal weaned pup playing with a piece of kelp. *Photo by author.*

FIG. 9.17. Northern elephant seal weaned pup playing with a Kleenex on bull's nose. *Photo by author.*

and forth while tossing a piece of dried kelp about with its teeth. Another group of weaners appeared to be fascinated by the white Kleenex tissue that was plastered against a sleeping bull's nose by the wind (Fig. 9.17). One weaner was especially attracted to the tissue and kept rotating its head, trying to grasp the fluttering tissue from the nose of the oblivious bull. Bright white objects may attract weaners, for many a weaner has eyed my white sneakers with keen interest. A friend of mine once complained that Guadalupe Island weaners took a fancy to his white tennis shoes one night while he was camping on the beach. He nervously awoke to find a gang of moonlit weaners surrounding his brand new white sneakers. Apparently they were not used to such enticing toys on this relatively uninhabited and remote island.

As representatives of a playful species ourselves, we easily find such behavior engaging and entertaining. Although we have learned a great deal in recent years, we still have much to learn from seals about their lives and their remarkable range of adaptations to diverse conditions. Perhaps the only way to truly understand a seal is to be one. Yet as mammals ourselves, we can certainly understand some of the problems seals have had to overcome to survive in the ocean and still be able to reproduce on land. When we put aside science for a moment and — for the pure pleasure of watching—observe a harbor seal gliding effortlessly among quiet kelp forests or playful young sea lions frolicking together and bodysurfing, our appreciation and respect for these marvelous mammals is renewed and deepened. We cannot help being enchanted by their beauty and ease and strength. And as warm-blooded land creatures who now and then venture beyond the surf and into the sea ourselves, we will no doubt continue to be intrigued by these captivating half-land and half-marine mammals that are so intimately at home in the ocean.

APPENDIX
Some Common Names
for Various Pinniped Species

Hawaiian monk seal (*Monachus schauinslandi*)
    Laysan monk seal
Caribbean monk seal (*Monachus tropicalis*)
    West Indian monk seal
Northern elephant seal (*Mirounga angustirostris*)
    northern sea elephant
Southern elephant seal (*Mirounga leonina*)
    southern sea elephant
Ross seal (*Ommatophoca rossi*)
    singing seal
    big-eyed seal
Leopard seal (*Hydrurga leptonyx*)
    sea leopard
Hooded seal (*Cystophora cristata*)
    crested seal
    bladdernose seal
Bearded seal (*Erignathus barbatus*)
    square flipper
    sea hare
    oogruk (Eskimo)
Grey seal (*Halichoerus grypus*)
    gray seal
    Atlantic seal
    hooknosed seal
    horse-head
Harp seal (*Phoca groenlandica*)
    Greenland seal
    saddleback seal
    saddle seal
    whitecoats (newborn pups)
    ragged jackets (partially molted pups)
    bedlamers (immature seals)

Ribbon seal (*Phoca fasciata*)
   banded seal
Largha seal (*Phoca largha*)
   spotted seal
Ringed seal (*Phoca hispida*)
   floe rat
   jar seal
   silver jar (molted pup)
Harbor seal (*Phoca vitulina*)
   common seal
Steller sea lion (*Eumetopias jubatus*)
   northern sea lion
   Steller's sea lion
Southern sea lion (*Otaria byronia*)
   South American sea lion
Australian sea lion (*Neophoca cinerea*)
   hair seal
   white-necked hair seal
   white-capped hair seal
   counselor seal
New Zealand sea lion (*Phocartos hookeri*)
   Hooker's sea lion
Northern fur seal (*Callorhinus ursinus*)
   Alaska fur seal
   Pribilof fur seal
Juan Fernández fur seal (*Arctocephalus philippii*)
   Philippi fur seal
South American fur seal (*Arctocephalus australis*)
   southern fur seal
   Falkland fur seal
New Zealand fur seal (*Arctocephalus forsteri*)
   western Australian fur seal
Antarctic fur seal (*Arctocephalus gazella*)
   Kerguelen fur seal
Subantarctic fur seal (*Arctocephalus tropicalis*)
   Amsterdam Island fur seal
South African fur seal (*Arctocephalus pusillus pusillus*)
   Cape fur seal
Australian fur seal (*A.p. doriferus*)
   Tasmanian fur seal
   Victorian fur seal
Walrus (*Odobenus rosmarus*)
   sea cow
   sea horse

GLOSSARY

Adaptation: Any characteristic that increases the probability that an organism will survive and reproduce more successfully than other members of the same species.

Adipose tissue: Fatty tissue.

Aerobic: A metabolic or cellular respiratory process that takes place in the presence of free oxygen.

Allomother: A female that assists the genetic parent in the care of its young.

Anaerobic: A metabolic or cellular respiratory process that takes place in the absence of free oxygen.

Atmosphere: A unit of pressure equivalent to 14.7 pounds per square inch (p.s.i.), the force exerted by air at sea level. One atmosphere is equal to a depth of 33 feet (29.4 p.s.i.); two atmospheres are equal to a depth of 66 feet (44.1 p.s.i.), etc.

Bends: A potentially fatal disorder that can occur when a scuba diver ascends too rapidly after spending a period of time at depth. Bends occur when nitrogen bubbles in blood and tissues under high pressure are released into joints and nerve tissue during a too rapid return to atmospheric pressure, causing pain, paralysis, and death unless quickly treated by gradual decompression. Also referred to as caisson disease or decompression sickness.

Biomass: The total mass of all organisms in a given region or at a given trophic level (a feeding stratum in a food chain containing organisms that occupy a similar functional position; e.g., T_1 = green plants).

Biphyletic: Originating from two separate ancestries.

Bottling: A posture sometimes assumed by a seal that is resting in the water: the face and nose are extended above the surface while the rest of the body remains submerged vertically under water.

347

Bradycardia: A decrease in heartbeat or an abnormally slow heartbeat.

Brown fat: Also known as brown adipose tissue (BAT). Specialized internal fat with a capacity for a high rate of oxygen consumption and heat production. Allows animals such as hibernating mammals and newborn seals to generate heat efficiently and rapidly to keep warm.

Carnivore: An organism whose diet consists of flesh or meat.

Chemoreception: A type of sensory reception that identifies substances from their chemical structure by means of taste and smell.

Circumantarctic region: The region surrounding the Antarctic.

Circumarctic region: The region surrounding the Arctic.

Circumpolar region: The region around the poles.

Close pack: Relatively solid ice that is not broken up by many leads (q.v.) of water.

Cognition: The mental process of thinking, reasoning, or learning.

Cohort: A particular age class. For example, all pups born during a certain breeding season at the same rookery represent a single cohort.

Countercurrent heat exchange: A versatile system of temperature regulation consisting of a network of veins and arteries in the extremities that allows an animal to retain or dissipate heat.

Delayed implantation: The attachment to the uterine wall of a fertilized egg whose development has been suspended at the blastocyst stage. After implantation the egg begins to grow again, but as a result of the delay birth is postponed by a matter of weeks or months.

Dialect: A geographic variation of a species' vocalizations.

Dominance hierarchy: A system of competition in which some members of a group predictably dominate other members by agonistic behavior. The highest-ranking individuals have greater access to resources (e.g., mates) than low-ranking individuals.

Echolocation: The production of sounds that allow an animal such as a dolphin to orient itself and locate objects by means of the returning sound waves or echoes.

Ecosystem: An ecological community and its physical environment considered as a whole.

El Niño Southern Oscillation (ENSO): Oceanographic events that occur in the eastern Pacific Ocean every few years in which prevailing current patterns change, water temperature increases, and upwelling declines, resulting in reduced food resources for pinnipeds and other marine mammals.

Estrous: Adjective describing estrus (q.v.) state of female.

Estrus: The period of sexual receptivity or "heat" in female mammals, usually coinciding with ovulation.

Estuary: The region where the ocean extends inland at the mouth of a freshwater river.

Evolution: The process of genetic change in organisms with each generation or of cumulative change in gene frequencies within populations in successive generations.

Fast ice: Ice that is attached to shore. A band of stable, shorefast ice usually forms along coastlines.

Fjord: A long narrow inlet of the sea between steep cliffs and bluffs.

Flaw zone: A lead, or lane of open water, that usually forms between the shorefast ice and pack ice, separating the stable fast ice from the moving pack.

Fresh ice: Ice that is newly formed.

Gastrolith: An ingested stone found in an animal's stomach.

Gestation: Period of pregnancy, or length of time elapsing between conception and birth.

Hemoglobin: In the red blood cells of vertebrates, the iron-containing pigment or blood protein that functions in oxygen transport.

Herbivore: An organism whose diet consists of plants.

Hummocks: Rounded hills of ice formed by wind-driven snow and ice fragments.

Hypermetropia: Farsightedness, in which vision is clear at a distance and blurry at close range. Because of a visual defect light rays are focused in back of the retina rather than on it.

Hyperosmotic: Having a higher osmotic pressure than the surrounding medium. Many marine invertebrates are hyperosmotic, the salt concentration in them being higher than that in seawater.

Hypo-osmotic: Having a lower osmotic pressure than the surrounding medium. Marine mammals are hypo-osmotic relative to the more concentrated, or salty, water in which they live.

Hypothermia: A condition in which an animal's body temperature falls to abnormally low levels. If not treated, it can lead to death.

Hypoxia: A lack of oxygen reaching body tissues, a condition that can result in suffocation.

Ice front: The constantly changing edge of moving pack ice.

Imprinting: A process that occurs early in an animal's life, often during a critical period following birth, in which it bonds to its parents or learns particular behaviors through association with parents or other animals. A mother also imprints on her newborn during this critical period.

Inguinal: Located in the groin.

Isosmotic: Having the same osmotic pressure as the surrounding medium. Some marine invertebrates are in osmotic equilibrium with seawater.

Krill: Small shrimplike marine crustaceans of the family Euphausiidae, which makes up a major portion of zooplankton. Especially plentiful in the Antarctic ocean, krill is the major food resource of baleen whales and crabeater seals.

K-selection: A relative concept in which natural selection favors competitive ability in predictable, stable environments with slow population growth. K-strategists tend to produce few offspring, which are intensively nurtured (cf. r-selection).

Lactation: The production of milk by female mammals to nurse young.

Lacustrine: Pertaining to, growing in, or living in a lake environment.

Laminar flow: Nonturbulent flow of a fluid in layers over an object or near a boundary.

Lanugo: The soft, fine hair covering most mammalian fetuses. In seals, the lanugo may be shed before birth or retained after the pup is born (for example, the newborn harp seal has white lanugo fur).

Leads: Channels of ice-free water that form within the ice pack.

Lek: An area where animals consistently gather for communal courtship displays.

Locomotion: Movement.

Mariposia: The practice of ingesting seawater.

Metabolic rate: The rate of energy expenditure in an animal. (Basal metabolism is the rate of energy expenditure when the animal is at rest.) It can be calculated by measuring either the amount of oxygen used or the amount of heat energy released.

Metabolism: All of the chemical and physical processes necessary for the maintenance of life within a cell or an entire organism, including the breakdown of molecules to release energy (catabolism) and the synthesis of complex molecules from simple molecules (anabolism).

Mitochondria: Microscopic bodies within the cytoplasm of every cell that contain enzymes used to convert food to usable energy; in these bodies aerobic (q.v.) respiration takes place.

Molt: The process of shedding an outer layer (e.g., fur or skin), which is replaced by new growth. In pinnipeds molting may be gradual (fur seals) or abrupt (elephant seals).

Monogamy: A breeding unit of one female and one male.

Monomorphism: A situation in which both sexes of the same species are similar in size and appearance.

Monophyletic: Originating from a single common ancestor.

Multiparous female: One that has produced more than one offspring.

Multiyear ice: Thickened ice that has remained unmelted through a second summer.

Myoglobin: A type of hemoglobin found in muscle fibers (*myo*, "muscle").

Myopia: Nearsightedness, in which vision is clear at close range and blurry at a distance. Because of a visual defect, distant objects are focused in front of the retina rather than on it.

Natality: Birthrate.

Nonshivering thermogenesis: A process in which an animal keeps warm not by shivering but by metabolizing brown fat (q.v.) to produce heat instead of storable energy.

Nulliparous female: One that has never produced offspring.

Open pack: Ice that is broken up to a high degree by leads (q.v.), or lanes of open water.

Osmoregulation: A process that keeps the osmotic concentration of body fluids constant despite fluctuations in the external aquatic environment.

Pack ice: Ice that is unattached to land. Pack ice is usually moving and shifting, although the interior, permanent pack ice can be stable.

Pagophilic: Literally "ice loving"; refers to seals that inhabit areas of ice.

Pancake ice: Large round plates of ice with upturned edges, formed by the action of swells on new ice crystals.

Pelage: The fur or hair covering a mammal.

Pelagic: Of or relating to open ocean waters as opposed to coastal waters.

Philopatry: The tendency of an animal to return to the rookery of its birth for reproduction and molting later in life.

Polygamy: A breeding system in which an animal of one sex mates with more than one member of the opposite sex. Such systems may be polyandrous (one female mates with several males) or polygynous.

Polygyny: A breeding system in which one male mates with more than one female.

Polynya: A large area of water in the pack ice that remains open throughout the year.

Precocial: Relatively well developed at birth, usually able to move about or swim soon after. (*Altricial,* in contrast, means comparatively helpless and undeveloped at birth.)

Pressure ridge: An often sizable pile of compressed rubble and ice fragments that extends above and below the surface of the ice, sometimes up to 40 feet in either direction.

Primiparous female: One that has produced only one offspring.

Promiscuity: A relatively unstructured breeding system in which no bonds are formed between mates and multiple matings are common (e.g., a lekking system).

Rete (pl. retia; also called *retia mirabilia*): A network of highly elaborated veins and arteries that function in countercurrent heat exchange and in marine mammals may serve as blood reservoirs during deep dives.

Reverse sexual dimorphism: A situation that exists when adult females are larger than adult males of the same species.

r-selection: A relative concept in which natural selection favors rapid population increase, often in unpredictable, fluctuating, or short-lived environments. r-strategists generally produce many offspring with little investment in any of them.

Serial monogamy: For pinnipeds, the mating of a single male with more than one female during the breeding season but with only one at a time. After mating with one female, the male may search for another available female to mate with.

Sexual bimaturism: The sexual maturation of one sex before that of the other sex in a particular species. In pinnipeds, females tend to mature earlier than males, the more competitive sex.

Sexual dimorphism: A situation that exists when adult males are larger than females and may exhibit distinctive secondary sexual characteristics, such as the hooded seal male's inflatable nasal sac.

Sexual maturity: The age at which an animal is physiologically capable of reproducing (in females, the age of
first mating rather than first birth).

Site fidelity: The tendency for an animal to return to the
site of its birth or to the same very specific area each
year for breeding.

Subantarctic region: The geographic region immediately
north of the Antarctic circle.

Subarctic region: The geographic region immediately south
of the Arctic circle.

Tachycardia: An extremely rapid heartbeat.

Tapetum lucidum: A specialized layer behind the retina
containing high levels of guanine crystals. This layer
enables the retina to reflect light like a mirror and
gives the eye a metallic appearance (like that of a cat's
eye, which seems to "glow" in the dark). The tapetum
gathers light and helps the animal to see better in
darkness.

Territory: An area occupied exclusively by one animal and
defended by agonistic behavior or advertisement.

Thermoregulation: The process of maintaining a constant
internal body temperature despite changes in the
outside environmental temperature.

Thigmotaxis: A tendency to rest in close bodily contact
with other animals.

Trachea: The "windpipe." In vertebrates, the respiratory
tube that extends from the pharynx into the thorax.

Ultrasonic: Beyond the range of sound frequencies audible
to the human ear (above about 20,000 hertz).

Vibrissa (pl. vibrissae): Whiskers; stiff hairs that usually
project from the face and serve as sensory receptors.

BIBLIOGRAPHY

Sources for personal communications cited in the text (pers. comm.) are listed at the end of the Bibliography. Multiple-author entries beginning with the same name are organized alphabetically.

Aguayo, A. 1971. The present status of the Juan Fernández fur seal. *K. Norske Vidensk. Selsk. Skr.* 1:1–4.

———. 1973. The Juan Fernández fur seal. In *Seals*. International Union for the Conservation of Nature (IUCN) Pub. New Ser. Supp. Paper. no. 39: 140–143.

Aguayo, A., and R. Maturana. 1973. Prescenia del lobo marino común *Otaria flavescens* en el litoral chileno. *Biol. Pesq. Chile, Santiago* (Chile) 6:45–75.

Ainley, D. G., R. P. Henderson, H. R. Huber, R. J. Boekelheide, S. G. Allen, and T. L. McElroy. 1985. Dynamics of white shark/pinniped interactions in the Gulf of the Farallons. In *Biology of the White Shark*. Memoirs of the Southern California Academy of Sciences, Los Angeles, vol. 9, 109–122.

Ainley, D. G., H. R. Huber, and K. M. Bailey. 1982. Population fluctuations of California sea lions and the Pacific whiting fishery off central California. *U.S. Natl. Mar. Fish. Serv. Bull.* 80:253–258.

Ainley, D. G., H. R. Huber, R. P. Henderson, and T. J. Lewis. 1977a. Studies of marine mammals at the Farallon Islands. Final Report to the U.S. Marine Mammal Commission, Washington, D.C., for contract MM4AC002. U.S. Dept. of Commerce, National Technical Information Service, Springfield, Va. PB-274076.

Ainley, D. G., H. R. Huber, R. P. Henderson, T. J. Lewis, and S. H. Morrell. 1977b. Studies of marine mammals at the Farallon Islands, California, 1975–1976. Final Report to the U.S. Marine Mammal Commission, Washington D.C., for contract MM5AC020. U.S. Dept. of Commerce, National Technical Information Service, Springfield, Va. PB-266249.

Ainley, D. G., H. R. Huber, S. H. Morrell, and R. H. LeValley. 1978. Studies at the Farallon Islands, 1977. Final Report to the U.S. Marine Mammal Commission, Washington, D.C., for contract MM6AC027.

Ainley, D. G., C. S. Strong, H. R. Huber, T. J. Lewis, and S. H. Morrell. 1981. Predation by sharks on pinnipeds at the Farallon Islands. *Fishery Bulletin* 78:941–945.

Alcorn, D. J. 1984. The Hawaiian monk seal on Laysan Island, 1982. NOAA Technical Memorandum, NMFS.

Alcorn, D. J., and A.K.H. Kam. 1986. Fatal attack on a Hawaiian monk seal (*Monachus schauinslandi*). *Marine Mammal Science* 2:313–315.

Allen, S. G. 1985. Mating behavior in the harbor seal. *Marine Mammal Science* 1:84–87.

Allen, S. G., D. G. Ainley, and G. W. Page. 1979. Assessment of harbor seal populations in Bolinas Lagoon, Marin County, California, 1978–1979. Report to the Marine Mammal Commission, Washington, D.C., for contract no. MM8AC012.

Almon, M., and D. Renouf. 1985. Sleep in harbor seals. In *Proceedings of the sixth biennial conference on the biology of marine mammals,* Nov. 22–26, Vancouver, British Columbia.

Ames, J. A., and G. V. Morejohn. 1980. Evidence of white shark, *Carcharodon carcharias,* attacks on sea otters, *Enhydra lutris. Calif. Fish and Game* 66: 196–209.

Amoroso, E. C., A. Goffin, G. Halley, L. H. Matthews, and D. J. Matthews. 1950. Lactation in the grey seal. *J. Physiol.* 113:4–5.

Amoroso, E. C., and J. H. Matthews. 1951. The growth of the grey seal (*Halichoerus grypus*) (Fabricius) from birth to weaning. *J. Anat.* (London) 85:426–428.

Amos, W., S. Anderson, and R. Hoezel. 1987. Marine mammals: Stock identity and paternity as determined by powerful new molecular analysis. In *Proceedings of the seventh biennial conference on the biology of marine mammals,* Dec. 5–9, Miami, Fla.

Andersen, H. T. 1966. Physiological adaptations in diving vertebrates. *Physiological Reviews* 46:212–243.

Anderson, S. S. 1977. The grey seal in Wales. *Nature in Wales* 15:114–123.

———. 1979. Cave breeding in another phocid seal, *Halichoerus grypus.* In *The Mediterranean monk seal,* ed. K. Ronald and R. Duguy, 151–155. Oxford: Pergamon.

———. 1981. Seals in Shetland waters. In *Proceedings of the Royal Society of Edinburgh* 80B:181–188.

Anderson, S. S., R. W. Burton, and C. F. Summers. 1975. Behavior of grey seals (*Halichoerus grypus*) during a breeding season at North Rona. *Journal of Zoology* (London) 177:179–195.

Anderson, S. S., and M. A. Fedak. 1985. Grey seal males: Energetic and behavioral links between size and sexual success. *Animal Behaviour* 33: 829–838.

Anderson, S. S., and J. Harwood. 1985. Time budgets and topography: Low energy reserves and terrain determine the breeding behavior of grey seals. *Animal Behaviour* 33:1343–1348.

Anderson, S. S., and A. D. Hawkins. 1978. Scaring seals by sound. *Mammal Rev.* 8:19–24.

Antonelis, G. A., Jr., C. H. Fiscus, and R. L. DeLong. 1984. Spring and summer prey of California sea lions, *Zalophus californianus,* at San Miguel Island, California, 1978–79. *Fishery Bulletin* 82:67–76.

Antonelis, G. A., Jr., M. S. Lowry, D. P. DeMaster, and C. H. Fiscus. 1987. Assessing northern elephant seal feeding habits by stomach lavage. *Marine Mammal Science* 3:308–322.

Antonelis, G. A., Jr., and A. E. York. 1985. Identification of individual male northern fur seals, *Callorhinus ursinus*, from their vocalizations. In *Proceedings of the sixth biennial conference on the biology of marine mammals*, Nov. 22–26, Vancouver, British Columbia.

Armstrong, T. E., B. Roberts, and C. Swithinbank. 1973. *Illustrated glossary of snow and ice*. 2d ed. Scott Polar Research Inst. Cambridge, England.

Arnason, U., and B. Widegren. 1986. Pinniped phylogeny enlightened by molecular hybridizations using highly repetitive DNA. *Molecular Biology and Evolution* 3:356–365.

Arseniev, V. A. 1941. The feeding of the ribbon seal (*Histriophoca fasciata* Zimm.). *Izv. Tikhookean. nauchno-issled. inst. rybn. khoz. okeanogr* 20:121–127.

Bailey, A. M., and I. H. Sorensen. 1962. Subantarctic Campbell Island. Proceedings no. 10. Denver Museum of Natural History, Denver, Colo.

Bailey, K. M., and D. G. Ainley. 1981/1982. The dynamics of California sea lion predation on Pacific hake. *Fisheries Research* (Amsterdam) 1:163–176.

Baker, M. C. 1974. Genetic structure of two populations of white-crowned sparrows with different song dialects. *Condor* 76:351–356.

———. 1975. Song dialects and genetic differences in white-crowned sparrows (*Zonotrichia leucophrys*). *Evolution* 29:226–241.

Balazs, G. H., and G. C. Whittow. 1979. First record of a tiger shark observed feeding on a Hawaiian monk seal. *Elepaio* 39:107–109.

Baptista, L. F. 1975. Song dialects and demes in sedentary populations of the white-crowned sparrow (*Zonotrichia leucophrys nuttalli*). Univ. of California Publ. Zool. 105:1–52.

———. 1977. Geographic variation in song and song dialects of montane white-crowned sparrows. *Condor* 82:267–284.

Baptista, L. F., and M. L. Morton. 1982. Song dialects and mate selection in montane white-crowned sparrows. *Auk* 99:537–547.

Baptista, L. F., and L. Petrinovich. 1986. Song development in the white-crowned sparrow: Social factors and sex differences. *Animal Behaviour* 34:1359–1371.

Barash, D. P. 1978. *Sociobiology and behavior*. New York: Elsevier.

Barber, R. T., and F. P. Chavez. 1983. Biological consequences of El Niño. *Science* 222:1203–1210.

Barlow, G. W. 1972. A paternal role for bulls of the Galápagos Islands sea lions. *Evolution* 26:307–310.

Bartholomew, G. A. 1959. Mother-young relations and the maturation of pup behavior in the Alaska fur seal. *Animal Behaviour* 7:163–171.

———. 1970. A model for the evolution of pinniped polygyny. *Evolution* 24:546–559.

Bartholomew, G. A., and N. E. Collias. 1962. The role of vocalization in the social behavior of the northern elephant seal. *Animal Behaviour* 10:7–14.

Bartholomew, G. A., and C. L. Hubbs. 1960. Winter population of pinnipeds about Guadalupe, San Benito, and Cedros islands, Baja California. *Journal of Mammalogy* 33:160–171.

Bartlett, D., and J. Bartlett. 1976. Patagonia's wild shore. *National Geographic* 149:312–317.

Bauer, G. B., A. Perry, N. Fuller, J. R. Dunn, J. Zoeger, and L. M. Herman. 1983. Biomagnetic studies of cetaceans. In *Proceedings of the fifth biennial conference on the biology of marine mammals,* Nov. 27–Dec. 1, Boston.

Beach, R. J., and S. I. Jeffries. 1981. Techniques and preliminary results in the capture and tagging of harbor seals (*Phoca vitulina richardsi*) in the Columbia River. In *Proceedings of the fourth biennial conference on the biology of marine mammals,* Dec. 14–18, San Francisco.

Beck, B., T. G. Smith, and A. W. Mansfield. 1970. Occurrence of the harbour seal, *Phoca vitulina* (Linnaeus), in the Thlewiaza River, N.W.T. *Canadian Field-Naturalist* 84:297–300.

Beck, B. B. 1980. *Animal tool behavior.* New York: Garland.

Beddington, J. R., R.J.H. Beverton, and D. M. Lavigne, eds. 1985. *Marine mammals and fisheries.* London: George Allen and Unwin.

Bekoff, M. 1972. The development of social interaction, play, and metacommunication in mammals: An ethological perspective. *Quarterly Review of Biology* 47:412–434.

Bel'kovich, V. M., and A. V. Yablokov. 1961. Among the walruses. *Priroda* (Moscow) 3:50–56.

Bell, R. D. 1979. Progress of fur seal pups mortality programme, 2V15 Bird Island 1978/79. British Antarctic Survey, Madingley Road, Cambridge, CB3 0ET, U.K.

Belopol'skii, L. O. 1939. On the migrations and ecology of reproduction of the Pacific walrus (*Odobenus rosmarus divergens* Illiger). *Zoologicheskii zhurnal* (Moscow) 18:762–774.

Bengtson, J. L., and D. B. Siniff. 1981. Reproductive aspects of female crabeater seals (*Lobodon carcinophagus*) along the Antarctic Peninsula. *Can. J. Zool.* 59:91–102.

Berta, A. 1987. Affinities of fossil enaliarctids and their role in pinniped phylogeny. In *Proceedings of the seventh biennial conference on the biology of marine mammals,* Dec. 5–9, Miami, Fla.

Berta, A., and G. S. Morgan. 1985. A new sea otter (Carnivora: Mustelidae) from the late Miocene and early Pliocene (Hemphilian) of North America. *J. Paleo.* 59:809–819.

Bertram, G.C.L. 1940. The biology of the Weddell and crabeater seals. Sci. Rept. British Graham Land Exped., 1934–37, 1:1–139.

Berzin, A. 1972. The sperm whale. Israel Program for Scientific Translations, U.S. Dept. of Commerce, National Technical Information Service, Springfield, Va.

Best, P. B. 1973. Seals and sealing in South and South-West Africa. *South Africa Shipping News and Fishing Industry Review* 28:49–57.

Best, P. B., M. A. Meyer, and R. W. Weeks. 1981. Interactions between a male elephant seal *Mirounga leonina* and Cape fur seals *Arctocephalus pusillus. South African Journal of Zoology* 16:59–66.

Best, R. C. 1977. Ecological aspects of polar bear predation. In *Proc. 1975 predator symp.,* ed. R. L. Phillips and C. Jonkel, 203–211. Univ. of Montana, Missoula.

Bester, M. N. 1975. The functional morphology of the kidney of the Cape fur seal, *Arctocephalus pusillus* (Schreber). Modogua Series 2, no. 4, 69–92.

———. 1977. Habitat selection, seasonal population changes, and behavior of the Amsterdam Island fur seal, *Arctocephalus tropicalis,* on Gough Island. D.Sc. diss., Univ. Pretoria, Republic of South Africa.

———. 1987. Subantarctic fur seal, *Arctocephalus tropicalis,* at Gough Island (Tristan Da Cunha Group). In *Status, biology, and ecology of fur seals,* ed. J. Croxall and R. L. Gentry, 57–60. Proceedings of an International Symposium and Workshop, Cambridge, England, Apr. 23–27, 1984. NOAA Technical Report NMFS 51.

Bester, M. N., and G.I.H. Kerley. 1983. Rearing of twin pups to weaning by subantarctic fur seal, *Arctocephalus tropicalis* female. *S. Afr. J. Wildl. Res.* 13:86–87.

Bester, M. N., and P. A. Laycock. 1985. Cephalopod prey of the fur seal, *Arctocephalus tropicalis* at Gough Island. In *Antarctic nutrient cycles and food webs,* ed. W. R. Siegfried, P. R. Condy, and R. M. Laws, 551–554. Proceedings of the Fourth SCAR Symposium on Antarctic Biology. Berlin: Springer-Verlag.

Bigg, M. A. 1969a. Clines in the pupping season of the harbour seal, *Phoca vitulina. J. Fish. Res. Board Can.* 26:449–455.

———. 1969b. The harbour seal in British Columbia. *Fish. Res. Board Canada Bull.* 172:33.

———. 1973. Adaptations in the breeding of the harbour seal, *Phoca vitulina. J. Reprod. Fert.,* Supp. 19:131–142.

———. 1981. Harbour seal—*Phoca vitulina* and *P. largha.* In *Handbook of marine mammals.* Vol. 2, *Seals,* ed. S. H. Ridgway and R. J. Harrison, 1–27. London: Academic Press.

Bigg, M. A., and H. D. Fisher. 1975. Effect of photoperiod on annual reproduction in female harbor seals. *Rapp. P.-v. Reun. Cons. Int. Explor. Mer* 169: 141–144.

Bishop, R. H. 1967. Reproduction, age determination, and behavior of the harbor seal, *Phoca vitulina* L. in the Gulf of Alaska. Master's thesis, Univ. of Alaska, College, Alaska.

Blix, A. S., H. J. Grav, and K. Ronald. 1975. Brown adipose tissue and the significance of the venous plexuses in pinnipeds. *Acta Physiologica Scandinavica* 94:133–135.

Blix, A. S., L. K. Miller, M. C. Keyes, H. J. Grav, and R. Elsner. 1979. Newborn mother fur seals (*Callorhinus ursinus*)—do they suffer from cold? *American Journal of Physiology* 236:322–327.

Blix, A. S., and J. D. Steen. 1979. Temperature regulation in newborn polar homeotherms. *Physiological Reviews* 59:285–304.

Boness, D. J. 1979. The social system of the grey seal (*Halichoerus grypus*) on Sable Island, Nova Scotia. Ph.D. diss., Dalhousie Univ., Halifax, Nova Scotia.

Boness, D. J., S. S. Anderson, and C. R. Cox. 1982. Functions of female aggression during the pupping and mating season of grey seals, *Halichoerus grypus* (Fabricius). *Can. J. Zool.* 60:2270–2278.

Boness, D. J., W. D. Bowen, and O. T. Oftedal. 1987. Spacing of hooded seals: Evidence of polygyny. In *Proceedings of the seventh biennial conference on the biology of marine mammals,* Dec. 5–9, Miami, Fla.

Boness, D. J., L. Dabek, K. Ono, and O. T. Oftedal. 1985. Female attendance behavior in California sea lions. In *Proceedings of the sixth biennial conference on the biology of marine mammals,* Nov. 22–26, Vancouver, British Columbia.

Boness, D. J., and H. James. 1979. Reproductive behavior of the grey seal (*Halichoerus grypus*) on Sable Island, Nova Scotia. *Journal of Zoology* (London) 188:477–500.

Bonnell, M. L., and R. K. Selander. 1974. Elephant seals: Genetic variation and near extinction. *Science* 184:908–909.

Bonner, W. N. 1968. The fur seal of South Georgia. Brit. Ant. Survey Sci. Rep. no. 56.

———. 1971. An aged seal (*Halichoerus grypus*). *Journal of Zoology* (London) 164:261–262.

———. 1972. The grey seal and common seal in European waters. *Oceanogr. Mar. Biol. Ann. Rev.* 10:461–507.

———. 1975. Population increase of grey seals at the Farne Islands. *Rapp. P.-v. Reun. Cons. Int. Explor. Mer* 169:366–370.

———. 1979. Largha seal. In *Mammals in the seas.* Vol. 2, *Pinniped species summaries and reports on sirenians.* Food and Agriculture Organization (FAO) Fisheries Series no. 5. Rome: United Nations FAO.

———. 1981a. Grey seal—*Halichoerus grypus.* In *Handbook of marine mammals.* Vol. 2, *Seals,* ed. S. H. Ridgway and R. J. Harrison, 111–144. London: Academic Press.

———. 1981b. Southern fur seals—*Arctocephalus.* In *Handbook of marine mammals.* Vol. 1, *The walrus, sea lions, fur seals, and sea otters,* ed. S. H. Ridgway and R. J. Harrison, 161–208. London: Academic Press.

———. 1982. *Seals and man: A study of interactions.* Washington Sea Grant Publication. Seattle: Univ. of Washington Press.

———. 1984. Lactation strategies in pinnipeds: Problems for a marine mammalian group. *Symp. Zool. Soc. Lond.* 51:253–272.

Bonner, W. N., and G. Hickling. 1971. The grey seals of the Farne Islands: Report for the period October 1969–July 1971. *Transactions of the Natural History Society of Northumbria* 17:141–162.

Bonner, W. N., and S. Hunter. 1982. Predatory interactions between Antarctic fur seals, Macaroni penguins, and giant petrels. *British Antarctic Survey Bulletin* 56:75–79.

Bonner, W. N., and T. S. McCann. 1982. Neck collars on fur seals, *Arctocephalus gazella,* at South Georgia. *British Antarctic Survey Bulletin* 57:73–77.

Bonner, W. N., and S. R. Witthames. 1974. Dispersal of common seals (*Phoca vitulina*) tagged in the Wash, East Anglia. *Journal of Zoology* (London) 174:528–531.

Boswall, J. 1972. The South American sea lion *Otaria byronia* as a predator on penguins. *Bulletin British Ornithologist's Club* 92:129–132.

Boulva, J. 1979. Mediterranean Monk Seal. In *Mammals in the seas.* Vol. 2, *Pinniped species summaries and reports on sirenians,* 95–100. Food and Agriculture Organization (FAO) Fisheries Series no. 5. Rome: United Nations FAO.

Boulva, J., and I. A. McLaren. 1979. Biology of the harbor seal, *Phoca vitulina,* in Eastern Canada. *Fish. Res. Board Canada Bull.* 200:1–24.

Bowen, W. D., D. J. Boness, and O. T. Oftedal. 1987. Mass transfer from mother to pup and subsequent mass loss by the weaned pup in the hooded seal, *Cystophora cristata. Can. J. Zool.* 65:1–8.

Bowen, W. D., O. T. Oftedal, and D. J. Boness. 1985. Birth to weaning in four days: Remarkable growth in the hooded seal, *Cystophora cristata. Can. J. Zool.* 63:2841–2846.

Boyd, I. L. 1983. Luteal regression, follicle growth, and the concentration of some plasma steroids during lactation in grey seals (*Halichoerus grypus*). *J. Reprod. Fert.* 69:157–164.

———. 1984. The relationship between body condition and the timing of implantation in pregnant grey seals (*Halichoerus grypus*). *Journal of Zoology* (London) 203:113–123.

———. 1985. Pregnancy and ovulation rates in grey seals (*Halichoerus grypus*) on the British coast. *Journal of Zoology* (London) 205:265–272.

Brodie, P., and B. Beck. 1983. Predation by sharks on the grey seal (*Halichoerus grypus*) in eastern Canada. *Canadian Journal of Fisheries and Aquatic Sciences* 40:267–271.

Brooks, J. W. 1953. The Pacific walrus and its importance to the Eskimo economy. In *Transactions of the eighteenth North American wildlife conference,* 503–510. Washington, D.C.: Wildlife Management Institute.

———. 1954. A contribution to the life history and ecology of the Pacific Walrus. Master's thesis, Univ. of Alaska, Fairbanks.

Brown, K. G. 1957. The leopard seal at Heard Island, 1951–54. Australian National Antarctic Research Expedition Interior Reports 16.

Brown, R., and J. Harvey. 1981. Movements and dive characteristics of free-ranging, radio-tagged harbor seals, *Phoca vitulina.* In *Proceedings of the fourth biennial conference on the biology of marine mammals,* Dec. 14–18, San Francisco.

Brownell, R. L., R. L. DeLong, and R. W. Schreiber. 1974. Pinniped populations at Islas de Guadalupe, San Benito, Cedros, and Natividad, Baja California, in 1968. *Journal of Mammalogy* 55:469–472.

Bruce, W. S., and W. E. Clarke. 1899. The mammalia and birds of Franz Josef Land, part 1: Mammalia. *Proc. R. Phys. Soc.* (Edinburgh) 14:78–86.

Bryden, M. M. 1968. Lactation and suckling in relation to early growth of the southern elephant seal, *Mirounga leonina* (L). *Aust. J. Zool.* 16:739–748.

———. 1978. Arteriovenous anastomoses in the skin of seals, part 3: The harp seal *Pagophilus groenlandicus* and the hooded seal *Cystophora cristata. Aquatic Mammals* 6:67–75.

Bryden, M. M., and W.J.L. Felts. 1974. Quantitative anatomical observations on the skeletal and muscular systems of four species of Antarctic seals. *J. Anat.* 118:589–600.

Buckley, J. L. 1958. The Pacific walrus: A review of current knowledge and suggested management needs. U. S. Fish and Wildlife Service, Spec. Sci. Rep.—Wildl. no. 41.

Bukhtiyarov, Y. A., K. J. Frost, and L. E. Lowry. 1981. New information on foods of the spotted seal, *Phoca largha,* in the Bering Sea in spring. In *Soviet-American cooperative research on marine mammals.* Vol. 1, *Pinnipeds,* ed. F. H. Fay and G. A. Fedoseev, 55–59. NOAA Technical Report NMFS 12.

Burns, J. J. 1965. The walrus in Alaska: Its ecology and management. Juneau: Alaska Dept. of Fish and Game.

———. 1967. The Pacific bearded seal. Juneau: Alaska Dept. of Fish and Game.

———. 1969. Marine Mammal Investigations. Juneau: Alaska Dept. of Fish and Game.

———. 1970. Remarks on the distribution and natural history of pagophilic pinnipeds in the Bering and Chukchi Seas. *Journal of Mammalogy* 51: 445–454.

———. 1971. Biology of the ribbon seal, *Phoca* (*Histriophoca*) *fasciata*, in the Bering Sea (abstract). *Proc. Twenty-second Alaska Sci. Conf.*, p. 135.

———. 1981a. Bearded seal—*Erignathus barbatus*. In *Handbook of marine mammals*. Vol. 2, *Seals,* ed. S. H. Ridgway and R. J. Harrison, 145–170. London: Academic Press.

———. 1981b. Ribbon seal—*Phoca fasciata*. In *Handbook of marine mammals*. Vol. 2, *Seals,* ed. S. H. Ridgway and R. J. Harrison, 89–109. London: Academic Press.

Burns, J. J., and F. H. Fay. 1970. Comparative morphology of the skull of the ribbon seal, *Histriophoca fasciata,* with remarks on systematics of Phocidae. *Journal of Zoology* (London) 161:363–394.

Burns, J. J., F. H. Fay, and G. A. Fedoseev. 1981a. Craniological analysis of harbor and spotted seals of the North Pacific region. In *Soviet-American cooperative research on marine mammals*. Vol. 1, *Pinnipeds,* ed. F. H. Fay and G. A. Fedoseev, 5–16. NOAA Technical Report NMFS 12.

Burns, J. J., and K. J. Frost. 1979. The natural history and ecology of the bearded seal, *Erignathus barbatus*. Final report outer continental shelf environmental assessment for contract 02.5.022.53. Alaska Dept. of Fish and Game, Fairbanks.

Burns, J. J., and V. N. Goltsev. 1981. Comparative biology of harbor seals, *Phoca vitulina* Linnaeus, 1788, of the Commander, Aleutian, and Pribilof Islands. In *Soviet-American cooperative research on marine mammals*. Vol. 1, *Pinnipeds,* ed. F. H. Fay and G. A. Fedoseev, 17–24. NOAA Technical Report NMFS 12.

Burns, J. J., and S. J. Harbo, Jr. 1972. An aerial census of ringed seals, northern coast of Alaska. *Arctic* 25:279–290.

Burns, J. J., G. C. Ray, F. H. Fay, and P. D. Shaughnessy. 1972. Adoption of a strange pup by the ice-inhabiting harbor seal, *Phoca vitulina largha. Journal of Mammalogy* 53:594–598.

Burns, J. J., L. H. Shapiro, and F. H. Fay. 1981b. Ice as marine mammal habitat in the Bering Sea. In *The eastern Bering Sea shelf: Oceanography and resources,* vol. 2, ed. D. W. Hood and J. A. Calder, 781–797. Seattle: Univ. of Washington Press.

Burton, R. W., S. S. Anderson, and C. F. Summers. 1975. Perinatal activities in the grey seal (*Halichoerus grypus*). *Journal of Zoology* (London) 177:197–201.

Busch, B. C. 1985. *The war against the seals: A history of the North American seal fishery*. Montreal, Quebec, and Kingston, Ontario: McGill-Queen's Univ. Press.

Butterworth, D. S., J.H.M. David, L. H. McQuaid, and S. S. Xulu. 1987. Modeling the population dynamics of the South African fur seal *Arctocephalus pusillus pusillus.* In *Status, biology, and ecology of fur seals,* ed. J. Croxall and R. L. Gentry, 141–164. Proceedings of an International Symposium and Workshop, Cambridge, England, Apr. 23–27, 1984. NOAA Technical Report NMFS 51.

Cabrera, A., and J. Yepes. 1940. *Mamíferos sudamericanos (vida, costumbres y descripción).* Buenos Aires: Campañía Argentina de Editores.

Calambokidis, J., K. Bowman, S. Carter, J. Cubbage, P. Dawson, T. Fleischner, J. Skidmore, and B. Taylor. 1978. Chlorinated hydrocarbon concentrations and the ecology and behavior of harbor seals in Washington State waters. Evergreen State College, Olympia, Wash.

Calvert, W., and I. Stirling. 1987. Walrus and polar bear interactions. In *Proceedings of the seventh biennial conference on the biology of marine mammals,* Dec. 5–9, Miami, Fla.

Cameron, A. W. 1967. Breeding behavior in a colony of western Atlantic gray seals. *Can. J. Zool.* 45:161–173.

Campagna, C. 1985. The breeding cycle of the southern sea lion, *Otaria byronia. Marine Mammal Science* 1:210–218.

———. 1987. The breeding behavior of the southern sea lion. Ph.D. diss., Univ. of California, Santa Cruz.

Campagna, C., and B. J. Le Boeuf. 1988. Reproductive behavior of southern sea lions. *Behaviour* 104:223–261.

———. Thermoregulatory behavior affects male mating behavior and mating success in southern sea lions. Under submission.

Campagna, C., B. J. Le Boeuf, and H. L. Cappozzo. 1988. Group raids: A mating strategy of male southern sea lions. *Behaviour* 105:224–249.

Carey, F. G., G. Gabrielson, J. W. Kanwisher, and O. Brazier. 1982. The white shark, *Carcharodon carcharias,* is warm-bodied. *Copeia,* 254–260.

Carl, G. C. 1964. Diving rhythm of the hair seal. *Victorian Naturalist* (British Columbia) 21:35–37.

Carrara, I. S. 1952. Lobos marinos, pinguinos y guaneras de las costas del litoral marítimo e islas adyacentes de la República Argentina. Universidad Nacional de la Plata, Fac. de Cienc. Veterinarias, Cátedra de Higiene e Industrias, Publ. Especial.

Carrick, R., S. E. Csordas, and S. E. Ingham. 1962a. Studies on the southern elephant seal, *Mirounga leonina* (L), part 4: Breeding and development. *Commonwealth Scientific and Industrial Research Organization (C.S.I.R.O.) Wildlife Research* 7:161–197.

Carrick, R., S. E. Csordas, S. E. Ingham, and K. Keith. 1962b. Studies on the southern elephant seal, *Mirounga leonina* (L), part 3: The annual cycle in relation to age and sex. *Commonwealth Scientific and Industrial Research Organization (C.S.I.R.O.) Wildlife Research* 7:119–160.

Carrick, R., and S. E. Ingham. 1962. Studies on the southern elephant seal, part 5: Population dynamics and utilization. *Commonwealth Scientific and Industrial Research Organization (C.S.I.R.O.) Wildlife Research* 7:190–206.

Castilla, J. C. 1981. Informe de actividades del Taller Científico sobre identificación de mandíbulas de cefalopodos (Plymouth, 1981). *Bol. Antart. Chil.* 1:23–25.

Chapskii, K. K. 1936. The walrus of the Kara Sea. *Trudy Vsesoiuz. arktich. inst.* (Leningrad) 67:1–124.

———. 1938. The bearded seal (*Erignathus barbatus,* Fabr.) of the Kara and Barents Seas. *Trans. Arctic Inst.* Leningrad. 123:7–70.

Charnov, E. L. 1982. *The theory of sex allocation.* Princeton, N.J.: Princeton Univ. Press.

Clarke, M. R., and N. Macleod. 1982a. Cephalopod remains in the stomachs of eight Weddell seals. *British Antarctic Survey Bulletin* 57:33–40.

———. 1982b. Cephalopods in the diet of elephant seals at Signy Island, South Orkney Islands. *British Antarctic Survey Bulletin* 57:27–31.

Clarke, M. R., and F. Trillmich. 1980. Cephalopods in the diet of fur seals of the Galápagos Islands. *Journal of Zoology* (London) 190:211–215.

Cleator, H., I. Stirling, and T. G. Smith. 1987. Geographical variation in the repertoire of the bearded seal. In *Proceedings of the seventh biennial conference on the biology of marine mammals,* Dec. 5–9, Miami, Fla.

Cole, L. C. 1954. The population consequences of life history phenomena. *Quarterly Review of Biology* 29:103–137.

Collett, R. 1881. On *Halichoerus grypus* and its breeding on the Fro Islands off Trondhjems-Fjord in Norway. *Proc. Zool. Soc.* (London) 380–387.

Conboy, M. E. 1975. Project "quickfind": A marine mammal system for object recovery. *Rapp. P.-v. Reun. Cons. Int. Explor. Mer* 169:487–500.

Condit, R. S., and B. J. Le Boeuf. 1984. Feeding habits and feeding grounds of the northern elephant seal. *Journal of Mammalogy* 65:281–290.

Condit, R. S., and C. L. Ortiz. 1985. Food consumption and utilization by northern elephant seals. In *Proceedings of the sixth biennial conference on the biology of marine mammals,* Nov. 22–26, Vancouver, British Columbia.

Condy, P. R. 1977a. Annual cycle of the elephant seal *Mirounga leonina* (Linn.) at Marion Island. *South African Journal of Zoology* 14:95–102.

———. 1977b. Whale observation in the pack ice off the Fimbul Ice Shelf, Antarctica. *South African J. Antarctic Res.* 7:7–9.

———. 1978. Distribution, abundance, and annual cycle of fur seals (*Arctocephalus* spp.) on the Prince Edward Islands. *S. Afr. J. Wildl. Res.* 8:159–168.

———. 1981. Annual food consumption, and seasonal fluctuations in biomass of seals at Marion Island. *Mammalia* (Paris) 45:21–30.

Condy, P. R., R. J. van Aarde, and M. N. Bester. 1978. The seasonal occurrence of killer whales (*Orcinus orca*) at Marion Island. *Journal of Zoology* (London) 184:449–464.

Cooper, C. F., and J. R. Jehl. 1980. Potential effects of space shuttle booms on the biota and geology of the California islands: Synthesis of research and recommendations. Center for Marine Studies, San Diego State Univ., San Diego, Calif. Technical Report 80-2.

Cooper, J. 1974. The predators of the jackass penguin *Spheniscus demerus. Bulletin British Ornithologist's Club* 94:21–24.

Costa, D. P. 1985. Lactation and bioenergetic strategies for rapid growth. Plenary talk presented at the Sixth Biennial Conference on the Biology of Marine Mammals, Nov. 22–26, Vancouver, British Columbia.

Costa, D. P., and R. L. Gentry. 1986. Free-ranging energetics of northern fur seal. In *Fur seals: Maternal strategies on land and at sea,* ed. R. L. Gentry and G. L. Kooyman, 79–101. Princeton, N.J.: Princeton Univ. Press.

Costa, D. P., B. J. Le Boeuf, A. C. Huntley, and C. L. Ortiz. 1986. The energetics of lactation in the northern elephant seal. *Journal of Zoology* (London) 209:21–33.

Costa, D. P., and C. L. Ortiz. 1983. The physiology of natural prolonged fasting in the northern elephant seal. IUPS, *Proceedings*, vol. 15, Twenty-ninth Congress, Sydney, Australia.

Costa, D. P., P. H. Thorson, J. G. Herpolsheimer, and J. P. Croxall. 1985. Reproductive bioenergetics of the Antarctic fur seal, *Arctocephalus gazella*. *Antarctic Journal of the United States* 20:176–177.

Cott, H. B. 1961. Scientific results of an inquiry into the ecology and economic status of the Nile crocodile (*Crocodilus niloticus*) in Uganda and northern Rhodesia. *Trans. Zool. Soc. Lond.* 29:211–356.

Coulson, J. C., and G. Hickling. 1964. The breeding biology of the grey seal, *Halichoerus grypus*. (Fab.), on the Farne Islands, Northumberland. *Journal of Animal Ecology* 33:485–512.

Cox, C. R., and B. J. Le Boeuf. 1977. Female incitation of male competition: A mechanism in sexual selection. *American Naturalist* 111:317–335.

Crawley, M. C., and R. Warneke. 1979. New Zealand fur seal. In *Mammals in the seas*. Vol. 2, *Pinniped species summaries and reports on sirenians*, 45–48. Food and Agriculture Organization (FAO) Fisheries Series no. 5. Rome: United Nations FAO.

Crawley, M. C., and G. J. Wilson. 1976. The natural history and behavior of the New Zealand fur seal (*Arctocephalus forsteri*). *Tuatata* 22:1–19.

Credle, V. R. 1987. Biogenic magnetite isolated from the dura of pygmy sperm whales. In *Proceedings of the seventh biennial conference on the biology of marine mammals*, Dec. 5–9, Miami, Fla.

Cressie, N.A.C., and P. D. Shaughnessy. 1987. Statistical methods for estimating numbers of Cape fur seal pups from aerial surveys. *Marine Mammal Science* 3:297–307.

Cresswell, F. P. 1985. Underwater locomotion in phocids: A descriptive and comparative study of the grey seal. In *Proceedings of the sixth biennial conference on the biology of marine mammals*, Nov. 22–26, Vancouver, British Columbia.

Croxall, J. P., and R. L. Gentry, eds. 1987. *Status, biology, and ecology of fur seals*. Proceedings of an International Symposium and Workshop, Cambridge, England, Apr. 23–27, 1984. NOAA Technical Report NMFS 51.

Croxall, J. P., and M. N. Pilcher. 1984. Characteristics of krill *Euphausia superba* eaten by Antarctic fur seals *Arctocephalus gazella* at South Georgia. *British Antarctic Survey Bulletin* 63:117–125.

Csordas, S. E. 1965. A few facts about the southern elephant seal. *Victorian Naturalist* 82:69–74.

Csordas, S. E., and S. E. Ingham. 1965. The New Zealand fur seal, *Arctocephalus forsteri* (Lesson), at Macquarie Island, 1949–64. *Commonwealth Scientific and Industrial Research Organization (C.S.I.R.O.) Wildlife Research* 10:83–89.

Curry-Lindahl, K. 1970. Breeding biology of the Baltic grey seal (*Halichoerus grypus*). *Der Zoologische Gärten* (Leipzig) 38:16–29.

———. 1975. Ecology and conservation of the grey seal *Halichoerus grypus*, common seal *Phoca vitulina*, and ringed seal *Pusa hispida* in the Baltic Sea. *Rapp. P.-v. Reun. Cons. Int. Explor. Mer* 169:527–532.

da Silva, J., and J. M. Terhune. 1985. Harbour seal grouping as an anti-predator strategy. In *Proceedings of the sixth biennial conference on the biology of marine mammals,* Nov. 22–26, Vancouver, British Columbia.

David, J.H.M. 1987a. Diet of the South African (Cape) fur seal (1974–1985) and an assessment of competition with fisheries in southern Africa. In *The Benguela and comparable ecosystems,* ed. A.I.L. Payne, J. A. Gulland, and K. H. Brink. *S. Afr. J. Mar. Sci.* 5.

———. 1987b. South African fur seal, *Arctocephalus pusillus pusillus.* In *Status, biology, and ecology of fur seals,* ed. J. Croxall and R. L. Gentry, 65–71. Proceedings of an International Symposium and Workshop, Cambridge, England, Apr. 23–27, 1984. NOAA Technical Report NMFS 51.

David, J.H.M., and R. W. Rand. 1986. Attendance behavior of South African fur seals. In *Fur seals: Maternal strategies on land and at sea,* ed. R. L. Gentry and G. L. Kooyman, 126–141. Princeton, N.J.: Princeton Univ. Press.

Davies, J. L. 1957. The geography of the gray seal. *Journal of Mammalogy* 38: 297–310.

Davydov, A. F., and A. R. Makarova. 1965. Changes in the temperature of the skin of the harp seal during ontogenesis as related to the degree of cooling. In *Morskie mlekopitaiushchie,* ed. E. N. Pavlovskii, 262–265. Moscow: Nauka. Translated by Trans-Bureau, Foreign Languages Division, Canada.

Dawson, W. R., G. A. Bartholomew, and A. F. Bennett. 1977. A reappraisal of the aquatic specializations of the Galápagos marine iguana (*Amblyrhynchus cristatus*). *Evolution* 31:891–897.

Dearborn, J. H. 1965. Food of Weddell seals at McMurdo Sound, Antarctica. *Journal of Mammalogy* 46:37–43.

de Jong, W. W. 1982. Eye lens proteins and vertebrate phylogeny. In *Macromolecular sequences in systematic and evolutionary biology,* ed. M. Goodman, 75–114. New York: Plenum.

Dellinger, T. 1987. Das Nahrungsspektrum der sympatrischen Galapagos-Seebären (*Arctocephalus galapagoensis*) und Galapagos-Seelöwen (*Zalophus californianus*) mit Versuchen zur Methodik der Kotanalyse. Diplomarbeit, Universität Konstanz.

DeLong, R. L. 1982. Population biology of northern fur seals at San Miguel Island, California. Ph.D. diss., Univ. of California, Berkeley.

DeLong, R. L., and G. A. Antonelis. 1985. Impact of the 1982–83 El Niño on the northern fur seal population at San Miguel Island. In *Proceedings of the sixth biennial conference on the biology of marine mammals,* Nov. 22–26, Vancouver, British Columbia.

DeLong, R. L., and R. L. Brownell. 1977. Hawaiian monk seal (*Monachus schauinslandi*) habitat and population survey in the Northwestern (Leeward) Hawaiian Islands, April 1977. Northwest and Alaska Fisheries Center Processed Report, National Marine Fisheries Service, Seattle, Wash.

DeMaster, D. P. 1978. Calculation of the average age of sexual maturity in marine mammals. *J. Fish. Res. Board Can.* 35:912–915.

Depocas, F., J. S. Hart, and H. D. Fisher. 1971. Sea water drinking and water flux in starved and fed harbor seals, *Phoca vitulina. Can. J. Physiol. Pharmacol.* 49:53–62.

Deutsch, C. J. 1985. Male-male competition in the Hawaiian monk seal. In

Proceedings of the sixth biennial conference on the biology of marine mammals, Nov. 22–26, Vancouver, British Columbia.

Deutsch, C. J. 1987. Behavioral aspects of reproductive effort in male northern elephant seals. Do alpha bulls try harder? In *Proceedings of the seventh biennial conference on the biology of marine mammals,* Dec. 5–9, Miami, Fla.

de Visscher, M. N. 1985. An integrated strategy for the conservation of the monk seal (*Monachus monachus*) in the Mediterranean. In *Proceedings of the sixth biennial conference on the biology of marine mammals,* Nov. 22–26, Vancouver, British Columbia.

Doidge, D. W. 1987. Rearing of twin offspring to weaning in Antarctic fur seals, *Arctocephalus gazella.* In *Status, biology, and ecology of fur seals,* ed. J. Croxall and R. L. Gentry, 107–111. Proceedings of an International Symposium and Workshop, Cambridge, England, Apr. 23–27, 1984. NOAA Technical Report NMFS 51.

Doidge, D. W., and J. P. Croxall. 1985. Diet and energy budget of the Antarctic fur seal *Arctocephalus gazella* at South Georgia. In *Antarctic nutrient cycles and food webs,* ed. W. R. Siegfried, P. R. Condy, and R. M. Laws, 543–550. Proceedings of the Fourth SCAR Symposium on Antarctic Biology. Berlin: Springer-Verlag.

Doidge, D. W., J. P. Croxall, and C. Ricketts. 1984. Growth rates of Antarctic fur seal *Arctocephalus gazella* pups at South Georgia. *Journal of Zoology* (London) 203:87–93.

Doidge, D. W., T. S. McCann, and J. P. Croxall. 1986. Attendance behavior of Antarctic fur seals. In *Fur seals: Maternal strategies on land and at sea,* ed. R. L. Gentry and G. L. Kooyman, 102–114. Princeton, N.J.: Princeton Univ. Press.

Eaton, R. L. 1979. A beluga whale imitates human speech. *Carnivore* 2:22–23.

Eibl-Eibesfeldt, I. 1967. Concepts of ethology and their significance in the study of human behavior. In *Early behavior: Comparative and developmental approaches,* ed. H. W. Stevenson, 127–146. New York: Wiley.

———. 1984a. The large iguanas of the Galápagos Islands. In *Key environments: Galápagos,* ed. R. Perry, 157–173. Oxford: Pergamon.

———. 1984b. The natural history of the Galápagos sea lion (*Zalophus californianus wollebaeki,* Sivertsen). In *Key environments: Galápagos,* ed. R. Perry, 207–214. Oxford: Pergamon.

Eisenberg, J. F. 1981. *The mammalian radiations: An analysis of trends in evolution, adaptation, and behavior.* Chicago: Univ. of Chicago Press.

Eley, T. J. 1978. A possible case of adoption in the Pacific walrus. *Murrelet* 59:77–78.

Elsner, R., D. D. Hammond, D. M. Denison, and R. Wyburn. 1977. Temperature regulation in the newborn Weddell seal *Leptonychotes weddelli.* In *Adaptations within Antarctic ecosystems,* ed. G. A. Llano, 531–540. Proceedings of the Third SCAR Symposium on Antarctic Biology, Aug. 26–30, 1974. Washington, D.C.: Smithsonian Institution.

Emlen, S. T., and L. W. Oring. 1977. Ecology, sexual selection, and the evolution of mating systems. *Science* 197:215–223.

Erickson, A. W., and R. J. Hofman. 1974. Antarctic seals. Antarctic Map Folio Series 18:4–12.

Erickson, A. W., D. B. Siniff, D. R. Cline, and R. J. Hofman. 1971. Distributional ecology of Antarctic seals. In *Symposium on Antarctic ice and water masses*, ed. G. Deacon, 55–76. Cambridge, England: Scientific Committee on Antarctic Research.

Estes, J. A. 1979. Exploitation of marine mammals: r-selection of K-strategists? *J. Fish. Res. Board Can.* 36:1009–1017.

———. (In press). Adaptations for aquatic living in carnivores. In *Carnivore behavior, ecology, and evolution*, ed. John Gittleman. Ithaca, N.Y.: Cornell Univ. Press.

Estes, J. A., and V. N. Goltsev. 1981. Abundance and distribution of the Pacific walrus: Results of the first Soviet-American joint aerial survey, autumn 1975. In *Soviet-American cooperative research on marine mammals*. Vol. 1, *Pinnipeds*, ed. F. H. Fay and G. A. Fedoseev. NOAA Technical Report NMFS 12.

Estes, J. A., R. J. Jameson, and A. M. Johnson. 1981. Food selection and some foraging tactics of sea otters. In *Worldwide furbearer conference proceedings*, ed. J. A. Chapman and D. Pursley, 606–641, Aug. 3–11, 1980. Frostburg, Md.

Estes, J. A., and J. F. Palmisano. 1974. Sea otters: Their role in structuring nearshore communities. *Science* 185:1058–1060.

Estes, R. D., and J. Goddard. 1967. Prey selection and hunting behavior of African wild dogs. *J. Wildl. Manage.* 31:52–70.

Evans, W. E., and J. Bastian. 1969. Marine mammal communication: Social and ecological factors. In *The biology of marine mammals*, ed. H. T. Anderson, 424–475. London: Academic Press.

Everitt, R. D., and H. W. Braham. 1980. Aerial survey of Pacific harbor seals in the southeastern Bering Sea. *Northwest Science* 54:281–288.

Fadely, B. S. 1987. Determination of seawater ingestion and food intake by water and sodium turnover in juvenile northern fur seals (*Callorhinus ursinus*). In *Proceedings of the seventh biennial conference on the biology of marine mammals*, Dec. 5–9, Miami, Fla.

Fagen, R. 1981. *Animal play behavior*. New York: Oxford Univ. Press.

Fancher, L. 1977. Diving behavior of the harbor seal, *Phoca vitulina*, in San Francisco Bay. In *Proceedings of the second biennial conference on the biology of marine mammals*, Dec. 12–15, San Diego, Calif.

Farentinos, R. C. 1971. Some observations on the play behavior of the Steller sea lion (*Eumetopias jubata*). *Z. Tierpsychol.* 28:428–438.

Fay, F. H. 1957. History and present status of the Pacific walrus population. In *Transactions of the twenty-second North American wildlife and natural resources conference*, 431–445, Mar. 4–6. Washington, D.C.: Wildlife Management Institute.

———. 1960. Carnivorous walrus and some arctic zoonoses. *Arctic* 13:111–122.

———. 1974a. Mammals and birds. In *Bering Sea oceanography: An update*, ed. Y. Takenouti and D. W. Hood, 133–138. Fairbanks: Univ. of Alaska Institute of Marine Science Report no. 75-2.

———. 1974b. The role of ice in the ecology of marine mammals of the Bering Sea. In *Oceanography of the Bering Sea*, ed. D. W. Hood and E. J. Kelley, 383–399. Fairbanks: Univ. of Alaska Institute of Marine Science.

———. 1981a. Marine mammals of the eastern Bering Sea shelf: An overview.

In *The eastern Bering Sea shelf: Oceanography and resources,* vol. 2, ed. D. W. Hood and J. A. Calder, 807–811. Seattle: Univ. of Washington Press.

———. 1981b. Walrus, *Odobenus rosmarus.* In *Handbook of marine mammals.* Vol 1, *The walrus, sea lions, fur seals, and sea otters,* ed. S. H. Ridgway and R. J. Harrison, 1–23. London: Academic Press.

———. 1982. Ecology and biology of the Pacific walrus, *Odobenus rosmarus divergens* Illiger. North American Fauna, no. 74. U.S. Dept. of the Interior, Fish and Wildlife Service, Washington, D.C.

Fay, F. H., Y. A. Bukhtiyarov, S. W. Stoker, and L. M. Shults. 1981a. Foods of the Pacific walrus in winter and spring in the Bering Sea. In *Soviet-American cooperative research on marine mammals.* Vol. 1, *Pinnipeds,* ed. F. H. Fay and G. A. Fedoseev, 81–88. NOAA Technical Report NMFS 12.

Fay, F. H., H. M. Feder, and S. W. Stoker. 1977. An estimate of the impact of the Pacific walrus population on its food resources in the Bering Sea. U.S. Dept. of Commerce, National Technical Information Service, Springfield, Va. PB-273-505.

Fay, F. H., B. P. Kelly, and J. L. Sease. 1989. Managing the exploitation of Pacific walruses: A tragedy of delayed response and poor communication. *Marine Mammal Science* 5:1–16.

Fay, F. H., and C. Ray. 1968. Influence of climate on the distribution of walruses, *Odobenus rosmarus* (Linnaeus), part 1: Evidence from thermo-regulatory behavior. *Zoologica* 53:1–18.

Fay, F. H., G. C. Ray, and A. A. Kibal'chich. 1981b. Time and location of mating and associated behavior of the Pacific walrus, *Odobenus rosmarus divergens* Illiger. In *Soviet-American cooperative research on marine mammals.* Vol. 1, *Pinnipeds,* ed. F. H. Fay and G. A. Fedoseev, 89–99. NOAA Technical Report NMFS 12.

Fedak, M. A., and S. S. Anderson. 1982. The energetics of lactation: Accurate measurements from a large wild mammal, the Grey seal (*Halichoerus grypus*). *Journal of Zoology* (London) 198:473–479.

Fedak, M. A., J. Kanwisher, B. McConnell, K. Nicholas, M. Pullen, and P. Thompson. 1985. Diving behavior and circulatory responses of free living common seals using VHF and acoustic telemetry. In *Proceedings of the sixth biennial conference on the biology of marine mammals,* Nov. 22–26, Vancouver, British Columbia.

Fedak, M. A., and T. S. McCann. 1987. Parental investment in southern elephant seals, *Mirounga leonina.* In *Proceedings of the seventh biennial conference on the biology of marine mammals,* Dec. 5–7, Miami, Fla.

Fedorova, Z. P., and Z. S. Yankina. 1963. The passage of Pacific Ocean water through the Bering Strait into the Chukchi Sea. *Okeanologiya* 3. In *Translations from "Okeanologiya."* Oxford: Pergamon.

Fedoseev, G. A. 1965. Food of the ringed seals. *Izv. Tikhookean. nauchno-issled. inst. rybn. khoz. okeanogr.* 59:216–223.

———. 1975. Ecotypes of the ringed seal (*Pusa hispida* Schreber, 1777) and their reproductive capabilities. *Rapp. P.-v. Reun. Cons. Int. Explor. Mer* 169: 156–160.

———. 1976. Principal population indicators of dynamics of numbers of seals of the family Phocidae. *Ekologiya* 5:62–70. Translated by Consultants Bureau, New York.

Fedoseev, G. A., and Y. A. Bukhtiyarov. 1972. The diet of seals of the Okhotsk Sea. In *Proceedings of the fifth all-union conference studies of marine mammals* (USSR), part 1, 110–112. Translated by F. H. Fay.

Feldkamp, S. D., R. L. DeLong, and G. A. Antonelis, Jr. 1983. Diving behavior of California sea lions (*Zalophus californianus*). In *Proceedings of the fifth biennial conference on the biology of marine mammals,* Nov. 27–Dec. 1, Boston.

Ferren, H., and R. Elsner. 1979. Diving physiology of the ringed seal: Adaptations and implications. In *Proceedings of the twenty-ninth Alaska science conference,* Aug. 15–17, 1978, 379–387, Fairbanks, Alaska.

Finch, V. A. 1966. Maternal behavior in the harbor seal. In *Proceedings of the third annual conference on biological sonar and diving mammals,* ed. C. E. Rice, 147–150. Menlo Park, Calif.: Stanford Research Institute Biological Sonar Laboratories.

Fink, B. D. 1959. Observations of porpoise predation on a school of Pacific sardines. *Calif. Fish and Game* 45:216–217.

Fiscus, C. H. 1978. Northern fur seal. In *Marine mammals of eastern North Pacific and Arctic waters,* ed. D. Haley, 152–159. Seattle, Wash.: Pacific Search Press.

Fiscus, C. H., and G. A. Baines. 1966. Food and feeding behavior of Steller and California sea lions. *Journal of Mammalogy* 47:195–200.

Fisher, H. D. 1952. The status of the harbour seal in British Columbia, with particular reference to the Skeena River. *Fish. Res. Board. Canada Bull.* 93:58.

Fleischer, L. A. 1978a. The distribution, abundance, and population characteristics of the Guadalupe fur seal, *Arctocephalus townsendi* (Merriam, 1897). Ph.D. diss., Univ. of Washington, Seattle.

———. 1978b. Guadalupe fur seal. In *Marine mammals of eastern North Pacific and Arctic waters,* ed. D. Haley, 160–165. Seattle, Wash.: Pacific Search Press.

———. 1987. Guadalupe fur seal, *Arctocephalus townsendi.* In *Status, biology, and ecology of fur seals,* ed. J. Croxall and R. L. Gentry, 43–48. Proceedings of an International Symposium and Workshop, Cambridge, England, Apr. 23–27, 1984. NOAA Technical Report NMFS 51.

Fobes, J. L., and C. C. Smock. 1981. Sensory capacities of marine mammals. *Psychol. Bull.* 89:288–307.

Fogden, S.C.L. 1968. Suckling behavior in the grey seal (*Halichoerus grypus*) and the northern elephant seal (*Mirounga angustirostris*). *Journal of Zoology* (London) 154:415–420.

———. 1971. Mother-young behavior at grey seal breeding beaches. *Journal of Zoology* (London) 164:61–92.

Fowler, C. W. 1982. Interactions of northern fur seals and commercial fisheries. In *Transactions of the forty-seventh North American wildlife and natural resources conference,* 278–292. Washington, D.C.: Wildlife Management Institute.

Francis, J. M. 1987. Inter-female aggression and spacing in the northern fur seal (*Callorhinus ursinus*) and California sea lion (*Zalophus californianus*). Ph.D. diss., Univ. of California, Santa Cruz.

Francis, J. M., and C. B. Heath. 1985. Duration of maternal care in the California sea lion—bias by sex. In *Proceedings of the sixth biennial conference on the biology of marine mammals,* Nov. 22–26, Vancouver, British Columbia.

Frankel, R. P., R. P. Blakemore, F. F. Torres de Araujo, D.M.S. Esquivel, and J. Dannon. 1981. Magnetotactic bacteria at the geomagnetic equator. *Science* 212:1269–1270.

Freuchen, P. 1935. Mammals, part 2: Field notes and biological observations. Report of the Fifth Thule Expedition, 1921–1924, vol. 2, 68–278. Copenhagen: Nordisk-Forlag.

Frisch, J., N. A. Øritsland, and J. Krog. 1974. Insulation of furs in water. *Comparative Biochemistry and Physiology* 47A:403–410.

Frost, K. J., and L. F. Lowry. 1980. Feeding of ribbon seals (*Phoca fasciata*) in the Bering Sea in spring. *Can. J. Zool.* 58:1601–1607.

———. 1981. Ringed, Baikal, and Caspian Seals. In *Handbook of marine mammals*. Vol. 2, *Seals,* ed. S. H. Ridgway and R. J. Harrison, 29–53. London: Academic Press.

Fukuyama, A. K., and J. S. Oliver. 1985. Sea star and walrus predation on bivalves in Norton Sound, Bering Sea, Alaska. *Ophelia* 24:17–36.

Gallo-R., J. P., and A. Ortega-O. 1986. The first report of *Zalophus californianus* in Acapulco, Mexico. *Marine Mammal Science* 2:158.

Gambell, R. 1973. Some effects of exploration in reproduction of whales. *J. Reprod. Fert. Supp.* 19:533–553.

Gaskin, D. E. 1972. *Whales, dolphins, and seals, with special reference to the New Zealand region.* London: Heinemann.

Geiger, A. C. 1985. Evaluation of seal harassment devices to protect salmon in gillnet fisheries. In *Proceedings of the sixth biennial conference on the biology of marine mammals,* Nov. 22–26, Vancouver, British Columbia.

Geist, U. 1971. Mountain sheep: a study in behavior and evolution. Chicago: Univ. of Chicago Press.

Gentry, R. L. 1967. Underwater auditory localization in the California sea lion (*Zalophus californianus*). *Journal of Auditory Research* 7:187–193.

———. 1970. Social behavior of the Steller sea lion. Ph.D. diss., Univ. of California, Santa Cruz.

———. 1973. Thermoregulatory behavior of eared seals. *Behaviour* 46:73–93.

———. 1974. The development of social behavior through play in the Steller sea lion. *American Zoologist* 14:391–403.

———. 1975. Comparative social behavior of eared seals. *Rapp. P.-v. Reun. Cons. Int. Explor. Mer* 169:189–194.

———. 1980. Set in their ways. *Oceans* 13:34–37.

———. 1981a. Northern fur seal—*Callorhinus ursinus.* In *Handbook of marine mammals.* Vol. 1, *The walrus, sea lions, fur seals, and sea otters,* ed. S. H. Ridgway and R. J. Harrison, 143–160. London: Academic Press.

———. 1981b. Seawater drinking in eared seals. *Comparative Biochemistry and Physiology* 68A:81–86.

Gentry, R. L., D. P. Costa, J. P. Croxall, J.H.M. David, R. W. Davis, G. L. Kooyman, P. Majluf, T. S. McCann, and F. Trillmich. 1986. Synthesis and conclusions. In *Fur seals: Maternal strategies on land and at sea,* ed. R. L. Gentry and G. L. Kooyman, 220–264. Princeton, N.J.: Princeton Univ. Press.

Gentry, R. L., and J. R. Holt. 1986. Attendance behavior of northern fur seals. In *Fur seals: Maternal strategies on land and at sea,* ed. R. L. Gentry and G. L. Kooyman, 41–60. Princeton, N.J.: Princeton Univ. Press.

Gentry, R. L., and J. H. Johnson. 1981. Predation by sea lions on northern fur seal neonates. *Mammalia* (Paris) 45:423–430.

Gentry, R. L., and G. L. Kooyman. 1986. Introduction. In *Fur seals: Maternal strategies on land and at sea,* ed. R. L. Gentry and G. L. Kooyman, 3–27. Princeton, N.J.: Princeton Univ. Press.

Gentry, R. L., G. L. Kooyman, and M. E. Goebel. 1986. Feeding and diving behavior of northern fur seals. In *Fur seals: Maternal strategies on land and at sea,* ed. R. L. Gentry and G. L. Kooyman, 61–78. Princeton, N.J.: Princeton Univ. Press.

Gentry, R. L., W. E. Roberts, and M. W. Cawthorn. 1987. Diving behavior of the Hooker's sea lion. In *Proceedings of the seventh biennial conference on the biology of marine mammals,* Dec. 5–9, Miami, Fla.

Geraci, J. R., and D. J. St. Aubin. 1980. Offshore petroleum resource development and marine mammals: A review and research recommendations. *Marine Fisheries Review* 42:1–12.

Gilbert, J. R., and A. W. Erickson. 1977. Distribution and abundance of seals in pack ice of the Pacific sector of the Southern Ocean. In *Adaptations within Antarctic ecosystems,* ed. G. A. Llano. Proceedings of the Third SCAR Symposium on Antarctic Biology, Aug. 26–30, 1974. Washington, D.C.: Smithsonian Institution.

Gilmartin, W. G. 1983. Recovery plan for the Hawaiian monk seal, *Monachus schauinslandi.* In cooperation with the Hawaiian Monk Seal Recovery Team, NOAA, NMFS, Southwest Region, Terminal Island, Calif.

Gilmartin, W. G., T. Gerrodette, D. J. Alcorn, T. C. Johanos, J. R. Henderson, R. L. Westlake, R. G. Forsyth, and L. D. Banish. 1987. The Hawaiian monk seal: Population status and current research and recovery activities. In *Proceedings of the seventh biennial conference on the biology of marine mammals,* Dec. 5–9, Miami, Fla.

Gisiner, R. C. 1985. Male territorial and reproductive behavior in the Steller sea lion, *Eumetopias jubatus.* Ph.D. diss., Univ. of California, Santa Cruz.

———. 1987. Reproductive life histories of male Steller sea lions: The importance of experience. In *Proceedings of the seventh biennial conference on the biology of marine mammals,* Dec. 5–9, Miami, Fla.

Goltsev, V. N. 1971. Feeding of the harbor seal (in Russian). *Ekologiya* 2:62–70. Available in translation from U.S. National Oceanic and Atmospheric Administration.

Goltsev, V. N., and G. A. Fedoseev. 1970. Dynamics of the age composition of rookeries and the replacement capacity of harbour seal populations (in Russian). *Izv. Tikhookean. nauchno-issled. inst. rybn. khoz. okeanogr.* 71:309–317. Fisheries Research Board of Canada Translation ser. no. 2401.

Goodall, J. 1986. *The chimpanzees of Gombe: Patterns of behavior.* Cambridge: Harvard Univ. Press, Belknap Press.

Gould, J. L., J. L. Kirschvink, and K. S. Deffeyes. 1978. Bees have magnetic remanence. *Science* 201:1026–1028.

Grav, H. J., and A. S. Blix. 1975. Brown adipose tissue—a factor in the survival of harp seal pups *Vitam impendere vero.* In *Depressed metabolism and cold thermogenesis,* ed. L. Jansky. Prague: Academia.

Grav, H. J., A. S. Blix, and A. Pasche. 1974. How do seal pups survive birth in Arctic winter? *Acta Physiologica Scandinavia* 92:427–429.

Griffin, D. R. 1984. *Animal thinking*. Cambridge: Harvard Univ. Press.

Guinee, L. N., K. Chu, and E. M. Dorsey. 1983. Changes over time in the songs of known individual humpback whales (*Megaptera novaeangliae*). In *Communication and behavior of whales,* ed. R. Payne. American Association for the Advancement of Science Symposium Series. Boulder, Colo.: Westview Press.

Gwynn, A. M. 1953. Notes on the fur seals at Macquarie Island and Heard Island. Australian National Antarctic Research Expedition (A.N.A.R.E.) Interim Report 4.

Hacker, E. S. 1986. Stomach content analysis of short-finned pilot whales (*Globicephala macrorhynchus*) and northern elephant seals (*Mirounga angustirostris*) from the southern California Bight. National Marine Fisheries Service, Southwest Fisheries Center Administrative Report LJ-86-08C.

Hamilton, C. L. 1981. The use of sound in harbor seal communication. Master's thesis, Humboldt State Univ., Arcadia, Calif.

Hamilton, J. E. 1934. The southern sea lion, *Otaria byronia* (de Blainville). *Discovery Reports* 8:269–318.

———. 1939. The leopard seal *Hydrurga leptonyx* (de Blainville). *Discovery Reports* 18:239–264.

———. 1946. Seals preying on birds. *Ibis,* Jan., 131–132.

Hamilton, W. D. 1964. The genetical theory of social behavior, parts 1 and 2. *Journal of Theoretical Biology* 7:1–52.

———. 1971. Geometry for the selfish herd. *Journal of Theoretical Biology* 31: 295–311.

Hammill, M. O., and T. G. Smith. 1987. Function of the ringed seal subnivean lair. In *Proceedings of the seventh biennial conference on the biology of marine mammals,* Dec. 5–9, Miami, Fla.

Hanggi, E. B., and R. J. Schusterman. 1987. Preferential social bonding in captive California sea lions. In *Proceedings of the seventh biennial conference on the biology of marine mammals,* Dec. 5–9, Miami, Fla.

Hansen, D. J. 1987. Shark predation and the decline of the northern fur seal. In *Proceedings of the seventh biennial conference on the biology of marine mammals,* Dec. 5–9, Miami, Fla.

Harrison, R. J. 1960. Reproduction and reproductive organs in common seals (*Phoca vitulina*) in the Wash, East Anglia. *Mammalia* (Paris) 24:372–385.

———. 1969. Reproduction and reproductive organs. In *The biology of marine mammals,* ed. H. T. Andersen, 253–348. London: Academic Press.

Harrison, R. J., and G. L. Kooyman. 1968. General physiology of the pinnipedia. In *The behavior and physiology of pinnipeds,* ed. R. J. Harrison, R. C. Hubbard, R. S. Peterson, C. E. Rice, and R. J. Schusterman, 211–296. New York: Appleton-Century-Crofts.

Hart, J. S., and L. Irving. 1959. The energetics of harbour seals in air and in water with special consideration of seasonal changes. *Can. J. Zool.* 37: 447–457.

Hartman, D. S. 1979. Ecology and Behavior of the Manatee (*Trichechus manatus*) in Florida. Special Publication, American Society of Mammalogy no. 5, 1–153.

Harwood, J. 1983. Interactions between marine mammals and fisheries. *Advances in Applied Biology* 8:189–214.

Harwood, J., and J. P. Croxall. 1988. The assessment of competition between seals and commercial fisheries in the North Sea and the Antarctic. *Marine Mammal Science* 4:13–33.

Harwood, J., and J. H. Prime. 1978. Some factors affecting the size of British grey seal populations. *Journal of Applied Ecology* 15:401–411.

Heath, C. B. 1985. The effects of environment on the breeding system of the California sea lion (*Zalophus californianus*). In *Proceedings of the sixth biennial conference on the biology of marine mammals,* Nov. 22–26, Vancouver, British Columbia.

Heath, C. B., and J. M. Francis. 1987. Mechanisms and consequences of mate choice in the California sea lion. In *Proceedings of the seventh biennial conference on the biology of marine mammals,* Dec. 5–9, Miami, Fla.

Hemmingsen, E. A., and E. L. Douglas. 1970. Ultraviolet radiation thresholds for corneal injury in Antarctic and temperate-zone animals. *Comparative Biochemistry and Physiology* 32:593–600.

Hendey, Q. B., and C. A. Repenning. 1972. A Pliocene phocid from South Africa. *Ann. S. Afr. Mus.* 59:71–98.

Heptner, V. G. 1976. *Mammals of the Soviet Union.* Vol. 2, *Pinnipeds and toothed whales* (in Russian). Moscow: Publishing House for Higher Schools.

Herman, L. M., D. G. Richards, and J. P. Wolz. 1984. Comprehension of sentences by bottlenosed dolphins. *Cognition* 16:129–219.

Herman, L. M., and W. N. Tavolga. 1980. The communication systems of cetaceans. In *Cetacean behavior: Mechanisms and functions,* ed. L. M. Herman, 149–210. New York: Wiley.

Hess, E. H. 1973. *Imprinting: Early experience and the developmental psychology of attachment.* New York: Van Nostrand Reinhold.

Hess, E. H., and S. Petrovich, eds. 1977. *Imprinting.* Stroudsburg, Pa.: Dowden, Hutchinson and Ross.

Hewer, H. R. 1957. A Hebridean breeding colony of grey seals, *Halichoerus grypus* (Fab.) with comparative notes on the grey seals of Ramsey Island, Pembrokeshire. *Proc. Zool. Soc.* (London) 128:23–64.

———. 1964. The determination of age, sexual maturity, longevity, and a life table in the grey seal *Halichoerus grypus. Proc. Zool. Soc.* (London) 132:593–624.

Hewer, H. R., and K. M. Backhouse. 1960. A preliminary account of a colony of grey seals, *Halichoerus grypus* (Fab.) in the southern Inner Hebrides. *Proc. Zool. Soc.* (London) 134:157–195.

Hickling, G. 1962. Grey seals and the Farne Islands. London: Routledge and Kegan Paul.

Higgins, L. V. 1987. Breeding behavior of the Australian sea lion, *Neophoca cinerea.* In *Proceedings of the seventh biennial conference on the biology of marine mammals,* Dec. 5–9, Miami, Fla.

Higgins, L. V., D. P. Costa, A. C. Huntley, and B. J. Le Boeuf. 1988. Behavior and physiological measurements of the maternal investment in the Steller sea lion, *Eumetopias jubatus. Marine Mammal Science* 4:44–58.

Hindell, M. A., and G. J. Little. 1988. Longevity, fertility, and philopatry of two female southern elephant seals (*Mirounga leonina*) at Macquarie Island. *Marine Mammal Science* 4:168–177.

Hofman, R. J., and W. N. Bonner. 1985. Conservation and protection of marine mammals: Past, present, and future. *Marine Mammal Science* 1: 109–127.

Hofman, R. J., R. A. Reichle, D. B. Siniff, and D. Muller-Schwarze. 1977. The leopard seal (*Hydrurga leptonyx*) at Palmer Station, Antarctica. In *Adaptations within Antarctic ecosystems*, ed. G. A. Llano, 769–782. Proceedings of the Third SCAR Symposium on Antarctic Biology, Aug. 26–30, 1974. Washington, D.C.: Smithsonian Institution.

Hoover, A. 1983. Behavior and ecology of harbor seals (*Phoca vitulina richardsi*) inhabiting glacial ice in Aialik Bay, Alaska. Master's thesis, Univ. of Alaska, Fairbanks.

Horning, D. S., Jr., and G. D. Fenwick. 1978. Leopard seals at the Snares Islands, New Zealand. *New Zealand Journal of Zoology* 5:171–172.

Huber, H. 1985. Age specific natality and the age of first reproduction in female northern elephant seals on the Farallon Islands, California. In *Proceedings of the sixth biennial conference on the biology of marine mammals*, Nov. 22–26, Vancouver, British Columbia.

Huntley, A. C. 1984. Relationships between metabolism, respiration, heart rate, and arousal states in the northern elephant seal. Ph.D. diss., Univ. of California, Santa Cruz.

Huntley, A. C., and D. P. Costa. 1983. Cessation of ventilation during sleep: A unique mode of energy conservation in the northern elephant seal. *Proc. Int. Union Physiol. Sci.* 15:203.

Huntley, A. C., D. P. Costa, and D. R. Rubin. 1984. The contribution of nasal countercurrent heat exchange to water balance in the northern elephant seal, *Mirounga angustirostris*. *J. Exp. Biol.* 113:447–454.

Huntley, A. C., D. P. Costa, G.A.J. Worthy, and M. A. Castellini, eds. 1987. Approaches to marine mammal energetics. Special Publications no. 1, Society for Marine Mammalogy, Lawrence, Kans.

Hyvarinen, H. 1987. Diving in darkness: Whiskers as sense organs of the ringed seal. In *Proceedings of the seventh biennial conference on the biology of marine mammals*, Dec. 5–9, Miami, Fla.

Imler, R. H., and H. R. Sarber. 1947. Harbor seals and sea lions in Alaska. U.S. Fish and Wildlife Service, Spec. Sci. Rep. 28.

Innes, H. S., D. M. Lavigne, W. M. Earle, and K. M. Kovacs. 1985. Feeding rates of pinnipeds and other carnivora. In *Proceedings of the sixth biennial conference on the biology of marine mammals*, Nov. 22–26, Vancouver, British Columbia.

———. 1987. Feeding rates of seals and whales. *Journal of Animal Ecology* 56:115–130.

Innes, H. S., and K. Ronald. 1985. Daily energy expenditure, heat increment of feeding, and postabsorptive, resting metabolic rates of ringed, harp, and grey seals. In *Proceedings of the sixth biennial conference on the biology of marine mammals*, Nov. 22–26, Vancouver, British Columbia.

Innes, S. R., E. A. Stewart, and D. M. Lavigne. 1978. Growth in northwest Atlantic harp seals, *Pagophilus groenlandicus*: Density dependence and recent changes in energy availability. Canadian Atlantic Fisheries Scientific Advisory Committee Working Paper 78/46.

Irving, L. 1966. Elective regulation of the circulation in diving animals. In *Whales, dolphins, and porpoises,* ed. K. S. Norris, 381–396. Berkeley: Univ. of California Press.

———. 1969. Temperature regulation in marine mammals. In *The biology of marine mammals,* ed. H. T. Andersen, 147–174. London: Academic Press.

———. 1973. Aquatic mammals. In *Comparative physiology of thermoregulation.* Vol. 3, *Special aspects of thermoregulation,* ed. G. C. Whittow, 47–96. New York: Academic Press.

Irving, L., K. C. Fisher, and F. C. McIntosh. 1935. The water balance of a marine mammal, the seal. *J. Comp. Physiol.* 6:387–391.

Irving, L., and J. S. Hart. 1957. The metabolism and insulation of seals as bare-skinned mammals in cold water. *Can. J. Zool.* 35:497–511.

Irving, L., L. G. Peyton, C. H. Bahn, and R. Peterson. 1962. Regulation of temperature in fur seals. *Physiological Zoology* 35:275–284.

Iverson, J. A., and J. Krog. 1973. Heat production and body surface area in seals and sea otters. *Norwegian J. Zoology* 21:51–54.

Jamieson, G. S., and H. D. Fisher. 1972. The pinniped eye: A review. In *Functional anatomy of marine mammals,* vol. 1, ed. R. J. Harrison. London: Academic Press.

Jannasch, H. W., R. E. Marquis, and A. M. Zimmerman. 1987. *Current perspectives in high pressure biology.* London: Academic Press.

Jarman, P. J. 1974. The social organization of antelope in relation to their ecology. *Behaviour* 58:215–267.

Jerison, H. J. 1973. *Evolution of the brain and intelligence.* New York: Academic Press.

Johanos, R., and A. Kam. 1986. The Hawaiian monk seal on Lisianski Island, 1983. U.S. Dept. of Commerce. NOAA Technical Memorandum, NMFS, Southwest Fisheries Center no. 58.

Johanos, T. C., and J. R. Henderson. 1986. Hawaiian monk seal reproduction and injuries on Lisianski Island, 1982. NOAA Technical Memorandum, NMFS, Southwest Fisheries Center no. 64.

Johnson, A. M., R. L. DeLong, C. H. Fiscus, and K. W. Kenyon. 1982. Population status of the Hawaiian monk seal (*Monachus schauinslandi*), 1978. *Journal of Mammalogy* 6:415–421.

Johnson, B. W., and P. A. Johnson. 1978. The Hawaiian monk seal on Laysan Island, 1977. Final Report to the U.S. Marine Mammal Commission, Washington, D.C., for Contract MM7AC009. U.S. Dept. of Commerce, National Technical Information Service, Springfield, Va. Report no. MMC-77/05.

———. 1984. Observations of the Hawaiian monk seal on Laysan Island from 1977–1980. NOAA Technical Memorandum, NMFS, Southwest Fisheries Center no. 49.

Johnson, M. L., C. H. Fiscus, B. T. Ostenson, and M. L. Barbour. 1966. Marine mammals. In *Environment of the Cape Thompson region, Alaska,* ed. N. J. Wilimovsky and J. N. Wolfe, 897–924. U.S. Atomic Energy Commission, Oak Ridge, Tenn.

Jouventin, P., and A. Cornet. 1980. The sociobiology of pinnipeds. In *Advances in the study of behavior,* vol. 2, 121–141. New York: Academic Press.

Kajimura, H. 1984. Opportunistic feeding of the northern fur seal, *Callorhinus ursinus,* in the eastern North Pacific Ocean and eastern Bering Sea. U.S. Dept. of Commerce, NOAA Technical Report NMFS SSRF-779.

Kapel, F. O. 1975. Recent research on seals and seal hunting in Greenland. *Rapp P.-v. Reun. Cons. Int. Explor. Mer* 169:462–478.

Kaufman, G. W., D. B. Siniff, and R. Reichle. 1975. Colony behavior of Weddell seals, *Leptonychotes weddelli,* at Hutton Cliffs, Antarctica. *Rapp. P.-v. Reun. Cons. Int. Explor. Mer* 169:228–246.

Kellogg, W. N. 1958. Echo ranging in the porpoise. *Science* 128:982–988.

Kelly, B. P., S. C. Amstrup, C. Gardner, and L. T. Quakenbush. 1987. Predation on ringed seals in the western Beaufort Sea. In *Proceedings of the seventh biennial conference on the biology of marine mammals,* Dec. 5–9, Miami, Fla.

Kenyon, K. W. 1952. Diving depths of the Steller sea lion and Alaska fur seal. *Journal of Mammalogy* 33:245–246.

———. 1966. Marine wildlife observations, Leeward Hawaiian Islands, Sept. 8–27, 1966. U.S. Fish and Wildlife Service.

———. 1969. *The sea otter in the eastern Pacific Ocean.* North American Fauna, no. 68. U.S. Dept. of the Interior, Bureau of Sport Fisheries and Wildlife.

———. 1972. Man versus the monk seal. *Journal of Mammalogy* 53:687–696.

———. 1973. Hawaiian monk seal (*Monachus schauinslandi*). International Union for the Conservation of Nature (IUCN) Pub. New Series Supp. Paper no. 39:88–97.

———. 1977. Caribbean monk seal extinct. *Journal of Mammalogy* 58:97–98.

———. 1981. Monk seals—*Monachus.* In *Handbook of marine mammals.* Vol. 2, *Seals,* ed. S. H. Ridgway and R. J. Harrison, 195–220. London: Academic Press.

Kenyon, K. W., and D. W. Rice. 1959. Life history of the Hawaiian monk seal. *Pacific Science* 13:215–252.

Kenyon, K. W., and F. Wilke. 1953. Migration of the northern fur seal, *Callorhinus ursinus. Journal of Mammalogy* 34:87–98.

Kerley, G.I.H. 1983. Comparison of seasonal haulout patterns of fur seals *Arctocephalus tropicalis* and *A. gazella* on subantarctic Marion Island. *S. Afr. J. Wildl. Res.* 13:71–77.

———. 1987. *Arctocephalus tropicalis* on the Prince Edward Islands. In *Status, biology, and ecology of fur seals,* ed. J. Croxall and R. L. Gentry, 61–64. Proceedings of an International Symposium and Workshop, Cambridge, England, Apr. 23–27, 1984. NOAA Technical Report NMFS 51.

Kerley, G.I.H., and T. J. Robinson. 1987. Skull morphometrics of male Antarctic and subantarctic fur seals *Arctocephalus gazella* and *A. tropicalis,* and their interspecific hybrids. In *Status, biology, and ecology of fur seals,* ed. J. Croxall and R. L. Gentry, 121–131. Proceedings of an International Symposium and Workshop, Cambridge, England, Apr. 23–27, 1984. NOAA Technical Report NMFS 51.

Keyes, M. C. 1968. The nutrition of pinnipeds. In *The behavior and physiology of pinnipeds,* ed. R. J. Harrison, R. C. Hubbard, R. S. Peterson, C. E. Rice, and R. J. Schusterman, 359–395. New York: Appleton-Century-Crofts.

Keyes, M. C., E. J. Barron, and A. J. Ross. 1971. Analysis of urine of the northern fur seal. *J. Am. Vet. Med. Assoc.* 159:567–570.

Khuzin, R. S., and A. V. Yablokov. 1963. Some functional aspects of the digestive tract of the hooded seal (*Cystophora cristata* Erxl.) in the period of milk feeding. *Zoologicheskii zhurnal* (Moscow) 42:1273–1275.

King, J. E. 1961. Notes on the pinnipeds from Japan described by Temminck in 1844. *Zoologische Mededelingen* (Leiden) 37:211–224.

———. 1964a. *Seals of the world.* London: British Museum of Natural History.

———. 1964b. Swallowing modifications in the Ross seal. *J. Anat.* (London) 99:206–207.

———. 1965. Giant epiphyses in a Ross seal. *Nature* (London) 205:515–516.

———. 1969. Some aspects of the anatomy of the Ross seal, *Ommatophoca rossi* (Pinnipedia: Phocidae). Brit. Ant. Survey Sci. Rep. no. 63.

———. 1983. *Seals of the world.* 2d ed. London: British Museum of Natural History; Ithaca, N.Y.: Cornell Univ. Press.

Kirschvink, J. L. 1982. Birds, bees, and magnetism. A new look at the old problem of magnetoreception. *Trends in Neurosciences* 5:160–167.

———. 1987. Geomagnetic sensitivity in marine animals: Possible magnetoreceptors and migration strategies. In *Proceedings of the seventh biennial conference on the biology of marine mammals,* Dec. 5–9, Miami, Fla.

Klinowska, M. 1985. Cetacean live strandings relate to geomagnetic topography. *Aquatic Mammals* 1:27–32.

———. 1986. Cetacean live stranding dates relate to geomagnetic disturbances. *Aquatic Mammals* 11:109–119.

Knudtson, P. M. 1973. Observations on the behavior of harbor seal pups (*Phoca vitulina richardsi*) in Humboldt Bay, California. Tenth Annual Conference on Biological Sonar and Diving Mammals.

Konishi, M. 1965. The role of auditory feedback in the control of vocalization in the white-crowned sparrow. *Z. Tierpsychol.* 22:770–783.

Kooyman, G. L. 1965a. Leopard seals of Cape Crozier. *Animals* 6:59–63.

———. 1965b. Techniques used in measuring diving capacities of Weddell seals. *Polar Record* 12:391–394.

———. 1966. Maximum diving capacities of the Weddell seal, *Leptonychotes weddelli. Science* 151:1553–1554.

———. 1968. An analysis of some behavioral and physiological characteristics related to diving in the Weddell seal. *Antarctic Research* 2:227–261.

———. 1969. The Weddell seal. *Scientific American* 221:100–106.

———. 1975. A comparison between day and night diving in the Weddell seal. *Journal of Mammalogy* 56:563–574.

———. 1981a. Crabeater seal, *Lobodon carcinophagus.* In *Handbook of marine mammals.* Vol. 2, *Seals,* ed. S. H. Ridgway and R. J. Harrison, 221–238. London: Academic Press.

———. 1981b. Leopard seal, *Hydrurga leptonyx.* In *Handbook of marine mammals.* Vol. 2, *Seals,* ed. S. H. Ridgway and R. J. Harrison, 261–274. London: Academic Press.

———. 1981c. Weddell seal: Consummate diver. Cambridge: Cambridge Univ. Press.

———. 1981d. Weddell seal, *Leptonychotes weddelli.* In *Handbook of marine mammals.* Vol. 2, *Seals,* ed. S. H. Ridgway and R. J. Harrison, 275–296. London: Academic Press.

Kooyman, G. L., M. A. Castellini, R. W. Davis, and R. A. Maue. 1983. Aerobic diving limits of immature Weddell seals. *J. Comp. Physiol.* 151:171–174.

Kooyman, G. L., and D. P. Costa. 1979. Effects of oiling on temperature regulation in sea otters. Outer Continental Shelf Energy Assessment Program Yearly Progress Report, NOAA.

Kooyman, G. L., R. W. Davis, and J. P. Croxall. 1986. Diving behavior of Antarctic fur seals. In *Fur seals: Maternal strategies on land and at sea,* ed. R. L. Gentry and G. L. Kooyman, 115–125. Princeton, N.J.: Princeton Univ. Press.

Kooyman, G. L., and C. M. Drabek. 1968. Observations on milk, blood, and urine components of the Weddell seal. *Physiological Zoology* 41:187–194.

Kooyman, G. L., and R. L. Gentry. 1986. Diving behavior of South African fur seals. In *Fur seals: Maternal strategies on land and at sea,* ed. R. L. Gentry and G. L. Kooyman, 142–152. Princeton, N.J.: Princeton Univ. Press.

Kooyman, G. L., D. H. Kerem, W. B. Campbell, and J. J. Wright. 1973. Pulmonary gas exchange in freely diving Weddell seals. *Respiration Physiology* 17:283–290.

Kooyman, G. L., J. P. Schroeder, D. M. Denison, D. D. Hammond, J. J. Wright, and W. P. Bergman. 1972. Blood nitrogen tensions of seals during simulated deep dives. *American Journal of Physiology* 223:1016–1020.

Kooyman, G. L., and F. Trillmich. 1986a. Diving behavior of Galápagos fur seals. In *Fur seals: Maternal strategies on land and at sea,* ed. R. L. Gentry and G. L. Kooyman, 186–195. Princeton, N.J.: Princeton Univ. Press.

———. 1986b. Diving behavior of Galápagos sea lions. In *Fur seals: Maternal strategies on land and at sea,* ed. R. L. Gentry and G. L. Kooyman, 209–219. Princeton, N.J.: Princeton Univ. Press.

Kooyman, G. L., E. A. Wahrenbrock, M. A. Castellini, R. W. Davis, and E. E. Sinnett. 1980. Aerobic and anaerobic metabolism during volunteer diving in Weddell seals: Evidence of preferred pathways from blood chemistry and behavior. *J. Comp. Physiol.* 138:335–346.

Kovacs, K. M. 1987. Maternal behavior and early behavioral ontogeny of harp seals, *Phoca groenlandica. Animal Behaviour* 35:844–855.

Kovacs, K. M., and D. M. Lavigne. 1986a. *Cystophora cristata. Mammalian Species* 258:1–9.

———. 1986b. Maternal investment and neonatal growth in phocid seals. *Journal of Animal Ecology* 55:1035–1051.

———. 1986c. Neonatal growth and organ allometry of northwest Atlantic harp seals (*Phoca groenlandica*). *Can. J. Zool.* 63:2793–2797.

Kozhov, M. 1963. *Lake Baikal and its life.* The Hague: Junk.

Krebs, J. R. 1974. Colonial nesting and social feeding as strategies for exploring food resources in the great blue heron (*Ardea herodias*). *Behaviour* 51:99–134.

Krebs, J. R., and D. E. Kroodsma. 1980. Repertoires and geographical variation in bird song. *Adv. Study Behav.* 11:143–177.

Kroodsma, D. E. 1983. The ecology of avian vocal learning. *BioScience* 33:165–171.

Krylov, V. I. 1966. The sexual maturation of Pacific walrus females. *Zoologicheskii zhurnal* (Moscow) 45:919–927.

———. 1968. Present condition of the Pacific walrus stocks and prospects of their rational exploitation. In *Lastonogie severnoi chasti Tikhogo okeana,* ed. V. A. Arsen'ev and K. I. Panin, 189–204. Moscow: Pishchevaia promyshlennost'.

———. 1969. The period of mating and pupping of the Pacific walrus. In *Marine mammals,* ed. V. A. Arsen'ev, B. A. Zenkovich, and K. K. Chapskii, 275–285. Moscow: Nauka.

———. 1971. On the food of the Pacific walrus (*Odobenus rosmarus divergens* Ill.). In *Issledovaniia morskikh mlekopitaiushchikh,* Trudy, vyp. 39, ed. E. S. Mil'chenko, T. M. Andreeva, and G. P. Burov, 110–116. Kaliningrad: Atlant. NIRO.

Kumlien, L. 1879. Contributions to the natural history of Arctic-America, made in connection with the Howgate Polar Expedition, 1877–78. *Bull. U.S. Nat. Mus.* 15:1–179.

Kvitek, R. G., and J. S. Oliver. 1986. Side-scan sonar estimates of the utilization of gray whale feeding grounds along Vancouver Island, Canada. *Continental Shelf Research* 6:639–654.

Laevastu, T., and H. A. Larkins. 1981. Marine fisheries ecosystem: Its quantitative evaluation and management. Farnham, England: Fishing News Books.

Lander, R. H. 1981. A life table and biomass estimate for northern fur seals. *Fisheries Research* (Amsterdam) 1:55–70.

Lander, R. H. and K. Kajimura. 1976. Status of northern fur seals. United Nations Food and Agriculture Organization, Advisory Committee on Maritime Resource Research, FAO Scientific Consultation on Marine Mammals, Aug. 31–Sept. 9, Bergen, Norway.

Lansing, A. 1959. *Endurance: Shackleton's incredible voyage.* New York: McGraw-Hill.

Lavigne, D. M. 1973. Visual sensitivity in seals. Ph.D. diss., Univ. of Guelph, Ontario.

———. 1979. Harp seal. In *Mammals in the seas.* Vol. 2, *Pinniped species summaries and reports on sirenians,* Food and Agriculture Organization (FAO) Fisheries Series, no. 5. Rome: United Nations FAO.

———. 1982a. Marine mammal-fishery interactions: A report from an International Union for the Conservation of Nature (IUCN) workshop. In *Transactions of the forty-seventh North American wildlife and natural resources conference,* 312–321. Washington, D.C.: Wildlife Management Institute.

———. 1982b. Pinniped thermoregulation: Comments on the "effects of cold on the evolution of pinniped breeding systems." *Evolution* 36:409–414.

Lavigne, D.M., C. D. Bernholz, and K. Ronald. 1977. Functional aspects of pinniped vision. In *Functional Anatomy of Marine Mammals,* vol. 3, ed. R. J. Harrison, 135–173. London: Academic Press.

Lavigne, D. M., S. Innes, G. A. Worthy, K. M. Kovacs, O. J. Schmitz, and J. P. Hickie. 1986. Metabolic rates of seals and whales. *Can. J. Zool.* 64:279–284.

Lavigne, D. M., and K. M. Kovacs. 1988. *Harps and hoods: Ice-breeding seals of the northwest Atlantic.* Ontario: Univ. of Waterloo Press.

Lavigne, D. M., and K. Ronald. 1972. The harp seal, *Pagophilus groenlandicus* (Erxleben, 1777). 23. Spectral Sensitivity. *Can. J. Zool.* 50:1197–1206.

———. 1975. Pinniped visual pigments. *Comparative Biochemistry and Physiology* 50B:325–329.

Lavigne, D. M., R.E.A. Stewart, and F. Fletcher. 1982. Changes in composition and energy content of harp seal milk during lactation. *Physiological Zoology* 55:1–9.

Laws, R. M. 1953a. The elephant seal (*Mirounga leonina* Linn.), part 1: Growth and age. Falkland Islands Dependencies Survey, London. Sci. Rep. no. 8, 1–62.

———. 1953b. A new method of age determination in mammals, with special reference to the elephant seal (*Mirounga leonina* Linn.). Falkland Islands Dependencies Survey, London. Sci. Rep. no. 2, 1–11.

———. 1956a. The elephant seal (*Mirounga leonina* Linn.), part 2: General, social and reproductive behavior. Falkland Islands Dependencies Survey, London. Sci. Rep. no. 13, 1–88.

———. 1956b. The elephant seal (*Mirounga leonina* Linn.), part 3: The physiology of reproduction. Falkland Islands Dependencies Survey, London. Sci. Rep. no. 15, 1–66.

———. 1957. On the growth rates of the leopard seal, *Hydrurga leptonyx* (de Blainville, 1820). *Saugetierkunde Mitteilungen* 5:49–55.

———. 1960. The southern elephant seal (*Mirounga leonina* Linn.) at South Georgia. *Norsk Hvalfangst-Tidende* 49:466–476 and 520–542.

———. 1962. Some effects of whaling on the southern stocks of baleen whales. In *The exploitation of natural animal populations,* ed. E. D. LeCren and M. W. Holdgate, 137–158. A Symposium of the British Ecological Society, Oxford.

———. 1977a. Seals and whales of the Southern Ocean. *Philosophical Transactions of the Royal Society of London* B279:81–96.

———. 1977b. The significance of vertebrates in the Antarctic marine ecosystem. In *Adaptations within Antarctic ecosystems,* ed. G. A. Llano, 411–438. Proceedings of the Third SCAR Symposium on Antarctic Biology. Washington, D.C.: Smithsonian Institution.

———. 1981. Biology of Antarctic seals. *Sci. Prog.* (Oxford) 67:377–397.

———. 1983. Antarctica: A convergence of life. *New Scientist* 99:608–616.

———. 1984. Seals. In *Antarctic ecology,* vol. 2, ed. R. M. Laws, 621–715. London: Academic Press.

Lea, R. N., and D. J. Miller. 1985. Shark attacks off the California and Oregon coasts: An update, 1980–84. In *Biology of the white shark.* Memoirs of the Southern California Academy of Sciences, Los Angeles, vol. 9, 136–150.

Leatherwood, S., and W. F. Samaris. 1974. Some observations of killer whales, *Orcinus orca,* attacking other marine animals. *Proc. South. Calif. Acad. Sci. Meet.,* Fullerton, Calif.

Le Boeuf, B. J. 1974. Male-male competition and reproductive success in elephant seals. *American Zoologist* 14:163–176.

———. 1977. Back from extation? *Pacific Discovery* 30:1–9.

———. 1978. Social behavior in some marine and terrestrial carnivores. In *Contrasts in behavior,* ed. E. S. Reese and F. J. Lighter, 251–279. New York: Wiley.

——. 1981. Elephant seals. In *The natural history of Año Nuevo,* ed. B. J. Le Boeuf and S. Kaza, 326–374. Pacific Grove, Calif.: Boxwood Press.

——. 1985. Mating systems on land, at sea, and in the air. In *Proceedings of the sixth biennial conference on the biology of marine mammals,* Nov. 22–26, Vancouver, British Columbia.

——. 1986. Sexual strategies of seals and walruses. *New Scientist* 1491:36–39.

——. (In press). *Negative effects of mating for female northern elephant seals: Advances in the study of behavior.* New York: Academic Press.

Le Boeuf, B. J., D. G. Ainley and T. J. Lewis. 1974. Elephant seals on the Farallones: Population structure of an incipient breeding colony. *Journal of Mammalogy* 55:370–385.

Le Boeuf, B. J., and M. Bonnell. 1980. Pinnipeds of the California islands: Abundance and distribution. In *Proceedings of a multidisciplinary symposium,* ed. D. Power, 475–493. Santa Barbara, Calif.: Santa Barbara Museum of Natural History.

Le Boeuf, B. J., and R. S. Condit. 1983. The high cost of living on the beach. *Pacific Discovery* 36:12–14.

Le Boeuf, B. J., D. P. Costa and A. C. Huntley. 1985. Pattern and depth of dives in female northern elephant seals. In *Proceedings of the sixth biennial conference on the biology of marine mammals,* Nov. 22–26, Vancouver, British Columbia.

Le Boeuf, B. J., D. P. Costa, A. C. Huntley, and S. D. Feldkamp. 1988. Continuous, deep diving in female northern elephant seals, *Mirounga angustirostris. Can. J. Zool.* 66:446–458.

Le Boeuf, B. J., D. P. Costa, A. C. Huntley, G. L. Kooyman, and R. W. Davis. 1986a. Pattern and depth of dives in northern elephant seals, *Mirounga angustirostris. Journal of Zoology* (London) 208:1–7.

Le Boeuf, B. J., K. W. Kenyon, and B. Villa-Ramirez. 1986b. The Caribbean monk seal is extinct. *Marine Mammal Science* 2:70–73.

Le Boeuf, B. J., Y. Naito, A. C. Huntley, and T. Asaga. (In press). Prolonged, continuous deep-diving by northern elephant seals. *Can. J. Zool.*

Le Boeuf, B. J., and C. L. Ortiz. 1977. Composition of elephant seal milk. *Journal of Mammalogy* 58:683–685.

Le Boeuf, B. J., and K. J. Panken. 1977. Elephant seals breeding on the mainland in California. In *Proceedings of the California Academy of Sciences* 41:267–280.

Le Boeuf, B. J., and R. S. Peterson. 1969. Dialects in elephant seals. *Science* 166:1654–1656.

Le Boeuf, B. J., and L. F. Petrinovich. 1974. Elephant seals: Interspecific comparisons of vocal and reproductive behavior. *Mammalia* (Paris) 38:16–32.

——. 1975. Elephant seal dialects: Are they reliable? *Rapp. P.-v. Reun. Cons. Int. Explor. Mer* 169:213–218.

Le Boeuf, B. J., and J. Reiter. 1988. Lifetime reproductive success in northern elephant seals. In *Reproductive success,* ed. T. H. Clutton-Brock, 344–362. Chicago: Univ. of Chicago Press.

Le Boeuf, B. J., M. L. Riedman, and R. S. Keyes. 1982. White shark predation on pinnipeds in California coastal waters. *Fishery Bulletin* 80:891–895.

Le Boeuf, B. J., R. J. Whiting, and R. F. Gantt. 1972. Perinatal behavior of northern elephant seal females and their young. *Behaviour* 43:121–156.

Lenfant, C., K. Johansen, and J. D. Torrance. 1970. Gas transport and oxygen storage capacity in some pinnipeds and the sea otter. *Respiration Physiology* 9:277–286.

Lewontin, R. C. 1965. Selection for colonizing ability. In *The genetics of colonizing species,* ed. H. G. Baker and G. L. Stebbins, 77–94. New York: Academic Press.

Lilly, J. C. 1962. Vocal behavior of the bottlenose dolphin. *Proceedings of the American Philosophical Society* 106:520–529.

———. 1965. Vocal mimicry in *Tursiops:* Ability to match numbers and durations of human vocal bursts. *Science* 147:300–301.

Limberger, D., F. Trillmich, G. L. Kooyman, and P. Majluf. 1983. Reproductive failure of fur seals in Galápagos and Peru in 1982–83. *Trop. Ocean-Atmos. Newsl.* no. 21, 16–17.

Lindsey, A. A. 1937. The Weddell seal in the Bay of Whales, Antarctica. *Journal of Mammalogy* 18:127–144.

Ling, J. K. 1968. The skin and hair of the southern elephant seal, *Mirounga leonina* (L.), part 3: Morphology of the adult integument. *Aust. J. Zool.* 16:629–645.

———. 1970. Pelage and molting in wild mammals with special reference to aquatic forms. *Quarterly Review of Biology* 45:16–54.

———. 1974. The integument of marine mammals. In *Functional anatomy of marine mammals,* vol. 2, ed. R. J. Harrison, 1–44. London: Academic Press.

———. 1987. New Zealand fur seal, *Arctocephalus forsteri,* in southern Australia. In *Status, biology, and ecology of fur seals,* ed. J. Croxall and R. L. Gentry, 53–55. Proceedings of an International Symposium and Workshop, Cambridge, England, Apr. 23–27, 1984. NOAA Technical Report NMFS 51.

Ling, J. K., and M. M. Bryden. 1981. Southern elephant seal — *Mirounga leonina.* In *Handbook of marine mammals.* Vol. 2, *Seals,* ed. S. H. Ridgway and R. J. Harrison, 297–327. London: Academic Press.

Ling, J. K., and G. E. Walker. 1976. Seal studies in South Australia: Progress report for the year 1975. *South Australian Naturalist* 50:59–68, 72.

———. 1977. Seal studies in South Australia. Progress report for the period January 1976 to March 1977. *South Australian Naturalist* 52:18–27.

———. 1978. An 18-month breeding cycle in the Australian sea lion? *Search* 9:464–465.

Lipps, J. H. 1980. Hunters among the ice floes. *Oceans* 13:45–47.

Loizos, C. 1966. Play in mammals. *Symp. Zool. Soc. Lond.* 18:1–9.

Lønø, O. 1970. The polar bear (*Ursus maritimus* Phipps) in the Svalbard area. *Norsk Polarinstitutt Skrifter* 149:1–103.

Lorenz, K. 1935. Der Kumpan in der Umwelt des Vögels. *J. Ornithol.* 83:213. Translated in K. Lorenz. 1970. *Studies in animal and human behavior,* vol. 1. Cambridge: Harvard Univ. Press.

Loughlin, T. R., and P. A. Livingston. 1986. Summary of joint research on the diets of northern fur seals and fish in the Bering Sea during 1985. Northwest and Alaska Fisheries Center, NMFS, NOAA. Seattle, Wash.

Loughlin, T. R., and R. Nelson, Jr. 1986. Incidental mortality of northern sea lions in Shelikof Strait, Alaska. *Marine Mammal Science* 2:14–33.

Loughlin, T. R., M. A. Perez, and R. L. Merrick. 1987. *Eumetopias jubatus. Mammalian Species* no. 283, 1–7.

Loughrey, A. G. 1959. Preliminary investigation of the Atlantic walrus, *Odobenus rosmarus rosmarus* (Linnaeus). Wildl. Mgt. Bull. no. 14. Canadian Wildlife Service, Ottawa, Ontario.

Lowell, W. R., and W. F. Flanigan, Jr. 1980. Marine mammal chemoreception. *Mammal Rev.* 10:53–59.

Lowenstein, J. M. 1985a. Marine mammal evolution: The molecular evidence. In *Proceedings of the sixth biennial conference on the biology of marine mammals,* Nov. 22–26, Vancouver, British Columbia.

———. 1985b. Molecular approaches to the identification of species. *American Scientist* 73:541–547.

———. 1986. The pinniped family tree puzzle. *Oceans* 19:72.

Lowry, L. F., and F. H. Fay. 1984. Seal eating by walruses in the Bering and Chukchi seas. *Polar Biol.* 3:11–18.

Lowry, L. F., K. J. Frost, and J. J. Burns. 1977. Trophic relationships among ice-inhabiting phocid seals. Annu. Rep. BLM/OCSEAP Contract 03-5-022-53.

———. 1978. Food of ringed seals and bowhead whales near Point Barrow, Alaska. *Canadian Field-Naturalist* 92:67–70.

———. 1980a. Feeding of bearded seals in the Bering and Chukchi seas and trophic interaction with Pacific walruses. *Arctic* 33:330–342.

———. 1980b. Variability in the diet of ringed seals, *Phoca hispida,* in Alaska. *Can. J. Zool.* 37:2254–2261.

Lyons, K. J., and J. A. Estes. 1985. Individual variation in diet and foraging strategy in the female California sea otter, *Enhydra lutris.* Paper presented at the Western Society of Naturalists, Sixty-sixth Annual Meeting, Dec. 27–30, Monterey, Calif.

McCann, T. S. 1980a. Population structure and social organization of southern elephant seals, *Mirounga leonina* (L.). *Biological Journal of the Linnaean Society* 14:133–150.

———. 1980b. Territoriality and breeding behavior of adult male Antarctic fur seals, *Arctocephalus gazella. Journal of Zoology* (London) 192:295–310.

———. 1981. Aggression and sexual activity of male southern elephant seals, *Mirounga leonina. Journal of Zoology* (London) 195:295–310.

———. 1987a. Female fur seal attendance behavior. In *Status, biology, and ecology of fur seals,* ed. J. Croxall and R. L. Gentry, 199–200. Proceedings of an International Symposium and Workshop, Cambridge, England, Apr. 23–27, 1984. NOAA Technical Report NMFS 51.

———. 1987b. Male fur seal tenure. In *Status, biology, and ecology of fur seals,* ed. J. Croxall and R. L. Gentry, 197–198. Proceedings of an International Symposium and Workshop, Cambridge, England, Apr. 23–27, 1984. NOAA Technical Report NMFS 51.

McCann, T. S., and D. W. Doidge. 1987. Antarctic fur seal, *Arctocephalus gazella.* In *Status, biology, and ecology of fur seals,* ed. J. Croxall and R. L. Gentry, 5–8. Proceedings of an International Symposium and Workshop, Cambridge, England, Apr. 23–27, 1984. NOAA Technical Report NMFS 51.

McCosker, J. E. 1985. White shark attack behavior: Observations of and speculations about predator and prey strategies. In *Biology of the white shark*. Memoirs of the Southern California Academy of Sciences, Los Angeles, vol. 9, 123–135.

McLaren, I. A. 1958. The biology of the ringed seal, *Phoca hispida,* in the eastern Canadian Arctic. *Fish. Res. Board Canada Bull.* no. 118.

———. 1960. Are the pinnipeds biphyletic? *Systematic Zoology* 9:18–28.

———. 1962. Population dynamics and exploitation of seals in the eastern Canadian Arctic. In *The exploitation of natural animal populations,* ed. E. D. LeCren and M. W. Holdgate, 168–183. Oxford: Blackwell Scientific Publications.

———. 1977. Report on the Davis Strait aerial survey 77-1 for Imperial Oil Ltd., Aquitaine Co. of Canada Ltd., and Canada Cities Services Ltd., Arctic Petroleum Operators Association, Project no. 134, Dec. 1977. McLaren Atlantic Limited, Dartmouth-Fredericton, Moncton.

McLaren, I. A., and T. G. Smith. 1985. Population ecology of seals: Retrospective and prospective views. *Marine Mammal Science* 1:54–83.

Majluf, P. 1987. South American fur seal, *Arctocephalus australis,* in Peru. In *Status, biology, and ecology of fur seals,* ed. J. Croxall and R. L. Gentry, 33–35. Proceedings of an International Symposium and Workshop, Cambridge, England, Apr. 23–27, 1984. NOAA Technical Report NMFS 51.

Majluf, P., M. Goebel, and G. L. Kooyman. 1985. Diving patterns in the South American fur seal at Porta San Juan, Peru. In *Proceedings of the sixth biennial conference on the biology of marine mammals,* Nov. 22–26, Vancouver, British Columbia.

Manning, T. H. 1961. Comments on "Carnivorous walrus and some arctic zoonoses." *Arctic* 14:76–77.

Mansfield, A. W. 1958a. The biology of the Atlantic walrus, *Odobenus rosmarus rosmarus* (Linnaeus) in the eastern Canadian Arctic. Fish. Res. Board Can., Manuscr. Rep. Ser. (Biol.) 653.

———. 1958b. The breeding behavior and reproductive cycle of the Weddell seal (*Leptonychotes weddelli* Lesson). Falkland Islands Dependencies Survey, London. Sci. Rep. no. 18.

———. 1966. The grey seal in Eastern Canadian waters. *Canadian Audubon Magazine,* Nov.–Dec., 161–166.

———. 1967. *Seals of arctic and eastern Canada,* 2d ed. *Fish Res. Board Canada Bull.* no. 137.

Mansfield, A. W., and B. Beck. 1977. The grey seal in eastern Canada. Dept. Fish. Env. Fisheries and Marine Service Tech. Rep. no. 704.

Marchessaux, D., and N. Muller. 1985. Le phoque moine, *Monachus monachus* distribution, stat et biologie sur la côte saharienne. Parc National, 50 avenue Gambetta, 83400 Hyères.

Markham, B. J. 1971. Catálogo de los anfibios, reptiles, aves y mamíferos de la provincia de Magallanes (Chile). Instituto de la Patagonia, Chile.

Marler, P. 1970. A comparative approach to vocal learning: Song development in white-crowned sparrows. *J. Comp. Physiol. Psychol.* 71:1–25.

Marler, P., and M. Tamura. 1962. Song "dialects" in three populations of white-crowned sparrows. *Condor* 64:368–377.

———. 1964. Culturally transmitted patterns of vocal behavior in sparrows. *Science* 146:1483–1486.

Marlow, B. J. 1972. Pup abduction in the Australian sea-lion, *Neophoca cinerea*. *Mammalia* (Paris) 36:161–165.

———. 1975. The comparative behavior of the Australasian sea lions *Neophoca cinerea* and *Phocartos hookeri* (Pinnipedia: Otariidae). *Mammalia* (Paris) 39:159–230.

Marlow, B. J., and J. E. King. 1974. Sea lions and fur seals of Australia and New Zealand—the growth of knowledge. *J. Aust. Mamm. Soc.* 1:117–136.

Marr, J. 1962. The natural history and geography of the Antarctic krill (*Euphausia superba* Dana). *Discovery Reports* 32:33–464.

Marsh, H., B. R. Gardner, and G. E. Heinsohn. 1981. Present-day hunting and distribution of dugongs in the Wellesley Islands (Queensland): Implications for conservation. *Biol. Conserv.* 19:255–268.

Marten, K., K. S. Norris, and P.W.B. Moore. 1987. Loud impulse sounds in bottlenose dolphin and killer whale predation and social behavior. In *Proceedings of the seventh biennial conference on the biology of marine mammals*, Dec. 5–9, Miami, Fla.

Mate, B. R. 1973. Population kinetics and related ecology of the northern sea lion, *Eumetopias jubatus*, and the California sea lion, *Zalophus californianus*, along the Oregon coast. Ph.D. diss., Univ. of Oregon, Eugene.

———. 1975. Annual migration of the sea lions *Eumetopias jubatus* and *Zalophus californianus* along the Oregon U.S. coast. *Rapp. P.-v. Reun. Cons. Int. Explor. Mer* 169:455–461.

———. 1978. California sea lion. In *Marine mammals of eastern North Pacific and Arctic waters*, ed. D. Haley, 172–177. Seattle, Wash.: Pacific Search Press.

———. 1980. Workshop on marine mammal-fisheries interactions in the northeastern Pacific. Washington, D.C.: U.S. Marine Mammal Commission. U.S. Dept. of Commerce, National Technical Information Service, Springfield, Va., no. PB80-175144.

Mathisen, O. A., R. T. Baade, and R. J. Lopp. 1962. Breeding habits, growth, and stomach contents of the Steller sea lion in Alaska. *Journal of Mammalogy* 43:469–477.

Matthews, L. H. 1952. *British Mammals*. London: Collins.

Mattlin, R. H. 1978. Population, biology, thermoregulation, and site preference of the New Zealand fur seal, *Arctocephalus forsteri* (Lesson, 1828), on the Open Bay Islands, New Zealand. Ph.D. diss., Univ. of Canterbury, Christchurch, New Zealand.

———. 1981. Pup growth of the New Zealand fur seal, *Arctocephalus forsteri* on the Open Bay Islands, New Zealand. *Journal of Zoology* (London) 193:305–314.

———. 1987. New Zealand fur seal, *Arctocephalus forsteri*, within the New Zealand region. In *Status, biology, and ecology of fur seals*, ed. J. Croxall and R. L. Gentry, 49–51. Proceedings of an International Symposium and Workshop, Cambridge, England, Apr. 23–27, 1984. NOAA Technical Report NMFS 51.

May, R. M., ed. 1986. Exploration of Marine Communities. Dahlem Workshop Reports. Life Sciences Research Report 32. Berlin, April 1984. Berlin: Springer-Verlag.

May, R. M., J. R. Beddington, C. W. Clark, S. J. Holt, and R. M. Laws. 1979. Management of multispecies fisheries. *Science* 205:267–277.

Maynard Smith, J. 1980. A new theory of sexual investment. *Behav. Ecol. Sociobiol.* 7:241–251.

Meyer-Holzapfel, M. 1956. Das Spiel bei Säugetieren. *Handb. Zool.*, Band 8, Teil 19:1–36.

Miller, D. G., and R. S. Collier. 1981. Shark attacks in California and Oregon, 1926–1979. *Calif. Fish and Game* 67:76–104.

Miller, E. H. 1975a. A comparative study of facial expressions of two species of pinnipeds. *Behaviour* 53:268–284.

———. 1975b. Social and evolutionary implications of territoriality in adult male New Zealand fur seals, *Arctocephalus forsteri* (Lesson, 1828), during the breeding season. *Rapp. P.-v. Reun. Cons. Int. Explor. Mer* 169:170–187.

———. 1975c. Walrus ethology, part 1: The social role of tusks and applications of multidimensional scaling. *Can. J. Zool.* 53:590–613.

———. 1985. Airborne acoustic communication in the Hawaiian monk seal, *Monachus schauinslandi.* In *Proceedings of the sixth biennial conference on the biology of marine mammals,* Nov. 22–26, Vancouver, British Columbia.

Miller, K., and L. Irving. 1975. Metabolism and temperature regulation in young harbor seals *Phoca vitulina richardsi* in water. *American Journal of Physiology* 229:506–511.

Miller, K., M. Rosenmann, and P. Morrison. 1976. Oxygen uptake and temperature regulation of young harbor seals (*Phoca vitulina richardsi*) in water. *Comparative Biochemistry and Physiology* 54A:105–107.

Mitchell, E. D. 1968. The Mio-Pliocene pinniped *Imagotaria. J. Fish. Res. Board Can.* 25:1843–1900.

Møhl, B. 1964. Preliminary studies on hearing in seals. *Videnskabelige Meddelelser fra Dansk naturhistorisk Forening* 127:283–294.

———. 1968. Auditory sensitivity of the common seal in air and water. *Journal of Auditory Research* 8:27–38.

Møhl, B., J. M. Terhune, and K. Ronald. 1975. Underwater calls of the harp seal, *Pagophilus groenlandicus. Rapp. P.-v. Reun. Cons. Int. Explor. Mer* 169:533–543.

Mohr, E. 1952. Die Robben der europäischen Gewässer. Monograph. Wildsäug. 12. Frankfurt am Main: Paul Schöps.

Molyneux, G. S., and M. M. Bryden. 1978. Arteriovenous anastomoses in the skin of seals, part 1: The Weddell seal *Leptonychotes weddelli* and the elephant seal *Mirounga leonina. Anat. Rec.* 191:239–252.

Montagna, W. 1967. Comparative anatomy and physiology of the skin. *Archives of Dermatol.* 96:357–363.

Moore, P. W. 1980. Cetacean obstacle avoidance. In *Animal sonar systems,* ed. R.-G. Busnel and J. F. Fish, 97–108. New York: Plenum.

Moore, P. W., and R. J. Schusterman. 1987. Audiometric assessment of northern fur seals, *Callorhinus ursinus. Marine Mammal Science* 3:31–53.

Mordvinov, Yu. E., and B. V. Kurbatov. 1972. Influence of hair cover on some species of true seals (*Phocidae*) on the total hydrodynamic resistance. *Zoologicheskii zhurnal* (Moscow) 51:242–247.

Morejohn, G. V. 1977. Feeding ecology of marine mammals in Monterey Bay,

California. In *Proceedings of the second biennial conference on the biology of marine mammals,* Dec. 12–15, San Diego, Calif.

Muizon, C. de, and Q. B. Hendey. 1980. Late Tertiary seals of the South Atlantic Ocean. *Ann. South Afr. Mus.* 82:91–128.

Müller-Schwarze, D., and C. Müller-Schwarze. 1971. Seeleoparden des Ross-Meeres sprangen uns an. *Das Tier* 11, *Jahrgang,* 38–43.

Murchison, E. A. 1980. Maximum detection range and range resolution in echolocating bottlenose porpoises, *Tursiops truncatus* (Montagu). Ph.D. diss., Univ. of California, Santa Cruz.

Murphy, E. C., and A. A. Hoover. 1981. Research study of the reactions of wildlife to boating activity along the Kenai Fjords coastline. Final report to the National Park Service, Anchorage, Alaska.

Mursaloglu, A. U. (In press). Pup-mother environmental relations in the Mediterranean monk seal, *Monachus monachus* (Hermann, 1877), on Turkish coasts. *Acta Zool. Fennica.*

Nachtigall, P. E. 1980. Odontocete echolocation performance on object size, shape, and material. In *Animal sonar systems,* ed. R.-G. Busnel and J. F. Fish. New York: Plenum.

Nansen, F. 1925. Hunting and adventure in the Arctic. New York: Duffield.

Naumov, S. P. 1933. *The seals of the USSR: The raw material of the marine mammal fishery.* Economically exploited animals of the USSR, series ed. N. A. Bobrinskii, Moscow.

Nelson, B. 1968. Galápagos: Islands of birds. London: Longman.

Nelson, C. H., and K. R. Johnson. 1987. Whales and walruses as tillers of the sea floor. *Scientific American* 256:112–117.

Nelson, R. K. 1969. Hunters of the northern ice. Chicago: Univ. of Chicago Press.

Newby, T. C. 1973. Observations on the breeding behavior of the harbour seal in the state of Washington. *Journal of Mammalogy* 54:540–543.

Ni, I-H., W. D. Bowen, G. B. Stenson, and E. L. Neary. 1987. Reproductive biology of male harp seals. In *Proceedings of the seventh biennial conference on the biology of marine mammals,* Dec. 5–9, Miami, Fla.

Niggol, K., C. H. Fiscus, and F. Wilke. 1959. Pelagic fur seal investigations: California-Oregon and Washington. U.S. Fish and Wildlife Service, Seattle, Wash.

Nikulin, P. G. 1941. The Chukchi walrus. *Izv. Tikhookean. nauchno-issled. inst. rybn khoz. okeanogr.* 20:21–59.

———. 1947. Biological characteristics of the shore herds of the walrus on the Chukchi Peninsula. *Izv. Tikhookean. nauchno-issled. inst. rybn khoz. okeanogr.* 25:226–228.

Norris, K. S., and B. Møhl. 1983. Can odontocetes debilitate prey with sound? *American Naturalist* 122:85–104.

Norris, K. S., and J. H. Prescott. 1961. *Observations on Pacific cetaceans of Californian and Mexican waters.* Univ. of California Publ. Zool. 63:291–402.

Norris, K. S., J. H. Prescott, P. V. Asa-Dorian, and P. Perkins. 1961. An experimental demonstration of echolocation behavior in the porpoise *Tursiops truncatus* (Montagu). *Biol. Bull.* 120:163–176.

North, A. W., J. P. Croxall, and D. W. Doidge. 1983. Fish prey of the Antarctic fur seal *Arctocephalus gazella* at South Georgia. *British Antarctic Survey Bulletin* 61:27–37.

Nottebohm, F. 1969. The song of the chingolo, *Zonotrichia capensis,* in Argentina: Description and evaluation of a system of dialects. *Condor* 71:299–315.

———. 1972. The origins of vocal learning. *American Naturalist* 106:116–140.

———. 1975. Continental patterns of song variability in *Zonotrichia capensis*: Some possible ecological correlates. *American Naturalist* 109:605–624.

Nowak, R. M., and J. L. Paradiso. 1983. *Walker's mammals of the world,* vol. 2. Baltimore, Md.: Johns Hopkins Univ. Press.

Nyholm, E. S. 1975. Observations on the walrus (*Odobenus rosmarus* L.) in Spitzbergen in 1971–1972. *Ann. Zool. Fennici* 12:193–196.

Odell, D. K. 1972. Studies on the biology of the California sea lion and the northern elephant seal on San Nicolas Island, California. Ph.D. diss., Univ. of California, Los Angeles.

———. 1974. Behavioral thermoregulation in the California sea lion. *Behavioral Biology* 10:231–237.

———. 1975. Breeding biology of the California sea lion, *Zalophus californianus. Rapp. P.-v. Reun. Cons. Int. Explor. Mer* 169:374–378.

Oftedal, O. T., D. J. Boness, and R. A. Tedman. 1987a. The behavior, physiology, and anatomy of lactation in Pinnipedia. In *Current Mammalogy,* vol. 1, ed. H. H. Genoways, 175–246. New York: Plenum.

Oftedal, O. T., S. J. Iverson, and D. J. Boness. 1985. Energy intake in relation to growth rate in pups of the California sea lion, *Zalophus californianus.* In *Proceedings of the sixth biennial conference on the biology of marine mammals,* Nov. 22–26, Vancouver, British Columbia.

———. 1987b. Milk and energy intake of suckling California sea lion *Zalophus californianus* pups in relation to sex, growth, and predicted maintenance requirements. *Physiological Zoology* 60:560–575.

Ognev, S. I. 1935. *Mammals of the USSR and adjacent countries.* Vol. 3, *Carnivora (Fissipedia and Pinnipedia)* (in Russian). Moscow: Acad. Sci. USSR. Translated by A. Birron and Z. S. Coles for the Israel Program for Scientific Translations, 1962. Available from U.S. Dept. of Commerce, Office of Technical Services, Washington, D.C.

O'Gorman, F. 1963. Observations on terrestrial locomotion in Antarctic seals. *Proc. Zool. Soc.* (London) 141:837–850.

Oliver, G. W. 1978. Navigation in mazes by a grey seal, *Halichoerus grypus* (Fabricius). *Behaviour* 67:97–114.

Oliver, J. S., and R. G. Kvitek. 1984. Side-scan sonar records and diver observations of the grey whale (*Eschrichtius robustus*) feeding grounds. *Biol. Bull.* 167:264–269.

Oliver, J. S., R. G. Kvitek, and P. N. Slattery. 1985. Walrus feeding disturbance: Scavenging habits and recolonization of the Bering Sea benthos. *J. Exp. Mar. Biol. Ecol.* 91:233–246.

Oliver, J. S., and P. N. Slattery. 1985. Destruction and opportunity on the sea floor: Effects of gray whale feeding. *Ecology* 66:1965–1975.

Oliver, J. S., P. N. Slattery, E. F. O'Connor, and L. F. Lowry. 1983a. Walrus, *Odobenus rosmarus,* feeding in the Bering Sea: A benthic perspective. *Fishery Bulletin* 81:501–512.

Oliver, J. S., P. N. Slattery, M. A. Silberstein, and E. F. O'Connor. 1983b. A comparison of gray whale, *Eschrichtius robustus*, feeding in the Bering Sea and Baja California. *Fishery Bulletin* 81:513–522.

———. 1984. Gray whale feeding on dense ampeliscid amphipod communities near Bamfield, British Columbia. *Can. J. Zool.* 62:41–49.

Ono, K. A., O. T. Oftedal, D. J. Boness, and S. J. Iverson. 1985. Effects of the 1983 El Niño on the mother-pup relationship in the California sea lion. In *Proceedings of the sixth biennial conference on the biology of marine mammals*, Nov. 22–26, Vancouver, British Columbia.

Øritsland, N. A. 1971. Wavelength-dependent solar heating of harp seals (*Pagophilus groenlandicus*). *Comparative Biochemistry and Physiology* 40A: 359–361.

Øritsland, N. A., and K. Ronald. 1975. Energetics in the free diving seal. *Rapp. P.-v. Reun. Cons. Int. Explor. Mer* 169:451–454.

Øritsland, T. 1960. Flyleting etter klapp-myss på fangstfeltet i Danmark-stredet. *Fauna* (Oslo) 13:153–162.

———. 1964. Klappmysshunnens forplantningsbiologi (The hooded seal female's reproductive biology). *Fisken og Havet* 1:1–15.

———. 1970a. Biology and population dynamics of Antarctic seals. In *Antarctic ecology*, vol. 1, ed. M. W. Holdgate, 361–366. London: Academic Press.

———. 1970b. Energetic significance of absorption of solar radiation in polar homeotherms. In *Antarctic ecology*, vol. 1, ed. M. W. Holdgate, 464–470. London: Academic Press.

———. 1970c. Sealing and seal research in the southwest Atlantic pack ice, Sept.–Oct. 1964. In *Antarctic ecology*, ed. M. W. Holdgate, 367–376. London: Academic Press.

———. 1975. Sexual maturity and reproductive performance of female hooded seals at Newfoundland. Research Bulletin, International Commission for Northwest Atlantic Fisheries (ICNAF), no. 11, 37–41.

———. 1977. Food consumption of seals in the Antarctic pack ice. In *Adaptations within Antarctic ecosystems*, ed. G. A. Llano, 749–768. Proceedings of the Third SCAR Symposium on Antarctic Biology. Washington, D.C.: Smithsonian Institution.

Øritsland, T., and T. Benjaminsen. 1975. Sex ratio, age composition, and mortality of hooded seals at Newfoundland. Research Bulletin, International Commission for Northwest Atlantic Fisheries (ICNAF), no. 11, 135–143.

Orr, R. T., and T. C. Poulter. 1965. The pinniped population of Año Nuevo Island, California. *Proceedings of the California Academy of Sciences* 32: 377–404.

Ortiz, C. L., D. Costa, and B. J. Le Boeuf. 1978. Water and energy flux in elephant seal pups fasting under natural conditions. *Physiological Zoology* 51:166–178.

Ortiz, C. L., B. J. Le Boeuf, and D. P. Costa. 1984. Milk intake of elephant seal pups: An index of parental investment. *American Naturalist* 124:416–422.

Pastukhov, V. D. 1969. Some results of observations on the Baikal seal under experimental conditions. Morskie mlekopitaiushchie 1:105–110. Translated by the National Marine Fisheries Service, no. 3544.

Paulbitski, P. A., and T. D. Maguire. 1972. Tagging harbor seals in San Francisco Bay (California, USA). In *Proceedings of the seventh annual conference on biological sonar and diving mammals,* 53–72. Menlo Park, Calif.: Stanford Research Institute Biological Sonar Laboratories.

Paulian, P. 1964. Contribution à l'étude de l'otarie de l'Ile Amsterdam. *Mammalia* (Paris) 28, supp. 1, 1–146.

Payne, K., P. Tyack, and R. Payne. 1983. Progressive changes in the songs of humpback whales (*Megaptera novaeangliae*): A detailed analysis of two seasons in Hawaii. In *Communication and behavior of whales,* ed. R. Payne. American Association for the Advancement of Science Selected Symposium Series. Boulder, Colo.: Westview Press.

Payne, M. R. 1977. Growth of a fur seal population. *Philosophical Transactions of the Royal Society of London* B279:67–79.

———. 1979a. Fur seals *Artocephalus tropicalis* and *A. gazella* crossing the Antarctic convergence at South Georgia. *Mammalia* (Paris) 43:93–98.

———. 1979b. Growth in the Antarctic fur seal *Arctocephalus gazella. Journal of Zoology* (London) 187:1–20.

Payne, R. S. 1978. Behaviors and vocalizations of humpback whales, *Megaptera* spp. Report on a workshop on problems related to humpback whales (*Megaptera novaeangliae*) in Hawaii. U.S. Marine Mammal Commission Report MMC-77/03.

Payne, R. S., and S. McVay. 1971. Songs of humpback whales. *Science* 173: 585–597.

Peaker, M., and J. A. Goode. 1978. The milk of the fur seal, *Arctocephalus tropicalis gazella,* in particular the composition of the aqueous phase. *Journal of Zoology* (London) 185:469–476. (According to Bonner 1984, these milk samples are actually from the southern elephant seal *Mirounga leonina.*)

Pederson, A. 1962. *Das Walross*. Wittenberg-Lutherstadt: A. Ziemsen.

Peek, F. W. 1971. Seasonal change in the breeding behavior of the male redwinged blackbird. *Wilson Bulletin* 83:383–395.

Pejoves, J. 1985. Location preference of mother-pup South American fur seal (*Arctocephalus australis* Z.) in a breeding colony. In *Proceedings of the sixth biennial conference on the biology of marine mammals,* Nov. 22–26, Vancouver, British Columbia.

Penney, R. L. 1969. The leopard seal—south polar predator. *Animal Kingdom* 72:2–7.

Penney, R. L., and G. Lowry. 1967. Leopard seal predation on Adélie penguins. *Ecology* 48:878–882.

Perez, M. A., and M. A. Bigg. 1986. Diet of northern fur seals, *Callorhinus ursinus,* off western North America. *Fishery Bulletin* 84:959–973.

Perkins, J. E. 1945. Biology at Little America III, the West Base of the United States Service Expedition, 1939–1941. *Proceedings of the American Philosophical Society* 89:270–284.

Perlov, A. S. 1971. The onset of sexual maturity in sea lions. *Proc. All Union Inst. Marine. Fish. Ocean* 80:174–187.

Pernia, S. D., A. Hill, and C. L. Ortiz. 1980. Urea turnover during prolonged fasting in the northern elephant seal. *Comparative Biochemistry and Physiology* 65B:731–734.

Perry, E., and D. Renouf. 1985. The vocalization of the harbour seal pup: A useful tool for identification and contact, especially in water. In *Proceedings of the sixth biennial conference on the biology of marine mammals,* Nov. 22–26, Vancouver, British Columbia.

Peterson, R. S. 1968. Social behavior in pinnipeds. In *The behavior and physiology of pinnipeds,* ed. R. J. Harrison, R. C. Hubbard, R. S. Peterson, C. E. Rice, and R. J. Schusterman, 3–53. New York: Appleton-Century-Crofts.

Peterson, R. S., and G. A. Bartholomew. 1967. The natural history and behavior of the California sea lion. Am. Soc. Mammal. Spec. Pub. no. 1.

Peterson, R. S., C. L. Hubbs, R. L. Gentry, and R. L. DeLong. 1968. The Guadalupe fur seal: Habitat, behavior, population size, and field identification. *Journal of Mammalogy* 49:665–675.

Peterson, R. S., and W. G. Reeder. 1966. Multiple births in the northern fur seal. *Z. Saugetierk.* 31:52–56.

Petrinovich, L. F. 1974. Individual recognition of pup vocalization by northern elephant seal mothers. *Z. Tierpsychol.* 34:308–312.

Pierson, M. O. 1978. A study of the population dynamics and breeding behavior of the Guadalupe fur seal, *Arctocephalus townsendi.* Ph.D. diss., Univ. of California, Santa Cruz.

———. 1987. Breeding behavior of the Guadalupe fur seal, *Arctocephalus townsendi.* In *Status, biology, and ecology of fur seals,* ed. J. Croxall and R. L. Gentry, 83–94. Proceedings of an International Symposium and Workshop, Cambridge, England, Apr. 23–27, 1984. NOAA Technical Report NMFS 51.

Pilson, E.E.Q. 1970. Water balance in California sea lions. *Physiological Zoology* 43:257–269.

Pilson, E.E.Q., and A. L. Kelly. 1962. Composition of the milk from *Zalophus californianus,* the California sea lion. *Science* 135:104–105.

Pitcher, K. W. 1977. Population productivity and food habits of harbor seals in the Prince William Sound–Copper River Delta area, Alaska. Final report to the U.S. Marine Mammal Commission, Washington, D.C., for Contract MMC-75/03. U.S. Dept. of Commerce, National Technical Information Service, Springfield, Va. PB-266935.

Pitcher, K. W., and D. C. Calkins. 1979. Biology of the harbor seal, *Phoca vitulina richardsi,* in the Gulf of Alaska. Final report submitted to Outer Continental Shelf Environmental Assessment Program (OCSEAP) by the Alaska Dept. of Fish and Game, Anchorage.

———. 1981. Reproductive biology of Steller sea lions in the Gulf of Alaska. *Journal of Mammalogy* 62:599–605.

Pitcher, K. W., and F. H. Fay. 1982. Feeding by Steller sea lions on harbor seals. *Murrelet* 63:70–71.

Platt, N. E. 1975. Infestation of cod (*Gadus morhua* L.) with larvae of codworm (*Terranova decipiens* Krabbe) and herringworm, *Anisakis* Sp. (*Nematoda Ascaridata*), in North Atlantic and Arctic waters. *Journal of Applied Ecology* 12:437–450.

Popov, L. A. 1958. Herd of walruses on Peschan Island. *Priroda* (Moscow) 9:102–103.

———. 1960a. Materials on the biology of reproduction in the walrus of the Laptev Sea. *Biull. Moskov. obshchestva ispytatelei prirody. Otdel biologicheskii* 65:25–30.

———. 1960b. The status of coastal herds of walruses in the Laptev Sea. *Okhrana Prirody Ozel.* (Moscow) 3:95–104.

———. 1979a. Ladoga seal. In *Mammals in the seas.* Vol. 2, *Pinniped species summaries and reports on sirenians.* Food and Agriculture Organization (FAO) Fisheries Series no. 5. Rome: United Nations FAO.

———. 1979b. Baikal seal. In *Mammals in the seas.* Vol. 2, *Pinniped species summaries and reports on sirenians.* Food and Agriculture Organization (FAO) Fisheries Series no. 5. Rome: United Nations FAO.

Popper, A. N. 1980. Sound emission and detection by delphinids. In *Cetacean behavior: Mechanisms and functions,* ed. L. M. Herman, 1–52. New York: Wiley.

Potelov, V. A. 1975a. Biological background for determining the abundance of bearded seals (*Erignathus barbatus*) and ringed seals (*Pusa hispida*). *Rapp. P.-v. Reun. Cons. Int. Explor. Mer* 169:553 (abstract).

———. 1975b. Reproduction of the bearded seal (*Erignathus barbatus*) in the Barents Sea. *Rapp. P.-v. Reun. Cons. Int. Explor. Mer* 169:554 (abstract).

Poulter, T. C. 1967. Systems of echolocation. In *Animal sonar systems,* ed. R.-G. Busnel, 157–186. Jouy-en-Josas, France: Laboratoire de Physiologie Acoustique.

———. 1968. Underwater vocalization and behavior of pinnipeds. In *The behavior and physiology of pinnipeds,* ed. R. J. Harrison, R. C. Hubbard, R. S. Peterson, C. E. Rice, and R. J. Schusterman, 69–84. New York: Appleton-Century-Crofts.

———. 1969. Sonar of penguins and fur seals. *Proceedings of the California Academy of Sciences* 36:363–380.

Poulter, T. C., and D. G. Del Carlo. 1971. Echo ranging signals: Sonar of the Steller sea lion, *Eumetopias jubata. Journal of Auditory Research* 11:43–52.

Poulter, T. C., T. C. Pinney, R. Jennings, and R. C. Hubbard. 1965. The rearing of Steller sea lion, *Eumetopias jubata. Journal of Auditory Research* 6:165–173.

Power, G., and J. Gregoire. 1978. Predation by fresh water seals on the fish community of Lower Seal Lake, Quebec. *J. Fish. Res. Board Can.* 35: 844–850.

Prime, J. H. 1978. Analysis of a sample of grey seal teeth from the Faroe Islands. ICES. C.M. 1978/N:5.

Radford, K. W., R. T. Orr, and C. L. Hubbs. 1965. Reestablishment of the northern elephant seal, *Mirounga angustirostris,* off central California. In *Proceedings of the California Academy of Sciences* 31:601–612.

Rae, B. B. 1960. Seals and Scottish fisheries. *Mar. Res. 1960,* no. 2.

———. 1968. The food of seals in Scottish waters. *Mar. Res. 1968,* no. 2.

———. 1973. Further observations on the food of seals. *Journal of Zoology* (London) 169:287–297.

Rae, B. B., and W. M. Shearer. 1965. Seal damage to salmon fisheries. *Mar. Res. 1965,* no. 2.

Rainey, W. E., J. M. Lowenstein, V. M. Sarich, and D. M. Magor. 1984. Sirenian molecular systematics—including the extinct Steller's sea cow (*Hydrodamalis gigas*). *Naturwissenschaften* 71:597–599.

Ralls, K. 1976. Mammals in which females are larger than males. *Quarterly Review of Biology* 51:245–276.

Ralls, K., P. Fiorelli, and S. Gish. 1985. Vocalizations and vocal mimicry in captive harbor seals, *Phoca vitulina. Can. J. Zool.* 63:1050–1056.

Rand, R. W. 1955. Reproduction of the Cape fur seal, *Arctocephalus pusillus* (Schreber). *Proc. Zool. Soc.* (London) 124:717–740.

———. 1956. The Cape fur seal, *Arctocephalus pusillus* (Schreber): Its general characteristics and moult. Union of South Africa, Dept. of Commerce and Industries, Div. Fish., Invest. Rep. no. 21, 1–52.

———. 1959. The Cape fur seal (*Arctocephalus pusillus*): Distribution, abundance, and feeding habits off the Southwestern coast of South Africa. Union of South Africa, Dept. of Commerce and Industries, Div. Fish., Invest. Rep. no. 34, 1–65.

———. 1967. The Cape fur seal (*Arctocephalus pusillus*), 3: General behavior on land and at sea. Republic of South Africa, Dept. of Commerce and Industries, Div. Sea. Fish., Invest. Rep. no. 60.

Rasa, O. A. 1971. Social interaction and object manipulation in weaned pups of the northern elephant seal, *Mirounga angustirostris. Z. Tierpsychol.* 29: 82–102.

Rasmussen, B. 1960. Om Klapmyssbestanden i det nordlige Atlanterhav (On the stock of hooded seals in the northern Atlantic). *Fisken og Havet* 1. Fisheries Research Board of Canada Translation Ser., no. 387.

Rasmussen, K. 1941. Alaskan Eskimo words, 1921–1924. Fifth Thule Expedition Report, vol. 3, no. 4. New York: Putnam.

Rausch, R. L. 1970. Trichinosis in the Arctic. In *Trichinosis in man and animals,* ed. S. E. Gould, 348–373. Springfield, Ill.: Charles C. Thomas.

Ray, C. 1963. Locomotion in pinnipeds. *Natural History* 72:10–21.

———. 1967. Social behavior and acoustics of the Weddell seal. *Antarctic Journal of the United States* 2:105–106.

Ray, C. E. 1976. Geography of phocid evolution. *Systematic Zoology* 25:391–406.

Ray, G. C. 1970. Population ecology of Antarctic seals. In *Antarctic ecology,* vol. 1, ed. M. W. Holdgate, 398–414. London: Academic Press.

———. 1981. Ross seal—*Ommatophoca rossi.* In *Handbook of marine mammals.* Vol. 2, *Seals,* ed. S. H. Ridgway and R. J. Harrison, 237–260. London: Academic Press.

Ray, G. C., and W. A. Watkins. 1975. Social function of underwater sounds in the walrus *Odobenus rosmarus. Rapp. P.-v. Reun. Cons. Int. Explor. Mer* 169: 524–526.

Rea, L. D. 1987. Post-absorptive metabolism of hooded seals, *Cystophora cristata.* In *Proceedings of the seventh biennial conference on the biology of marine mammals,* Dec. 5–9, Miami, Fla.

Reeves, R. R. 1978. Atlantic walrus (*Odobenus rosmarus rosmarus*): A literature survey and status report. Wildlife Research Report 10. U.S. Fish and Wildlife Service, Washington, D.C.

Reeves, R. R., and J. K. Ling. 1981. Hooded seal—*Cystophora cristata.* In *Handbook of marine mammals.* Vol. 2, *Seals,* ed. S. H. Ridgway and R. J. Harrison, 171–194. London: Academic Press.

Reiss, M. J. 1982. Functional aspects of reproduction: Some theoretical considerations. Ph.D. diss., Cambridge Univ.

Reiter, J. 1984. Studies of female competition and reproductive success in the northern elephant seal. Ph.D. diss., Univ. of California, Santa Cruz.

Reiter, J., K. J. Panken, and B. J. Le Boeuf. 1981. Female competition and reproductive success in northern elephant seals. *Animal Behaviour* 29: 670–687.

Reiter, J., N. L. Stinson, and B. J. Le Boeuf. 1978. Northern elephant seal development: The transition from weaning to nutritional independence. *Behav. Ecol. Sociobiol.* 3:337–367.

Renouf, D. 1981. Evidence for echolocation in harbour seals. In *Proceedings of the fourth biennial conference on the biology of marine mammals*, Dec. 14–18, San Francisco.

———. 1984. The vocalization of the harbor seal pup and its role in the maintenance of contact with the mother. *Journal of Zoology* (London) 202:583–590.

Renouf, D., and M. B. Davis. 1982. Evidence that seals may use echolocation. *Nature* 300:635–637.

Renouf, D., and J. Lawson. 1985. Harbour seal play. In *Proceedings of the sixth biennial conference on the biology of marine mammals*, Nov. 22–26, Vancouver, British Columbia.

Repenning, C. A. 1972. Underwater hearing in seals: Functional morphology. In *Functional anatomy of marine mammals*, vol. 1, ed. R. J. Harrison, 307–331. London: Academic Press.

———. 1976. Adaptive evolution of sea lions and walruses. *Systematic Zoology* 25:375–390.

———. 1980. Warm-blooded life in cold ocean currents. *Oceans* 13:18–24.

Repenning, C. A., R. S. Peterson, and C. L. Hubbs. 1971. Contributions to the systematics of the southern fur seals, with particular reference to the Juan Fernández and Guadalupe species. In *Antarctic Pinnipedia*, ed. W. H. Burt, 1–34. Antarctic Research Series, vol. 18. Washington D.C.: American Geophysical Union.

Repenning, C. A., and C. E. Ray. 1977. The origin of the Hawaiian monk seal. *Proc. Biol. Soc. Wash.* 89:667–688.

Repenning, C. A., C. E. Ray, and D. Grigorescu. 1979. Pinniped biogeography. In *Historical biogeography, plate tectonics, and the changing environment*, ed. J. Gray and A. J. Boucot, 357–369. Corvallis: Oregon State Univ. Press.

Repenning, C. A., and R. H. Tedford. 1977. Otarioid seals of the Neogene. U.S. Dept. of the Interior Geol. Surv. Professional Paper 992.

Rice, D. W. 1960. Population dynamics of the Hawaiian monk seal. *Journal of Mammalogy* 41:376–385.

———. 1968. Stomach contents and feeding behavior of killer whales in the eastern North Pacific. *Norsk Hvalfangst-Tidende* 57:35–38.

Richards, D. G., J. P. Wolz, and L. M. Herman. 1984. Vocal mimicry of computer-generated sounds by a bottlenosed dolphin *Tursiops truncatus*. *J. Comp. Psychol.* 98:10–28.

Ridgway, S. H. 1972. Homeostasis in the aquatic environment. In *Mammals of the sea: Biology and medicine*, ed. S. H. Ridgway, 590–747. Springfield, Ill.: Charles C. Thomas.

———. 1973. Control mechanisms in diving dolphins and seals. Ph.D. diss., Univ. of Cambridge, England.

———. 1985. The bends problem: Dolphins, seals, and nitrogen. In *Proceedings of the sixth biennial conference on the biology of marine mammals,* Nov. 22–26, Vancouver, British Columbia.

Ridgway, S. H., J. R. Geraci, and W. Medway. 1975a. Diseases of pinnipeds. *Rapp. P.-v. Reun. Cons. Int. Explor. Mer* 169:327–337.

Ridgway, S. H., R. J. Harrison, and P. L. Joyce. 1975b. Sleep and cardiac rhythm in the gray seal. *Science* 187:553–555.

Riedman, M. L. 1982. The evolution of alloparental care and adoption in mammals and birds. *Quarterly Review of Biology* 57:405–435.

———. 1983. The difficult art of mothering in an elephant seal rookery. *Pacific Discovery* 36:4–11.

Riedman, M. L., and J. A. Estes. 1987. A review of the history, distribution, and foraging ecology of sea otters. In *The community ecology of sea otters,* ed. G. R. VanBlaricom and J. A. Estes, 4–21. Ecological Studies, vol. 65. Berlin: Springer-Verlag.

———. 1988. Predation on seabirds by sea otters. *Can. J. Zool.* 66:1396–1402.

———. (In review). The biology of the sea otter: A review.

Riedman, M. L., and B. J. Le Boeuf. 1982. Mother-pup separation and adoption in northern elephant seals. *Behav. Ecol. Sociobiol.* 11:203–215.

Riedman, M. L., and C. L. Ortiz. 1979. Changes in milk composition during lactation in the northern elephant seal. *Physiological Zoology* 52:240–249.

Riedman, M. L., M. Staedler, and J. A. Estes. 1988. Behavioral and biological studies on the California sea otter in the northern part of its range. Research plan, Monterey Bay Aquarium, Monterey, Calif.

Roest, A. I. 1964. A ribbon seal from California. *Journal of Mammalogy* 45:416–420.

Romanenko, E. V., V. E. Sokolov, and N. M. Kalinichenko. 1973. Hydrodynamic properties of the hair of *Phoca sibirica. Zoologicheskii zhurnal* (Moscow) 52:1537–1542.

Ronald, K., and J. L. Dougan. 1982. The ice lover: Biology of the harp seal (*Phoca groenlandica*). *Science* 215:928–933.

Ronald, K., and P. J. Healey. 1981. Harp seal—*Phoca groenlandica.* In *Handbook of marine mammals.* Vol. 2, *Seals,* ed. S. H. Ridgway and R. J. Harrison, 55–87. London: Academic Press.

Ronald, K., E. Johnson, M. E. Foster, and D. Vander Pol. 1970. The harp seal, *Pagophilus groenlandicus* (Erxleben, 1777), part 1: Methods of handling, moult, and diseases in captivity. *Can. J. Zool.* 48:1035–1040.

Ross, G.J.B. 1972. Nuzzling behavior in captive Cape fur seals. *Internat. Zoo Yearbook* 12:183–184.

Rostain, J. C., C. Lamaire, and R. Naquet. 1986. Deep-diving neurological problems. In *Comparative physiology of environmental adaptation,* ed. P. de Jours. Proceedings of the Eighth ESCP Conference, Strasbourg, France. Basel, Switzerland: Karger.

Roux, J.-P. 1987a. Recolonization processes in the subantarctic fur seal, *Arctocephalus tropicalis,* on Amsterdam Island. In *Status, biology, and ecology of fur seals,* ed. J. Croxall and R. L. Gentry, 189–194. Proceedings of an International Symposium and Workshop, Cambridge, England, Apr. 23–27, 1984. NOAA Technical Report NMFS 51.

———. 1987b. Subantarctic fur seal, *Arctocephalus tropicalis,* in French subantarctic territories. In *Status, biology, and ecology of fur seals,* ed. J. Croxall and R. L. Gentry, 79–81. Proceedings of an International Symposium and Workshop, Cambridge, England, Apr. 23–27, 1984. NOAA Technical Report NMFS 51.

Roux, J.-P., and A. D. Hes. 1984. The seasonal haul-out cycle of the fur seal, *Arctocephalus tropicalis* (Gray, 1872) on Amsterdam Island. *Mammalia* (Paris) 48:377–389.

Roux, J.-P., and P. Jouventin. 1987. Behavioral cues to individual recognition in the subantarctic fur seal, *Arctocephalus tropicalis.* In *Status, biology, and ecology of fur seals,* ed. J. Croxall and R. L. Gentry, 95–102. Proceedings of an International Symposium and Workshop, Cambridge, England, Apr. 23–27, 1984. NOAA Technical Report NMFS 51.

Rumyantsev, V., and L. Khiras'kin. 1978. (Caspian seals and wolves). *Okhota i okhotnich'e khoziaistvo* 2:22–24.

Ryder, R. A. 1957. Avian-pinniped feeding associations. *Condor* 59:68–69.

Sandegren, F. E. 1970. Breeding and maternal behavior of the Steller sea lion (*Eumetopias jubatus*) in Alaska. Master's thesis, Univ. of Alaska, College, Alaska.

———. 1976. Courtship display, agonistic behavior, and social dynamics in the Steller sea lion (*Eumetopias jubatus*). *Behaviour* 57:159–171.

Sapin-Jaloustre, J. 1952. Les phoques de Terre Adélie. *Mammalia* (Paris) 16:179–212.

Sarich, V. 1969. Pinniped phylogeny. *Systematic Zoology* 18:416–422.

Scammon, C. M. 1874. The marine mammals of the north-western coast of North America, described and illustrated, together with an account of the American whale-fishery. San Francisco: J. H. Carmany and Co.

Scheffer, V. B. 1950. Growth layers on the teeth of pinnipedia as an indication of age. *Science* 112:309–311.

———. 1958. *Seals, sea lions, and walruses: A review of the pinnipedia.* Stanford, Calif.: Stanford Univ. Press.

———. 1974. February birth of Mexican harbor seals. *Murrelet* 55:44.

———. 1976. *A natural history of marine mammals.* New York: Scribner.

Scheffer, V. B., C. H. Fiscus, and E. I. Todd. 1984. History of scientific study and management of the Alaska fur seal, *Callorhinus ursinus,* 1786–1964. U.S. Dept. of Commerce, NOAA Technical Report, NMFS SSRF-780.

Schevill, W. E. 1956. Evidence for echolocation by cetaceans. *Deep-Sea Res.* 3:153–154.

———. 1964. Underwater sounds of cetaceans. In *Marine bioacoustics,* vol. 1, ed. W. N. Tavolga, 307–316. Oxford: Pergamon.

———. 1968. Sea lion echo ranging? *Journal of the Acoustical Society of America* 43:1458–1459.

Schevill, W. E., and W. A. Watkins. 1965. Underwater calls of *Leptonychotes* (Weddell seal). *Zoologica* 50:45–46.

———. 1971. Directionality of the sound beam in *Leptonychotes weddelli* (Mammalia: Pinnipedia). In *Antarctic pinnipedia,* ed. W. H. Burt, 163–168. Antarctic Research Series, vol. 18. Washington, D.C.: American Geophysical Union.

Schevill, W. E., W. A. Watkins, and C. Ray. 1966. Analysis of underwater *Odobenus* calls with remarks on the development and function of the pharyngeal pouches. *Zoologica* 51:103–106.

Schlatter, R. P. 1976. Penetración del lobo marino común, *Otaria flavescens* Shaw, en el Río Valdivia y afluentes. *Medio ambiente* 2:86–90.

Scholander, P. J. 1940. Experimental investigations on the respiratory functions in diving mammals and birds. *Hvalrådets Skrifter* 22:1–131.

Schusterman, R. J. 1967. Attention shift and errorless reversal learning by the California sea lion. *Science* 156:833–835.

———. 1968. Experimental studies of pinniped behavior. In *The behavior and physiology of pinnipeds,* ed. R. J. Harrison, R. C. Hubbard, R. S. Peterson, C. E. Rice, and R. J. Schusterman, 87–171. New York: Appleton-Century-Crofts.

———. 1972. Visual acuity in pinnipeds. In *The behavior of marine animals.* Vol. 2, *Vertebrates,* ed. H. E. Winn and B. L. Olla, 469–494. New York: Plenum.

———. 1974. Auditory sensitivity of a California sea lion to airborne sound. *Journal of the Acoustical Society of America* 756:1248–1251.

———. 1975. Pinniped sensory perception. *Rapp. P.-v. Reun. Cons. Int. Explor. Mer* 169:165–168.

———. 1978. Vocal communication in pinnipeds. In *Behavior of captive wild animals,* ed. H. Markowitz and F. J. Stevens, 247–308. Chicago: Nelson-Hall.

———. 1980. Behavioral methodology in echolocation by marine animals. In *Animal sonar systems,* ed. R.-G. Busnel and J. F. Fish. New York: Plenum.

———. 1981a. Behavioral capabilities of seals and sea lions: A review of their hearing, visual, learning, and diving skills. *Psychological Record* 31:125–143.

———. 1981b. Steller sea lion—*Eumetopias jubatus.* In *Handbook of marine mammals.* Vol. 1, *The walrus, sea lions, fur seals, and sea otters,* ed. S. H. Ridgway and R. J. Harrison, 119–141. London: Academic Press.

———. 1985. Imprinting in California sea lions. In *Proceedings of the sixth biennial conference on the biology of marine mammals,* Nov. 22–26, Vancouver, British Columbia.

Schusterman, R. J., and R. F. Balliet. 1969. Underwater barking by male sea lions (*Zalophus californianus*). *Nature* (London) 222:1179–1181.

———. 1971. Aerial and underwater activity in the California sea lion, *Zalophus californianus,* as a function of luminance. *Ann. N.Y. Acad. Sci.* 188:37–47.

Schusterman, R. J., R. F. Balliet, and J. Nixon. 1972. Underwater audiogram of the California sea lion by the conditioned vocalization technique. *Journal of Experimental Animal Behavior* 17:339–350.

Schusterman, R. J., and R. G. Dawson. 1968. Barking, dominance, and territoriality in male sea lions. *Science* 160:434–436.

Schusterman, R. J., and R. L. Gentry. 1971. Development of a fatted male phenomenon in California sea lions. *Developmental Psychobiology* 4:333–338.

Schusterman, R. J., and R. Gisiner. 1988. Animal language research: Marine mammals re-enter the controversy. *Psychological Record* 38:311–348.

Schusterman, R. J., and K. Krieger. 1984. California sea lions are capable of semantic comprehension. *Psychological Record* 34:3–23.

———. 1986. Artificial language comprehension and size transposition by a California sea lion (*Zalophus californianus*). *J. Comp. Psychol.* 100:348–355.

Schusterman, R. J., and P.W.B. Moore. 1978. Underwater audiogram of the northern fur seal *Callorhinus ursinus*. *Journal of the Acoustical Society of America* 64, Supp. 1, 586.

———. 1980. Noise disturbance and audibility in pinnipeds. *Journal of the Acoustical Society of America* 70, Supp. 1, 583.

Scronce, B. L., and S. H. Ridgway. 1980. Gray seal, *Halichoerus grypus:* Echolocation not demonstrated. In *Animal sonar systems,* ed. R.-G. Busnel and J. F. Fish, 991–994. New York: Plenum.

Sergeant, D. E. 1951. The status of the common seal (*Phoca vitulina* L.) on the East Anglian coast. *J. Mar. Biol. Ass.* (U.K.) 29:707–717.

———. 1965. Migrations of harp seals *Pagophilus groenlandicus* (Erxleben) in the northwest Atlantic. *J. Fish. Res. Board Can.* 23:433–464.

———. 1966. Reproductive rates of harp seals, *Pagophilus groenlandicus* (Erxleben). *J. Fish. Res. Board Can.* 23:757–766.

———. 1970. Migration and orientation in harp seals. In *Proceedings of the seventh annual conference on biological sonar and diving mammals,* 123–131. Menlo Park, Calif.: Stanford Research Institute Biological Sonar Laboratories.

———. 1973a. Environment and reproduction in seals. *J. Reprod. Fert.* Supp. 19:555–561.

———. 1973b. Feeding, growth, and productivity of northwest Atlantic harp seal (*Pagophilus groenlandicus*). *J. Fish. Res. Board Can.* 30:17–29.

———. 1974. A rediscovered whelping population of hooded seals *Cystophora cristata* Erxleben and its possible relationship to other populations. *Polarforschung* 44:1–7.

———. 1976. History and present status of populations of harp and hooded seals. Scientific Consultation on Marine Mammals, Aug. 31–Sept. 9, Bergen, Norway.

———. 1977. Research on hooded seals in the western North Atlantic in 1977. International Commission for Northwest Atlantic Fisheries Res. Doc. 77/XI/57.

Sergeant, D. E., K. Ronald, J. Boulva, and F. Berkes. 1978. The recent status of *Monachus monachus,* the Mediterranean monk seal. *Biol. Conser.* 14:259–287.

Shaughnessy, P. D. 1979. Cape (South African) fur seals. In *Mammals in the seas.* Vol. 2, *Pinniped species summaries and reports on sirenians,* 37–40. Food and Agriculture Organization (FAO) Fisheries Series no. 5. Rome: United Nations FAO.

———. 1982. The status of seals in South Africa and Namibia. In *Mammals in the seas.* Vol. 4, *Small cetaceans, seals, sirenians, and otters,* 383–410. Food and Agriculture Organization (FAO) Fisheries Series no. 5. Rome: United Nations FAO.

———. 1987. Population size of Cape fur seal *Arctocephalus pusillus* I. From aerial photography. Investigational reports, Sea Fisheries Research Institute, South Africa.

Shaughnessy, P. D., and P. B. Best. 1976. A discrete population model for the South African fur seal, *Arctocephalus pusillus pusillus*. Scientific Consultation on Marine Mammals, Aug. 31–Sept. 9, Bergen, Norway.

Shaughnessy, P. D., and F. H. Fay. 1977. A review of the taxonomy and nomenclature of North Pacific harbor seals. *Journal of Zoology* (London) 182:385–419.

Shaughnessy, P. D., and L. Fletcher. 1987. Fur seal, *Arctocephalus* spp., at Macquarie Island. In *Status, biology, and ecology of fur seals,* ed. J. Croxall and R. L. Gentry, 177–188. Proceedings of an International Symposium and Workshop, Cambridge, England, Apr. 23–27, 1984. NOAA Technical Report NMFS 51.

Shaughnessy, P. D., and K. R. Kerry. 1989. Crabeater seals *Lobodon carcinophagus* during the breeding season: Observations on five groups near Enderby Land, Antarctica. *Marine Mammal Science* 5:68–77.

Shaughnessy, P. D., and A.I.L. Payne. 1979. Incidental mortality of cape fur seals during trawl fishing activities in South African waters. *Fisheries Bulletin* (South Africa) 12:20–25.

Shaughnessy, P. D., and R. M. Warneke. 1987. Australian fur seal, *Arctocephalus pusillus doriferus.* In *Status, biology, and ecology of fur seals,* ed. J. Croxall and R. L. Gentry, 73–77. Proceedings of an International Symposium and Workshop, Cambridge, England, Apr. 23–27, 1984. NOAA Technical Report NMFS 51.

Shepeleva, V. K. 1973. Adaptation of seals to life in the arctic. In *Morphology and ecology of marine mammals,* ed. K. K. Chapskii and V. E. Sokolov, 1–58. New York: Wiley. Translated from Russian by H. Mills for the Israel Program for Scientific Translations, Jerusalem.

Shustov, A. P. 1965a. The food of the ribbon seal in the Bering Sea. *Izv. Tikhookean. nauchno-issled. inst. rybn. khoz. okeanogr.* 59:178–183.

———. 1965b. Some biological features and reproductive rates of the ribbon seal (*Histriophoca fasciata*) in the Bering Sea. *Izv. Tikhookean. nauchno-issled. inst. rybn. khoz. okeanogr.* 59:183–192.

———. 1969. Relative indices and possible causes of mortality of Bering Sea ribbon seals. In *Morskie mlekopitaiushchie,* ed. V. A. Arsen'ev, B. A. Zenkovich, and K. K. Chapskii, 83–92. Moscow: Nauka.

Siebenaller, J., and G. N. Somero. 1978. Pressure adaptive differences in lactate dehydrogenases of congeneric fishes living at different depths. *Science* 201:255–257.

Sinha, A. A., and A. W. Erickson. 1972. Ultrastructure of the placenta of Antarctic seals during the first third of pregnancy. *American Journal of Anatomy* 141:317–327.

Siniff, D. B. 1982. Seal population dynamics and ecology. *Journal of the Royal Society of New Zealand* 11:317–327.

Siniff, D. B., and J. L. Bengtson. 1977. Observations and hypotheses concerning the interactions among crabeater seals, leopard seals, and killer whales. *Journal of Mammalogy* 58:414–416.

Siniff, D. B., R. Reichle, G. Kaufman, and D. Kuehn. 1971a. Population dynamics and behavior of Weddell seals at McMurdo Station. *Antarctic Journal of the United States* 6:97–98.

Siniff, D. B., I. Stirling, J. L. Bengtson, and R. A. Reichle. 1979. Social and reproductive behavior of crabeater seals (*Lobodon carcinophagus*) during the austral spring. *Can. J. Zool.* 57:2243–2255.

Siniff, D. B., and S. Stone. 1985. The role of the leopard seal in the tropho-dynamics of the Antarctic marine ecosystem. In *Antarctic nutrient cycles and food webs,* ed. W. R. Siegfried, P. R. Condy, and R. M. Laws, 555–560. Proceedings of the Fourth SCAR Symposium on Antarctic Biology. Berlin: Springer-Verlag.

Siniff, D. B., S. Stone, D. Reichle, and T. Smith. 1980. Aspects of leopard seals (*Hydrurga leptonyx*) in the Antarctic Peninsula pack ice. *Antarctic Journal of the United States* 15:160.

Siniff, D. B., J. R. Tester, and V. B. Kuechle. 1971b. Some observations on the activity patterns of Weddell seals are recorded by telemetry. In *Antarctic pinnipedia,* ed. W. H. Burt, 173–180. Antarctic Research Series, vol. 18. Washington, D.C.: American Geophysical Union.

Slijper, E. J. 1962. *Whales.* Ithaca, N.Y.: Cornell Univ. Press.

Smith, E. A. 1966. A review of the world's grey seal population. *Journal of Zoology* (London) 150:463–489.

———. 1968. Adoptive suckling in the grey seal. *Nature* (London) 217: 762–763.

Smith, T. G. 1973a. Management research on the Eskimo's ringed seal. *Canadian Geographical Journal* 86:118–125.

———. 1973b. Population dynamics of the ringed seal in the Canadian eastern Arctic. *Fish. Res. Board Canada Bull.* 181.

———. 1976. Predation of ringed seal pups (*Phoca hispida*) by the arctic fox (*Alopex lagopus*). *Can J. Zool.* 54:1610–1616.

———. 1980. Polar bear predation of ringed and bearded seals in the land-fast sea ice habitat. *Can. J. Zool.* 58:2201–2209.

Smith, T. G., D. B. Siniff, R. Reichle, and S. Stone. 1981. Coordinated behavior of killer whales (*Orcinus orca*) hunting a crabeater seal (*Lobodon carcinophagus*). *Can. J. Zool.* 59:1185–1189.

Sokolov, V. E. 1982. *Mammal Skin.* Berkeley: Univ. of California Press.

Sonafrank, N., R. Elsner, and D. Wartzok. 1983. Under-ice navigation by the spotted seal, *Phoca largha.* In *Proceedings of the fifth biennial conference on the biology of marine mammals,* Nov. 27–Dec. 1, Boston.

Spalding, D. J. 1964. Comparative feeding habits of the fur seal, sea lion, and harbour seal on the British Columbia coast. *Fish. Res. Board Canada Bull.* 146.

Spotte, S. 1982. The incidence of twinning in pinnipeds. *Can. J. Zool.* 60: 2226–2233.

Springer, S. 1967. Social organization of shark populations. In *Sharks, skates, and rays,* ed. P. W. Gilbert, R. F. Mathewson, and D. P. Ralls, 149–174. Baltimore, Md.: Johns Hopkins Univ. Press.

Squire, J. L., Jr. 1967. Observations of basking sharks and great white sharks in Monterey Bay, 1948–1950. *Copeia* 1:247–250.

Steiger, G. H., J. Calambokidis, J. C. Cubbage, D. C. Gribble, D. E. Skilling, and A. W. Smith. 1985. Comparative mortality, pathology, and microbiology of harbor seal neonates at different sites in Puget Sound, Washington. In *Proceedings of the sixth biennial conference on the biology of marine mammals,* Nov. 22–26, Vancouver, British Columbia.

Stein, J. L., and S. L. Jeffries. 1985. Some aspects of the pupping phenology of harbor seals, *Phoca vitulina,* in Grays Harbor, Washington. In *Proceedings of the sixth biennial conference on the biology of marine mammals,* Nov. 22–26, Vancouver, British Columbia.

Stephens, R. J., I. J. Beebe, and T. C. Poulter. 1973. Innervation of the vibrissae of the California sea lion, *Zalophus californianus. Anat. Rec.* 176:421–442.

Stewart, B. S., and P. K. Yochem. 1985. Feeding habits of harbor seals (*Phoca vitulina richardsi*) at San Nicolas Island, California, 1980–1985. In *Proceedings of the sixth biennial conference on the biology of marine mammals,* Nov. 22–26, Vancouver, British Columbia.

Stewart, B. S., P. K. Yochem, R. L. DeLong, and G. A. Antonelis, Jr. 1987. Interactions between Guadalupe fur seals and California sea lions at San Nicolas and San Miguel Islands, California. In *Status, biology, and ecology of fur seals,* ed. J. Croxall and R. L. Gentry, 103–106. Proceedings of an International Symposium and Workshop, Cambridge, England, Apr. 23–27, 1984. NOAA Technical Report NMFS 51.

Stewart, R.E.A. 1983. Behavioral and energetic aspects of reproductive effort in female harp seals, *Phoca groenlandicus.* Ph.D. diss., Univ. of Guelph, Ontario.

Stewart, R.E.A, and D. M. Lavigne. 1980. Neonatal growth of northwest Atlantic harp seals, *Pagophilus groenlandicus. Journal of Mammalogy* 61: 670–680.

Stirling, I. 1969a. Ecology of the Weddell seal in McMurdo Sound, Antarctica. *Ecology* 50:573–586.

———. 1969b. Toothwear as a mortality factor in the Weddell seal, *Leptonychotes weddelli. Journal of Mammalogy* 50:559–565.

———. 1971a. Studies on the behavior of the South Australian fur seal, *Arctocephalus forsteri* (Lesson), part 1: Annual cycle, postures, and calls and adult males during the breeding season. *Aust. J. Zool.* 19:243–266.

———. 1971b. Studies on the behavior of the South Australian fur seal, *Arctocephalus forsteri* (Lesson), part 2: Adult females and pups. *Aust. J. Zool.* 19:267–273.

———. 1971c. Variation in sex ratio of newborn seals during the pupping season. *Journal of Mammalogy* 52:842–844.

———. 1972. Observations on the Australian sea lion *Neophoca cinerea* (Peron). *Aust. J. Zool.* 20:271–279.

———. 1973. Vocalization in the ringed seal (*Phoca hispida*). *J. Fish. Res. Board Can.* 30:1592–1594.

———. 1974a. Mid-summer observations on the behavior of wild polar bears (*Ursus maritimus*). *Can. J. Zool.* 52:1191–1198.

———. 1974b. Movements of Weddell seals in McMurdo Sound, Antarctica. *Aust. J. Zool.* 22:39–43.

———. 1975a. Adoptive suckling in pinnipeds. *Journal of the Australian Mammal Society* 1:389–391.

———. 1975b. Factors affecting the evolution of social behavior in the pinnipedia. *Rapp. P.-v. Reun. Cons. Int. Explor. Mer* 169:205–212.

———. 1977. Adaptations of Weddell and ringed seals to exploit polar fast ice habitat in the presence or absence of land predators. In *Adaptations within Antarctic ecosystems,* ed. G. A. Llano, 741–748. Proceedings of the Third SCAR Symposium on Antarctic Biology, Aug. 26–30, 1974. Washington, D.C.: Smithsonian Institution.

———. 1983. The evolution of mating systems in pinnipeds. In *Recent advances in the study of mammalian behavior,* ed. J. F. Eisenburg and D. G. Kleiman, 489–527. Special Publication, American Society of Mammalogy no. 7.

Stirling, I., and W. R. Archibald. 1977. Aspects of predation of seals by polar bears. *J. Fish. Res. Board Can.* 34:1126–1129.

———. 1979. Bearded seal. In *Mammals in the seas.* Vol. 2, *Pinniped species summaries and reports on sirenians.* Food and Agriculture Organization (FAO) Fisheries Series no. 5. Rome: United Nations FAO.

Stirling, I., W. R. Archibald, and D. DeMaster. 1977. Distribution and abundance of seals in the eastern Beaufort Sea. *J. Fish Res. Board Can.* 34:976–988.

Stirling, I., and W. Calvert. 1979. Ringed seal. In *Mammals in the seas.* Vol. 2, *Pinniped species summaries and reports on sirenians.* Food and Agriculture Organization (FAO) Fisheries Series no. 5. Rome: United Nations FAO.

Stirling, I., and G. L. Kooyman. 1971. The crabeater seal (*Lobodon carcinophagus*) in McMurdo Sound, Antarctica, and the origin of mummified seals. *Journal of Mammalogy* 52:175–180.

Stirling, I., and J. Roux. 1987. Fur seal vocalizations. In *Status, biology, and ecology of fur seals,* ed. J. Croxall and R. L. Gentry, 201–202. Proceedings of an International Symposium and Workshop, Cambridge, England, Apr. 23–27, 1984. NOAA Technical Report NMFS 51.

Stirling, I., and E. D. Rudolph. 1968. Inland record of a live crabeater seal in Antarctica. *Journal of Mammalogy* 49:161–162.

Stirling, I., and R. M. Warneke. 1971. Implications of a comparison of the airborne vocalizations and some aspects of the behavior of the two Australian fur seals, *Arctocephalus* spp., on the evolution and present taxonomy of the genus. *Aust. J. Zool.* 19:227–241.

Stone, H. S., and D. B. Siniff. 1981. Leopard seal feeding and food consumption in Antarctic spring and summer. In *Proceedings of the fourth biennial conference on the biology of marine mammals,* Dec. 14–18, San Francisco.

Street, R. J. 1964. Feeding habits of the New Zealand fur seal, *Arctocephalus forsteri.* New Zealand Marine Dept. of Fisheries Technical Report no. 9.

Strombom, D. B. 1981. Marine mammal-fishery interactions in the northeast Pacific. Master's thesis, Univ. of Washington, Seattle.

Sullivan, R. M. 1981. Aquatic displays and interactions in harbor seals, *Phoca vitulina,* with comments on mating systems. *Journal of Mammalogy* 62: 825–831.

Summers, C. F. 1974. The grey seal (*Halichoerus grypus*) in Cornwall and the Isles of Scilly. *Biol Conserv.* 6:285–291.

Summers, C. F., W. N. Bonner, and J. Van Haaften. 1978. Changes in the seal populations of the North Sea. *Rapp. P.-v. Reun. Cons. Int. Explor. Mer* 172: 278–285.

Taggart, S. J., and C. J. Zabel. 1987. The grouping behavior of walruses (*Odobenus rosmarus*): An evolutionary perspective. In *Proceedings of the seventh biennial conference on the biology of marine mammals*, Dec. 5–9, Miami, Fla.

Tarasoff, F. J., A. Bisaillon, J. Pierard, and A. P. Whitt. 1972. Locomotory patterns and external morphology of the river otter, sea otter, and harp seal (Mammalia). *Can. J. Zool.* 50:915–929.

Tarasoff, F. J., and H. D. Fisher. 1970. Anatomy of the hind flippers of two species of seals with reference to thermoregulation. *Can. J. Zool.* 48: 821–829.

Tarasoff, F. J., and D. P. Toews. 1972. The osmotic and ironic regulatory capacities of the kidney of the harbor seal, *Phoca vitulina. J. Comp. Physiol.* 81:121–132.

Tavolga, M. C., and E. S. Essapian. 1957. The behavior of the bottlenosed dolphin (*Tursiops truncatus*): Mating, pregnancy, parturition, and mother-infant behavior. *Zoologica* 42:11–31.

Tavrovskii, V. A. 1971. *Mammals of the Yakutia* (in Russian). Moscow: Nauka.

Taylor, F.H.C, M. Fujinaga, and F. Wilke. 1955. Distribution and food habits of the fur seals of the North Pacific Ocean. Report of cooperative investigations by governments of Canada, Japan, and the United States of America, Feb.–July 1952. Washington, D.C.: U.S. Fish and Wildlife Service.

Taylor, L. R., and G. Naftel. 1978. Preliminary investigations of shark predation on the Hawaiian monk seal at Pearl and Hermes Reef and French Frigate Shoals. Final Report to the U.S. Marine Mammal Commission, Washington, D.C., for Contract 7AC011. U.S. Dept. of Commerce, National Technical Information Service, Springfield, Va. PB285-626.

Tedford, R. H. 1976. Relationships of pinnipeds to other carnivores (Mammalia). *Systematic Zoology* 25:363–374.

Terhune, J. M., and K. Ronald. 1971. The harp seal, *Pagophilus groenlandicus* (Erxleben, 1977), part 10: The air audiogram. *Can. J. Zool.* 49:385–390.

———. 1972. The harp seal, *Pagophilus groenlandicus* (Erxleben, 1977), part 3: The underwater audiogram. *Can J. Zool.* 50:565–569.

———. 1985. Distant and near range functions of harp seal underwater calls. In *Proceedings of the sixth biennial conference on the biology of marine mammals*, Nov. 22–26, Vancouver, British Columbia.

Testa, J. W. 1985. Patterns in Weddell seal population parameters. In *Proceedings of the sixth biennial conference on the biology of marine mammals*, Nov. 22–26, Vancouver, British Columbia.

Thomas, J. A., 1979. Quantitative analysis of the vocal repertoire of Weddell seals (*Leptonychotes weddelli*) in McMurdo Sound, Antarctica. Ph.D. diss., Univ. of Minnesota, Minneapolis.

Thomas, J. A., D. P. DeMaster, S. Stone, and D. Andrioshek. 1980. Observations of a newborn Ross seal pup (*Ommatophoca rossi*) near the Antarctic Peninsula. *Can. J. Zool.* 58:2156–2158.

Thomas, J. A., S. R. Fisher, W. E. Evans, and F. T. Awbrey. 1983. Ultrasonic vocalizations of leopard seals (*Hydrurga leptonyx*). *Antarctic Journal of the United States* 17:186.

Thomas, J. A., and V. B. Kuechle. 1982. Quantitative analysis of seal

(*Leptonychotes weddelli*) underwater vocalizations at McMurdo Sound, Antarctica. *Journal of the Acoustical Society of America* 72:1730–1738.

Thomas, J. A., and I. Stirling. 1984. Geographic variation in the underwater vocalizations of Weddell seals (*Leptonychotes weddelli*) from Palmer Peninsula and McMurdo Sound, Antarctica. *Can. J. Zool.* 61:2203–2214.

Thomson, D. 1954. *The people of the sea.* London: Paladin Books.

Thompson, D., P. S. Hammond, J. Gordon, K. S. Nicholas, and M. A. Fedak. 1987. Diving behavior of grey seals (*Halichoerus grypus*) in the North Sea. In *Proceedings of the seventh biennial conference on the biology of marine mammals,* Dec. 5–9, Miami, Fla.

Tikhomirov, E. A. 1959. The feeding of the sea lion on warm-blooded animals (in Russian). *Izv. Tikhookean. nauchno-issled. inst. rybn. khoz. okeanogr.* 47:185. Translated by the National Marine Fisheries Service, Northwest and Alaska Fisheries Center, Marine Mammal Division, Seattle, Wash.

———. 1961. Distribution and migration of seals in waters of the Far East. In *Transactions of the conference on ecology and hunting of marine mammals,* ed. E. H. Pavlovskii and S. K. Kleinenberg, 199–200. Moscow: Akademiia nauk SSSR, Ikhtiologicheskaia komissiia. Translated by the U.S. Fish and Wildlife Service.

———. 1964a. Distribution and hunting of the sea lion in the Bering Sea and adjacent parts of the Pacific (in Russian). *Trudy vsesoiuz. nauchno-issled. inst. morskogo rybn. khoziaistva i okeanografii* 53 (*Izv. Tikhookean. nauchno-issled. inst. rybn. khoz. okeanogr.*) 52:281–285. Translation in *Soviet fisheries investigations in the northeast Pacific,* part 3, 281–285; translated by the Israel Program for Scientific Translation, Jerusalem; available from the National Technical Information Service, Springfield, Va., no. TT67-510215.

———. 1964b. On the distribution and biology of pinnipeds of the Bering Sea. In *Sovetskie rybnye issledovaniia na Severnovostochnoi chasti Tikhogo okeana,* ed. P. A. Moiseev, A. G. Kaganovskii, and I. V. Kisevetter, 277–285. Moscow: Pishchevaia promyshlennost'.

———. 1966. Certain data on the distribution and biology of the harbor seal in the Sea of Okhotsk during the summer-autumn period and hunting it (in Russian). *Izv. Tikhookean. nauchno-issled. inst. rybn. khoz. okeanogr.* 58: 105–115. Translated by the Bureau of Commercial Fisheries, Washington, D.C.

———. 1968. Body growth and development of reproductive organs of the North Pacific phocids. In *Pinnipeds of the North Pacific,* ed. U. A. Arsen'ev and K. I. Panin. Moscow. Translated from Russian by the Israel Program for Scientific Translation, Jerusalem, 1971.

———. 1975. Biology of the ice forms of seals in the Pacific section on the Antarctic. *Rapp. P.-v. Reun. Cons. Int. Explor. Mer* 169:409–412.

Tikhomirov, E. A., and G. M. Kosygin. 1966. Prospects for commercial sealing in the Bering Sea. *Rybnoe khoziaistvo* 8:25–28.

Tomilin, A. G., and A. A. Kibal'chich. 1975. Walruses of the Wrangell Island region. *Zoologicheskii zhurnal* (Moscow) 54:266–272.

Torres, N. D. 1987. Juan Fernandez fur seal, *Arctocephalus philippi*. In *Status, biology, and ecology of fur seals,* ed. J. Croxall and R. L. Gentry, 37–41. Proceedings of an International Symposium and Workshop, Cambridge, England, Apr. 23–27, 1984. NOAA Technical Report NMFS 51.

Tricas, T. C. 1985. Feeding ethology of the white shark, *Carcharodon carcharias*. In *Biology of the White Shark*. Memoirs of the Southern California Academy of Sciences, Los Angeles, vol. 9, 81–91.

Tricas, T. C., and J. E. McCosker. 1984. Predatory behavior of the white shark, *Carcharodon carcharias*, with notes on its biology. *Proceedings of the California Academy of Sciences* 43:221–238.

Trillmich, F. 1979. Galápagos sea lions and fur seals. Noticias de Galápagos 29:8–14.

———. 1981. Mutual mother-pup recognition in Galápagos fur seals and sea lions: Cues used and functional significance. *Behaviour* 78:21–42.

———. 1984. The natural history of the Galápagos fur seal, *Arctocephalus galapagoensis*, Heller. In *Key environments: Galápagos*, ed. R. Perry, 215–223. Oxford: Pergamon.

———. 1986a. Attendance behavior of Galápagos fur seals. In *Fur seals: Maternal strategies on land and at sea*, ed. R. L. Gentry and G. L. Kooyman, 168–185. Princeton, N.J.: Princeton Univ. Press.

———. 1986b. Attendance behavior of Galápagos sea lions. In *Fur seals: Maternal strategies on land and at sea*, ed. R. L. Gentry and G. L. Kooyman, 196–208. Princeton, N.J.: Princeton Univ. Press.

———. 1986c. Maternal investment and sex-allocation in the Galápagos fur seal, *Arctocephalus galapagoensis*. *Behav. Ecol. Sociobiol.* 19:157–164.

———. 1987a. Behavior of fur seals. In *Status, biology, and ecology of fur seals*, ed. J. Croxall and R. L. Gentry, 195–196. Proceedings of an International Symposium and Workshop, Cambridge, England, Apr. 23–27, 1984. NOAA Technical Report NMFS 51.

———. 1987b. Galápagos fur seal, *Arctocephalus galapagoensis*. In *Status, biology, and ecology of fur seals*, ed. J. Croxall and R. L. Gentry, 23–27. Proceedings of an International Symposium and Workshop, Cambridge, England, Apr. 23–27, 1984. NOAA Technical Report NMFS 51.

———. 1987c. Seals under the sun. *Natural History* 96:42–49.

Trillmich, F., G. L. Kooyman, P. Majluf, and M. Sanchez-Grinan. 1986. Attendance and diving behavior of South American fur seals during El Niño in 1983. In *Fur seals: Maternal strategies on land and at sea*, ed. R. L. Gentry and G. L. Kooyman, 153–167. Princeton, N.J.: Princeton Univ. Press.

Trillmich, F., and E. Lechner. 1986. Milk of the Galápagos fur seal and sea lion, with a comparison of the milk of eared seals (Otariidae). *Journal of Zoology* (London) 209:271–277.

Trillmich, F., and D. Limberger. 1985. Drastic effects of El Niño on Galápagos pinnipeds. *Oecologia* (Berlin) 67:19–22.

Trillmich, F., and P. Majluf. 1981. First observations on colony structure, behavior, and vocal repertoire of the South American fur seal (*Arctocephalus australis*, Zimmermann, 1783) in Peru. *Z. Saugetierk.* 46:310–322.

Trillmich, F., and W. Mohren. 1981. Effects of the lunar cycle on the Galápagos fur seal, *Arctocephalus galapagoensis*. *Oecologia* (Berlin) 48:85–92.

Trivers, R. L. 1972. Parental investment and sexual selection. In *Sexual selection and the descent of man, 1871–1971*, ed. B. Campbell, 136–179. Chicago: Aldine.

————. 1985. *Social evolution*. Menlo Park, Calif.: Benjamin-Cummings Publishing Co.

Trivers, R. L., and D. E. Willard. 1973. Natural selection of parental ability to vary the sex ratio of offspring. *Science* 179:90–92.

Troughton, E. 1951. *Furred animals of Australia*. Sydney: Angus and Robertson.

Tsalkin, V. I. 1937. Materials on the biology of the walrus of the Franz Josef Archipelago. *Biull. Moskov. obshchestva ispytatelei prirody. Otdel biologicheskii* 46:43–51.

Uchiyama, K. 1965. California sea lion twins at Tokuyama Zoo. *Internat. Zoo Yearbook* 5:11.

VanBlaricom, G. R., and J. A. Estes. 1987. The community ecology of sea otters. Ecological Studies, vol. 65. Berlin: Springer-Verlag.

Vardy, P. H., and M. M. Bryden. 1981. The kidney of *Leptonychotes weddelli* (Pinnipedia: Phocidae) with some observations on the kidneys of two other southern phocid seals. *Journal of Morphology* 167:13–34.

Vaz-Ferreira, R. 1950. Observaciones sobre la Isla de Lobos. *Revista Fac. Hum. y Ciencias Montevideo* 5:145–176.

————. 1975a. Behavior of the southern sea lion, *Otaria flavescens* (Shaw) in the Uruguayan islands. *Rapp. P.-v. Reun. Cons. Int. Explor. Mer* 169:219–227.

————. 1975b. Factors affecting numbers of sea lions and fur seals on the Uruguayan islands. *Rapp. P.-v. Reun. Cons. Int. Explor. Mer* 169:257–262.

————. 1979. South American fur seal. In *Mammals in the seas*. Vol. 2, *Pinniped species summaries and reports on sirenians*, 34–36. Food and Agriculture Organization (FAO) Fisheries Series no. 5. Rome: United Nations FAO.

————. 1981. South American sea lion—*Otaria flavescens*. In *Handbook of marine mammals*. Vol. 1, *The walrus, sea lions, fur seals, and sea otters*, ed. S. H. Ridgway and R. J. Harrison, 39–65. London: Academic Press.

————. 1982a. *Arctocephalus australis* Zimmerman, South American fur seal. In *Mammals in the seas*. Vol. 4, *Small cetaceans, seals, sirenians, and otters*, 497–508. Food and Agriculture Organization (FAO) Fisheries Series no. 5. Rome: United Nations FAO.

————. 1982b. *Otaria flavescens* (Shaw), South American sea lion. In *Mammals in the seas*. Vol. 4, *Small cetaceans, seals, sirenians, and otters*, 477–495. Food and Agriculture Organization (FAO) Fisheries Series no. 5. Rome: United Nations FAO.

————. 1985. Male-male competition in South American sea lion. In *Proceedings of the sixth biennial conference on the biology of marine mammals*, Nov. 22–26, Vancouver, British Columbia.

————. 1987. Ecology, behavior, and survival of the South American fur seal in Uruguay. In *Status, biology, and ecology of fur seals*, ed. J. Croxall and R. L. Gentry, 165–175. Proceedings of an International Symposium and Workshop, Cambridge, England, Apr. 23–27, 1984. NOAA Technical Report NMFS 51.

Vaz-Ferreira, R., and E. Palerm. 1961. Efectos de los cambios meteorológicos sobre agrupaciones terrestres de Pinnipedios. *Revista Fac. Hum. y Ciencias Montevideo* 19:281–293.

Vaz-Ferreira, R., and A. Ponce de Leon. 1987. South American fur seal, *Arctocephalus australis*, in Uruguay. In *Status, biology, and ecology of fur seals*, ed. J. Croxall and R. L. Gentry, 29–32. Proceedings of an International Symposium and Workshop, Cambridge, England, Apr. 23–27, 1984. NOAA Technical Report NMFS 51.

Venables, U. M., and L.S.V. Venables. 1957. Mating behavior of the seal *Phoca vitulina* in Shetland. *Proc. Zool. Soc.* (London) 128:387–396.

Vibe, C. 1950. The marine mammals and the marine fauna in the Thule District (Northwest Greenland), with observations on ice conditions in 1939–41. *Medd. om. Grønl.* 150:1–115.

Vladimirov, V. A. 1987. Age-specific reproductive behavior of northern fur seals on the Commander Islands. In *Status, biology, and ecology of fur seals,* ed. J. Croxall and R. L. Gentry, 113–120. Proceedings of an International Symposium and Workshop, Cambridge, England, Apr. 23–27, 1984. NOAA Technical Report NMFS 51.

Vorozhtsov, G. A., V. D. Rumyantsev, G. A. Sklarova, and L. S. Khuraskin. 1972. *Trudy VINRO* 89:19–29. Translated by the National Marine Fisheries Service, no. 3108.

Walcott, C., J. L. Gould, and J. L. Kirschvink. 1979. Pigeons have magnets. *Science* 205:1027–1029.

Walker, G. E., and J. K. Ling. 1981a. Australian sea lion—*Neophoca cinerea.* In *Handbook of marine mammals.* Vol. 1, *The walrus, sea lions, fur seals, and sea otters,* ed. S. H. Ridgway and R. J. Harrison, 99–118. London: Academic Press.

———. 1981b. New Zealand sea lion—*Phocartos hookeri.* In *Handbook of marine mammals.* Vol. 1, *The walrus, sea lions, fur seals, and sea otters,* ed. S. H. Ridgway and R. J. Harrison, 25–38. London: Academic Press.

Warneke, R. M. 1979. Australian fur seal. In *Mammals in the seas.* Vol. 2, *Pinniped species summaries and reports on sirenians,* 41–44. Food and Agriculture Organization (FAO) Fisheries Series no. 5. Rome: United Nations FAO.

———. 1982. The distribution and abundance of seals in the Australasian region, with summaries of biology and current research. In *Mammals in the seas.* Vol. 4, *Small cetaceans, seals, sirenians, and otters,* 431–475. Food and Agriculture Organization (FAO) Fisheries Series no. 5. Rome: United Nations FAO.

Warneke, R. M., and P. D. Shaughnessy. 1985. *Arctocephalus pusillus.* The South African and Australian fur seal: Taxonomy, evolution, biogeography, and life history. In *Studies of sea mammals in south latitudes,* ed. J. K. Ling and M. M. Bryden, 53–77. Adelaide: South Australian Museum.

Wartzok, D., R. Elsner, B. P. Kelly, G. Mimken, and L. T. Quakenbush. 1987. Visual, acoustic, and vibrissae contributions to under-ice navigation by ringed seals, *Phoca hispida.* In *Proceedings of the seventh biennial conference on the biology of marine mammals,* Dec. 5–9, Miami, Fla.

Wartzok, D., and M. G. McCormick. 1978. Color discrimination by a Bering Sea spotted seal, *Phoca largha. Vision Res.* 18:781–784.

Watkins, W. A., and W. E. Schevill. 1979. Distinctive characteristics of underwater calls of the harp seal, *Phoca groenlandica,* during the breeding season. *Journal of the Acoustical Society of America* 66:983–988.

Watkins, W. A., and D. Wartzok. 1985. Sensory biophysics of marine mammals. *Marine Mammal Science* 1:219–260.

Wendell, F. A., R. A. Hardy, and J. A. Ames. 1985. Assessment of the accidental take of sea otters, *Enhydra lutris,* in gill and trammel nets. California Dept. of Fish and Game, Marine Resources Branch.

Werner, T. K., and T. W. Sherry. 1987. Behavioral feeding specialization in *Pinaroloxias inornata,* the "Darwin's Finch" of Cocos Island, Costa Rica. *Proceedings of the National Academy of Science* 84:5506–5510.

West, L. 1986a. Interindividual variation in prey selection by the snail *Nucella (Thais) emarginata. Ecology* 67:798–809.

West, L. 1986b. Prey selection by the tropical marine snail *Thais melones:* A study of the effects of interindividual variation and foraging experience on growth and gonad development. Ph.D. diss., Oregon State Univ., Corvallis.

Whittow, G. C. 1978. Thermoregulatory behavior of the Hawaiian monk seal (*Monachus schauinslandi*). *Pacific Science* 32:47–60.

———. 1987. Thermoregulatory adaptations in marine mammals: Interacting effects of exercise and body mass. A review. *Marine Mammal Science* 3: 220–241.

Whittow, G. C., C. A. Ohata, and P. T. Matsuura. 1971. Behavioral control of body temperature in the unrestrained California sea lion. *Comm. Behav. Biol.* 6:87–91.

Wickham, L. L., L. H. Cornell, and R. Elsner. 1985. Red cell morphology and blood viscosity: Possible adaptations of phocid seals. In *Proceedings of the sixth biennial conference on the biology of marine mammals,* Nov. 22–26, Vancouver, British Columbia.

Wiley, R. H. 1974. Evolution of social organization and life history patterns among grouse: Tetraonidae. *Quarterly Review of Biology* 49:201–227.

Wilke, F. 1954. Seals of northern Hokkaido. *Journal of Mammalogy* 35:218–224.

Williams, G. C. 1966. *Adaptation and natural selection.* Princeton, N.J.: Princeton Univ. Press.

Williams, T. D., D. D. Allen, J. M. Groff, and R. L. Glass. (In press). An analysis of California sea otter (*Enhydra lutris*) pelage and integument. *Marine Mammal Science.*

Williams, T. M. 1985. Hydrodynamic, energetic, and biochemical trends in swimming for marine mammals—sea otters to killer whales. In *Proceedings of the sixth biennial conference on the biology of marine mammals,* Nov. 22–26, Vancouver, British Columbia.

Wilson, E. A. 1907. *Lobodon carcinophagus.* The white Antarctic or crabeating seal. In *Mammalia (whales and seals): National Antarctic Expedition, 1901–1904.* Natural History, Zoology 2: ix–xii, 1–41. London: British Museum of Natural History.

Wilson, E. O. 1975. *Sociobiology.* Cambridge: Harvard Univ. Press, Belknap Press.

Wilson, G. J. 1974. The distribution, abundance, and population characteristics of the New Zealand fur seal (*Arctocephalus forsteri*). Master's thesis, University of Canterbury, Christchurch, New Zealand.

———. 1979. Hooker's sea lions in southern New Zealand. *New Zealand Journal of Marine and Freshwater Research* 13:373–375.

———. 1981. Distribution and abundance of the New Zealand fur seal, *Arctocephalus forsteri.* New Zealand Fish. Res. Div. Occas. Publ. 20.

Wilson, J. 1975. Killers in the surf. *Audubon* 77:2–5.

Wilson, S. 1974. Juvenile play of the common seal, *Phoca vitulina vitulina*, with comparative notes on the grey seal *Halichoerus grypus*. *Behaviour* 48:37–60.

Wilson, S. C. 1975. Attempted mating between a male grey seal and female harbor seals. *Journal of Mammalogy* 56:531–534.

Wilson, S. C., and D. G. Kleiman. 1974. Eliciting play: A comparative study (*Octodon, Octodontomys, Pediolagus, Phoca, Choeropsis, Ailuropoda*). *American Zoologist* 14:341–370.

Winn, H. E., and J. Schneider. 1977. Communication in sirenians, sea otters, and pinnipeds. In *How animals communicate*, ed. T. A. Sebeok, 809–840. Bloomington: Indiana Univ. Press.

Winn, H. E., and L. K. Winn. 1978. The song of the humpback whale, *Megaptera novaeangliae*, in the East Indies. *Mar. Biol.* 47:97–114.

Wirtz, W. O., II. 1968. Reproduction, growth and development, and juvenile mortality in the Hawaiian monk seal. *Journal of Mammalogy* 49:229–238.

Wood Jones, F. 1925. The eared seals of South Australia. *Rec. South Aust. Mus.* 3:9–16.

Würsig, B., and M. Würsig. 1980. Behavior and ecology of dusky dolphin, *Lagenorhynchus obscurus*, in the South Atlantic. *Fishery Bulletin* 77:871–890.

Wyss, A. R. 1987. Pinnipeds: How many origins and why the controversy? In *Proceedings of the seventh biennial conference on the biology of marine mammals*, Dec. 5–9, Miami, Fla.

Yaldwyn, J. C. 1958. Decapod crustacea from subantarctic seal and shag stomachs. *Records of the Dominion Museum* 3:121–127.

York, A. E. 1979. Analysis of pregnancy rates in female fur seals in the combined United States-Canada pelagic collections, 1958–74. In *Preliminary analysis of pelagic fur seal data collected by the United States and Canada during 1958–74*, ed. H. Kajimura, R. H. Landers, M. A. Perez, A. E. York, and M. A. Bigg, 50–122. Unpubl. MS. National Marine Mammal Lab., U.S. National Marine Fisheries Service, Seattle, Wash.

———. 1983. Average age at first reproduction of the northern fur seal (*Callorhinus ursinus*). *Can. J. Fish. Aquat. Sci.* 40:121–127.

Yurkowski, M. 1985. Lipid content of selected tissues from mother-pup harp seal (*Phoca groenlandica*) pairs during lactation, and of pups during fasting. In *Proceedings of the sixth biennial conference on the biology of marine mammals*, Nov. 22–26, Vancouver, British Columbia.

Zabel, C. J., S. J. Taggert, and B. P. Kelley. 1987. A comparison of aggressive behavior in female and male walruses (*Odobenus rosmarus*) during the fall migration. In *Proceedings of the seventh biennial conference on the biology of marine mammals*, Dec. 5–9, Miami, Fla.

Zenkovich, B. A. 1938. On the grampus or killer whale, *Grampus orca* Linn. *Priroda* (Moscow) 4:109–112.

List of Personal Communications

Alan Baldridge
Hopkins Marine Station
Stanford University
Pacific Grove, Calif. 93950

Don Bowen
Chief, Marine Fish Division
Dept. of Fisheries and Oceans
P.O. Box 1006
Dartmouth, Nova Scotia B2Y 4A2,
    Canada

Bob Byrne
Washington State Dept. of Wildlife
16018 Mill Creek Blvd.
Mill Creek, Wash. 98012

Charles Deutsch
Institute of Marine Sciences
University of California
Santa Cruz, Calif. 95064

James Estes
National Ecology Research Center
U.S. Fish and Wildlife Service
Institute of Marine Sciences
University of California
Santa Cruz, Calif. 95064

Roger Gentry
National Marine Mammal
    Laboratory
National Marine Fisheries Service,
    NWAFC
7600 Sand Point Way N.E., Bldg. 4
Seattle, Wash. 98115-0070

Tim Gerrodette
National Marine Fisheries Service
Southwest Fisheries Center
Honolulu Laboratory
2570 Dole Street
Honolulu, Hawaii 96822-2396

Bob Gisiner
Long Marine Lab
100 Shaffer Road
Santa Cruz, Calif. 95060

Anne Hoover
Pacific Rim Research
P.O. Box 509
Haines, Alaska 99827

Julie Hymer
Husbandry Dept.
Monterey Bay Aquarium
886 Cannery Row
Monterey, Calif. 93940

Frans Lanting
714A Riverside St.
Santa Cruz, Calif. 95060

David Lavigne
Dept. of Zoology
College of Biological Science
University of Guelph
Guelph, Ontario N1G 2W1, Canada

Burney Le Boeuf
Institute of Marine Sciences
272 Applied Sciences Building
University of California
Santa Cruz, Calif. 95064

John Ling
Head, Division of Natural Science
The South Australian Museum
North Terrace
Adelaide, South Australia 5000,
    Australia

Roger Luckenbach
Santa Catalina School
Mark Thomas Drive
Monterey, Calif. 93940

T. Seamus McCann
British Antarctic Survey
Natural Environment Research
    Council
High Cross, Madingley Road
Cambridge CB3 OET
United Kingdom

Leo Ortiz
Institute of Marine Sciences
University of California
Santa Cruz, Calif. 95064

Mark Pierson
Minerals Management Service
U.S. Dept. of the Interior
1340 West 6th Street
Los Angeles, Calif. 90017

Joanne Reiter
Institute of Marine Sciences
272 Applied Sciences Building
University of California
Santa Cruz, Calif. 95064

Ron Schusterman
Dept. of Psychology
California State University
Hayward, Calif. 94542

Michelle Staedler
Research Dept.
Monterey Bay Aquarium
886 Cannery Row
Monterey, Calif. 93940

Debbie Vanderbrook
California Marine Mammal Center
Golden Gate National Recreation
    Area
Sausalito, Calif. 94965

INDEX

The scientific name is given in parentheses after the common name of each species.

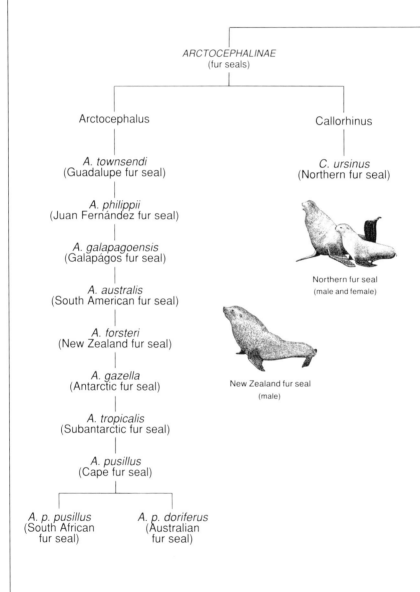

ARCTOCEPHALINAE
(fur seals)

Arctocephalus

Callorhinus

A. townsendi
(Guadalupe fur seal)

C. ursinus
(Northern fur seal)

A. philippii
(Juan Fernández fur seal)

A. galapagoensis
(Galápagos fur seal)

A. australis
(South American fur seal)

A. forsteri
(New Zealand fur seal)

A. gazella
(Antarctic fur seal)

A. tropicalis
(Subantarctic fur seal)

A. pusillus
(Cape fur seal)

A. p. pusillus
(South African
fur seal)

A. p. doriferus
(Australian
fur seal)

Northern fur seal
(male and female)

New Zealand fur seal
(male)

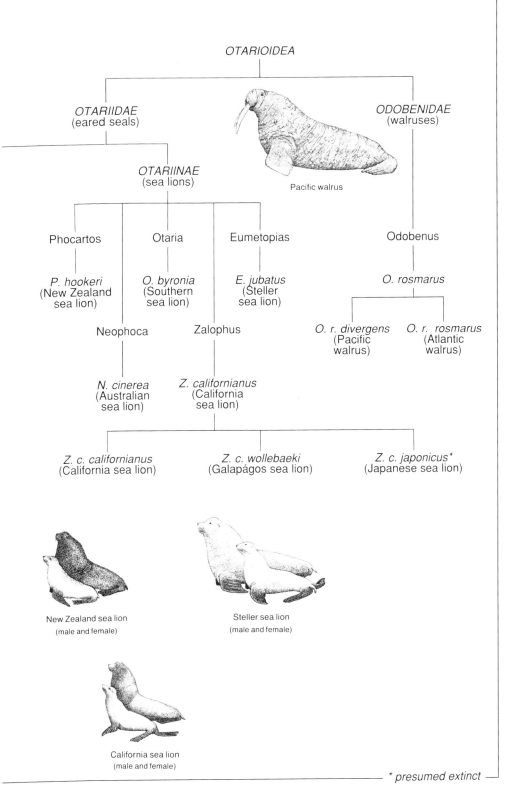

OTARIOIDEA

OTARIIDAE
(eared seals)

ODOBENIDAE
(walruses)

OTARIINAE
(sea lions)

Pacific walrus

Phocartos

Otaria

Eumetopias

Odobenus

P. hookeri
(New Zealand
sea lion)

O. byronia
(Southern
sea lion)

E. jubatus
(Steller
sea lion)

O. rosmarus

Neophoca

Zalophus

O. r. divergens
(Pacific
walrus)

O. r. rosmarus
(Atlantic
walrus)

N. cinerea
(Australian
sea lion)

Z. californianus
(California
sea lion)

Z. c. californianus
(California sea lion)

Z. c. wollebaeki
(Galapágos sea lion)

Z. c. japonicus*
(Japanese sea lion)

New Zealand sea lion
(male and female)

Steller sea lion
(male and female)

California sea lion
(male and female)

* presumed extinct